Pesticide Residues in Foods

CHEMICAL ANALYSIS

A SERIES OF MONOGRAPHS ON ANALYTICAL CHEMISTRY
AND ITS APPLICATIONS

Editor
J. D. WINEFORDNER

VOLUME 151

A WILEY-INTERSCIENCE PUBLICATION

JOHN WILEY & SONS, INC.

New York / Chichester / Weinheim / Brisbane / Singapore / Toronto

Pesticide Residues in Foods

Methods, Techniques, and Regulations

W. GEORGE FONG

Chemical Residue Laboratory
Florida Department of Agriculture
Tallahassee, Florida

H. ANSON MOYE

Food and Environmental Toxicology Laboratory
Food Science and Human Nutrition Department
University of Florida
Gainesville, Florida

JAMES N. SEIBER

University Center for Environmental Science and Engineering
University of Nevada
Reno, Nevada

JOHN P. TOTH

Food and Environmental Toxicology Laboratory
Food Science and Human Nutrition Department
University of Florida
Gainesville, Florida

A WILEY-INTERSCIENCE PUBLICATION

JOHN WILEY & SONS, INC.

New York / Chichester / Weinheim / Brisbane / Singapore / Toronto

This book is printed on acid-free paper. ♾

Published simultaneously in Canada.

Library of Congress Cataloging-in-Publication Data:

Pesticide residues in foods : methods, techniques, and regulations / W. George Fong ... [et al.].
p. cm. — (Chemical analysis ; v. 151)
ISBN 0-471-57400-7 (cloth : alk. paper)
1. Pesticide residues in food. I. Fong, W. George. II. Series.
TX571.P4P486 1999
363. 19′2—dc21 98-27512

Printed in the United States of America

10 9 8 7 6 5 4 3 2 1

CONTENTS

PREFACE

This book is intended to provide an overview of current and emerging analytical methodology for pesticide residues in food. Pesticide residue analysts are increasingly challenged to conduct analyses with more speed, sensitivity, accuracy, and precision than ever before. In order to meet these challenges, the analytical tools now available must be continually refined and improved, and new innovations fostered and put into practice.

Pesticide regulations also discussed in this book and new analytical method developments are more intertwined than one might expect. The advancement of analytical technology reveals trace amounts of residues that could not be detected previously. The finding of residues, no matter how small, fuels the public's perception of toxic contamination and the pressure for more government regulations. On the other hand, new government regulations usually require the development of more sensitive detection methods.

It is not the purpose of this book to uncouple the two. We rather include both the available regulations and the state-of-the-art technology for supporting these regulations. By presenting both areas based on sound science, it is the authors' hope to foster regulations and scientific advancement where it is truly needed to safeguard health and well-being.

CHEMICAL ANALYSIS

A SERIES OF MONOGRAPHS ON ANALYTICAL CHEMISTRY AND ITS APPLICATIONS

J. D. Winefordner, *Series Editor*

Vol. 50. **Analytical Laser Spectroscopy**. Edited by Nicolo Omenetto

Vol. 51. **Trace Element Analysis of Geological Materials**. By Roger D. Reeves and Robert R. Brooks

Vol. 52. **Chemical Analysis by Microwave Rotational Spectroscopy**. By Ravi Varma and Lawrence W. Hrubesh

Vol. 53. **Information Theory as Applied to Chemical Analysis**. By Karel Eckschlager and Vladimir Štěpánek

Vol. 54. **Applied Infrared Spectroscopy: Fundamentals, Techniques, and Analytical Problem-Solving**. By A. Lee Smith

Vol. 55. **Archaeological Chemistry**. By Zvi Goffer

Vol. 56. **Immobilized Enzymes in Analytical and Clinical Chemistry**. By P. W. Carr and L. D. Bowers

Vol. 57. **Photoacoustics and Photoacoustic Spectroscopy**. By Allan Rosencwaig

Vol. 58. **Analysis of Pesticide Residues**. Edited by H. Anson Moye

Vol. 59. **Affinity Chromatography**. By William H. Scouten

Vol. 60. **Quality Control in Analytical Chemistry**. *Second Edition*. By G. Kateman and L. Buydens

Vol. 61. **Direct Characterization of Fineparticles**. By Brian H. Kaye

Vol. 62. **Flow Injection Analysis**. By J. Ruzicka and E. H. Hansen

Vol. 63. **Applied Electron Spectroscopy for Chemical Analysis**. Edited by Hassan Windawi and Floyd Ho

Vol. 64. **Analytical Aspects of Environmental Chemistry**. Edited by David F. S. Natusch and Philip K. Hopke

Vol. 65. **The Interpretation of Analytical Chemical Data by the Use of Cluster Analysis**. By D. Luc Massart and Leonard Kaufman

Vol. 66. **Solid Phase Biochemistry: Analytical and Synthetic Aspects**. Edited by William H. Scouten

Vol. 67. **An Introduction to Photoelectron Spectroscopy**. By Pradip K. Ghosh

Vol. 68. **Room Temperature Phosphorimetry for Chemical Analysis**. By Tuan Vo-Dinh

Vol. 69. **Potentiometry and Potentiometric Titrations**. By E. P. Serjeant

Vol. 70. **Design and Application of Process Analyzer Systems**. By Paul E. Mix

Vol. 71. **Analysis of Organic and Biological Surfaces.** Edited by Patrick Echlin

Vol. 72. **Small Bore Liquid Chromatography Columns: Their Properties and Uses**. Edited by Raymond P. W. Scott

Vol. 73. **Modern Methods of Particle Size Analysis**. Edited by Howard G. Barth

Vol. 74. **Auger Electron Spectroscopy**. By Michael Thompson, M. D. Baker, Alec Christie, and J. F. Tyson

Vol. 75. **Spot Test Analysis: Clinical, Environmental, Forensic and Geochemical Applications**. By Ervin Jungreis

Vol. 76. **Receptor Modeling in Environmental Chemistry**. By Philip K. Hopke

Vol. 77. **Molecular Luminescence Spectroscopy: Methods and Applications** (*in three parts*). Edited by Stephen G. Schulman

Vol. 78. **Inorganic Chromatographic Analysis**. By John C. McDonald

Vol. 79. **Analytical Solution Calorimetry**. Edited by J. K. Grime

Vol. 80. **Selected Methods of Trace Metal Analysis: Biological and Environmental Samples**. By Jon C. Van Loon

Vol. 81. **The Analysis of Extraterrestrial Materials**. By Isidore Adler

Vol. 82. **Chemometrics.** By Muhammad A. Sharaf. Deborah L. Illman, and Bruce Kowalski

Vol. 83. **Fourier Transform Infrared Spectrometry**. By Peter R. Griffiths and James A. de Haseth

Vol. 84. **Trace Analysis: Spectroscopic Methods for Molecules**. Edited by Gary Christian and James B. Callis

Vol. 85. **Ultratrace Analysis of Pharmaceuticals and Other Compounds of Interest**. Edited by S. Ahuja

Vol. 86. **Secondary Ion Mass Spectrometry: Basic Concepts, Instrumental Aspects, Applications and Trends.** By A. Benninghoven, F. G. Rüdenauer, and H. W. Werner

Vol. 87. **Analytical Applications of Lasers**. Edited by Edward H. Piepmeier

Vol. 88. **Applied Geochemical Analysis** By C. O. Ingamells and F. F. Pitard

Vol. 89. **Detectors for Liquid Chromatography**. Edited by Edward S. Yeung

Vol. 90. **Inductively Coupled Plasma Emission Spectroscopy: Part I: Methodology, Instrumentation, and Performance; Part II: Applications and Fundamentals**. Edited by J. M. Boumans

Vol. 91. **Applications of New Mass Spectrometry Techniques in Pesticide Chemistry**. Edited by Joseph Rosen

Vol. 92. **X-Ray Absorption: Principles, Applications, Techniques of EXAFS, SEXAFS, and XANES**. Edited by D. C. Konnigsberger

Vol. 93. **Quantitative Structure-Chromatographic Retention Relationships**. By Roman Kaliszan

Vol. 94. **Laser Remote Chemical Analysis**. Edited by Raymond M. Measures

Vol. 95. **Inorganic Mass Spectrometry**. Edited by F. Adams, R. Gijbels, and R. Van Grieken

Vol. 96. **Kinetic Aspects of Analytical Chemistry**. By Horacio A. Mottola

Vol. 97. **Two-Dimensional NMR Spectroscopy**. By Jan Schraml and Jon M. Bellama

Vol. 98. **High Performance Liquid Chromatography**. Edited by Phyllis R. Brown and Richard A. Hartwick

Vol. 99. **X-Ray Fluorescence Spectrometry**. By Ron Jenkins

Vol. 100. **Analytical Aspects of Drug Testing**. Edited by Dale G. Deutsch

Vol. 101. **Chemical Analysis of Polycyclic Aromatic Compounds**. Edited by Tuan Vo-Dinh

Vol. 102. **Quadrupole Storage Mass Spectrometry**. By Raymond E. March and Richard J. Hughes

Vol. 103. **Determination of Molecular Weight**. Edited by Anthony R. Cooper

Vol. 104. **Selectivity and Detectability Optimizations in HPLC**. By Satinder Ahuja

Vol. 105. **Laser Microanalysis**. By Lieselotte Moenke-Blankenburg

Vol. 106. **Clinical Chemistry**. Edited by E. Howard Taylor

Vol. 107. **Multielement Detection Systems for Spectrochemical Analysis**. By Kenneth W. Busch and Marianna A. Busch

Vol. 108. **Planer Chromatography in the Life Science**. Edited by Joseph C. Touchstone

Vol. 109. **Fluorometric Analysis in Biomedical Chemistry: Trends and Techniques including HPLC Applications**. By Norio Ichinose, George Schwedt, Frank Michael Schnepel, and Kyoko Adochi

Vol. 110. **An Introduction to Laboratory Automation**. By Victor Cerdá and Guillermo Ramis

Vol. 111. **Gas Chromatography: Biochemical, Biomedical, and Clinical Applications**. Edited by Ray E. Clement

Vol. 112. **The Analytical Chemistry of Silicones**. Edited by A. Lee Smith

Vol. 113. **Modern Methods of Polymer Characterization**. Edited by Howard G. Barth and Jimmy W. Mays

Vol. 114. **Analytical Raman Spectroscopy**. Edited by Jeannette Graselli and Bernard J. Bulkin

Vol. 115. **Trace and Ultratrace Analysis by HPLC**. By Satinder Ahuja

Vol. 116. **Radiochemistry and Nuclear Methods of Analysis**. By William D. Ehmann and Diane E. Vance

Vol. 117. **Applications of Fluorescence in Immunoassays**. By Ilkka Hemmila

Vol. 118. **Principles and Practice of Spectroscopic Calibration**. By Howard Mark

Vol. 119. **Activation Spectrometry in Chemical Analysis**. By S. J. Parry

Vol. 120. **Remote Sensing by Fourier Transform Spectrometry**. By Reinhard Beer

Vol. 121. **Detectors for Capillary Chromatography**. Edited by Herbert H. Hill and Dennis McMinn

Vol. 122. **Photochemical Vapor Deposition**. By J. G. Eden

Vol. 123. **Statistical Methods in Analytical Chemistry**. By Peter C. Meier and Richard Zund

Vol. 124. **Laser Ionization Mass Analysis**. Edited by Akos Vertes, Renaat Gijbels, and Fred Adams

Vol. 125. **Physics and Chemistry of Solid State Sensor Devices**. By Andreas Mandelis and Constantinos Christofides

Vol. 126. **Electroanalytical Stripping Methods**. By Khjena Z. Brainina and E. Neyman

Vol. 127. **Air Monitoring by Spectroscopic Techniques**. Edited by Markus W. Sigrist

Vol. 128. **Information Theory in Analytical Chemistry**. By Karel Eckschlager and Klaus Danzer

Vol. 129. **Flame Chemiluminescence Analysis by Molecular Emission Cavity Detection**. Edited by David Stiles, Anthony Calokerinos, and Alan Townshend

Vol. 130. **Hydride Generation Atomic Absorption Spectrometry**. Edited by Jiri Dedina and Dimiter L. Tsalev

Vol. 131. **Selective Detectors: Environmental, Industrial, and Biomedical Applications**. Edited by Robert E. Sievers

Vol. 132. **High-Speed Countercurrent Chromatography**. Edited by Yoichiro Ito and Walter D. Conway

Vol. 133. **Particle-Induced X-Ray Emission Spectrometry**. By Sven A. E. Johansson, John L. Campbell, and Klas G. Malmqvist

Vol. 134. **Photothermal Spectroscopy Methods for Chemical Analysis**. By Stephen Bialkowski

Vol. 135. **Element Speciation in Bioinorganic Chemistry**. Edited by Sergio Caroli

Vol. 136. **Laser-Enhanced Ionization and Spectrometry**. Edited by John C. Travis and Gregory C. Turk

Vol. 137. **Fluorescence Imaging Spectroscopy and Microscopy**. Edited by Xue Feng Wang and Brian Herman

Vol. 138. **Introduction to X-Ray Powder Diffractometry**. By Ron Jenkins and Robert L. Snyder

Vol. 139. **Modern Techniques in Electroanalysis**. Edited by Peter Vanýsek

Vol. 140. **Total-Reflection X-Ray Fluorescence Analysis**. By Reinhold Klockenkämper

Vol. 141. **Spot Test Analysis: Clinical, Enviromental, Forensic, and Geochemical Applications**. *Second Edition*. By Ervin Jungreis

Vol. 142. **The Impact of Stereochemistry on Drug Development and Use**. Edited by Hassan Y. Aboul-Enein and Irving W. Wainer

Vol. 143. **Macrocyclic Compounds in Analytical Chemistry**. Edited by Yury A. Zolotov

Vol. 144. **Surface-Launched Acoustic Wave Sensors: Chemical Sensing and Thin-Film Characterization**. By Micheal Thompson and David Stone

Vol. 145. **Modern Isotope Ratio Mass Spectrometry**. By I. T. Platzner

Vol. 146. **High Performance Capillary Electrophoresis: Theory, Techniques, and Applications**. Edited by Morteza G. Khaledi

Vol. 147. **Solid-Phase Extraction: Principles and Practice**. By E. M. Thurman and M. S. Mills

Vol. 148. **Commercial Biosensors: Applications to Clinical, Bioprocess, and Environmental Samples**. Edited by Graham Ramsay

Vol. 149. **A Practical Guide to Graphite Furnace Atomic Absorption Spectrometry**. By David J. Butcher and Joseph Sneddon

Vol. 150. **Principles of Chemical and Biological Sensors**. Edited by Dermot Diamond

Vol. 151. **Pesticide Residues in Foods: Methods, Technologies, and Regulations.** By W. George Fong, H. Anson Moye, James N. Seiber, and John P. Toth

Pesticide Residues in Foods

CHAPTER

1

THE ANALYTICAL APPROACH

JAMES N. SEIBER

ORIGIN OF PESTICIDE RESIDUES

A residue will result when a crop or food animal (commodity) or medium of the environment (air, water, soils, wildlife, etc.) is treated with a chemical, or exposed unintentionally to it by drift, in irrigation water, in feed, or by other means. The chemical will typically be detectable in the commodity or medium at the time of treatment or exposure, and also for some period of time afterward. The magnitude of the residue at any point in time will depend on the treatment or exposure level and on the rate at which the residue dissipates from the commodity. Dissipation rates are a function of the chemical's physical and chemical properties, and also of the environment in which the chemical resides. For example, an insecticide sprayed to protect apples may dissipate by a number of processes, including

1. Volatilization to the atmosphere (influenced by the insecticide's volatility or vapor pressure and the temperature and wind movement in the orchard).
2. Washing off by rainfall or overhead irrigation (a function of the chemical's water solubility and the frequency and intensity of waterfall).
3. Chemical degradation (influenced by the molecular makeup of the insecticide and by such factors as sunlight, moisture, and temperature).
4. Growth dilution (as the apples become larger, the chemical residue *concentration* will decrease even without physical or chemical dissipation of residue from the fruit).
5. Metabolism and/or excretion in the case of animals and some plants. The parent chemical may be converted to one or more degradation

Pesticide Residues in Foods: Methods, Techniques, and Regulations, by W. G. Fong, H. A. Moye, J. N. Seiber, and J. P. Toth. Chemical Analysis Series, Vol. 151
ISBN 0-471-57400-7

products that will then constitute the components of the remaining residue. If the degradation processes result in formation of a product or products that are more toxic, or more prone to contaminate the food chain or some other segment of the biosphere than the parent, *activation* is said to have occurred (1,2). If the products are less toxic than the parent, *deactivation* or *detoxification* has occurred. From the regulatory viewpoint, breakdown products of toxicological significance are included in the tolerance for pesticides with food and feed uses and thus must be included in analyses.

The relationship between the theoretical initial residue, actual initial residue, and residue that exists at different times after exposure ends, extending beyond the point at which detection ends, is shown in Figure 1.1 for a hypothetical pesticide applied to foliage. The actual residue on the crop when treatment ends is generally far less than calculated on the basis of application rate and the total weight of crop in the treated field because (1) some of the chemical drifts or evaporates from the bulk of the sprayed

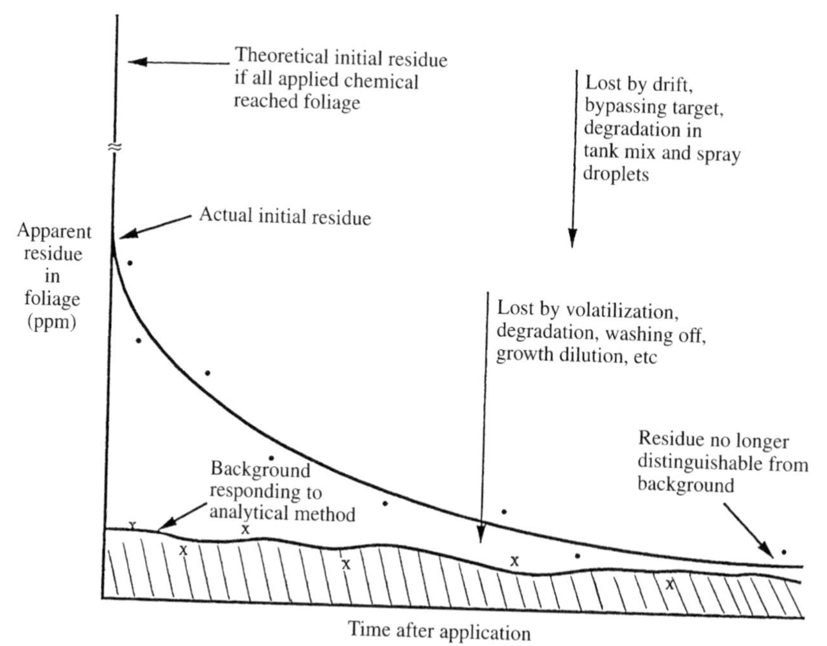

Figure 1.1. Hypothetical pesticide residue decline or dissipation curve, showing the relationship between actual residues and the background due to matrix-derived interferences that behave like the analyte. The limit of detection is reached when actual residues can no longer be distinguished from background (far right of diagram).

droplet cloud and moves off target, beyond the zone of the field; (2) some of the chemical bypasses the target (foliage in this example) and winds up in a nontarget component of the field (soil, in the case of a chemical applied to foliage); and/or (3) some of the chemical breaks down, in the spray tank, spray droplets, or target surface literally within minutes of contact. An example of scenario 3 is for merphos (tributyl phosphorotrithioite), which forms Def (tributyl phosphorotrithioate) and simple sulfur-containing byproducts such as butyl mercaptan and dibutyl disulfide in the tank and in freshly sprayed deposits on the target crop, cotton (3). Another example is for metam sodium, which releases methyl isothiocyanate when treated to soil for nematode control.

The actual residue at the end of treatment then undergoes decline or dissipation by physical (volatilization, washing off) or chemical processes (oxidation, photolysis, etc.) until it is below the limit of detection of the method. We can surmise that decline continues, but the method is insufficiently sensitive to distinguish it.

Dissipation proceeds at an overall rate that is a composite of the rate constants of the individual processes (volatilization, degradation, etc.) (4). The typical result is that concentrations of overall residue (parent plus products) decrease with time after exposure or treatment ends. Since most individual dissipation processes follow first-order kinetics, overall dissipation or decline is also first-order. This has profound ramifications because first-order decline is logarithmic in nature and a concentration–time plot is asymptotic with respect to the time coordinate. Thus, residue concentrations will approach zero with time but never, in fact, cease to exist entirely. Stated simply, a commodity treated with or exposed to a pesticide can theoretically never totally rid itself of all traces of residue. However, our ability to detect residues is limited by the background response of other chemicals in the matrix, the analytical sensitivity of measuring instrumentation, and other factors. Thus, at some point in time after treatment or exposure, the commodity will no longer have a *detectable* residue present. This point is reached when the residue concentration is below the limit of detection (LOD) of the analytical method employed. If a different method, with a lower LOD, is employed, detectable residues may be observed for a longer period of time after treatment or exposure. Figure 1.1 illustrates the concept of dissipation of residues to a point at which the residue remaining is no longer distinguishable from the background contributed by the substrate's matrix.

LIMIT OF DETECTION

The limit of detection (LOD) is the lowest concentration of the analyte that can be determined to be different, with a high degree of confidence, from the blank or background (5). LOD is set by the background that is carried through the method and sensed by the analytical instrumentation. It mimics the analyte by producing a signal that is sensed as though it were an analyte. For example, when using a UV/visible or fluorescence detection system, any matrix-derived coextractives, or solvent impurities, which lead to absorption or emission of light in the same wavelength region as the analyte's absorption or emission bands is a potential interference. In order to confirm that the absorption or emission intensity that we wish to measure is due to the analyte and not background, blank determinations are run, from which the difference between readings for the treated sample (samples that are expected to have analyte measurably present) and the background is calculated. The LOD, the lowest measurable signal ascribable to analyte, is generally set at 2 times the background as a minimum:

$$\text{Apparent LOD} = 2 \times \text{background response}$$

Thus, the analyte signal must be twice the background response (as a minimum) before we can reliably report it. LOD has approximately the same meaning as *method detection limit* (MDL) (6), except that, as implied, each method will have a distinct LOD for the same analyte in the same matrix. Thus, if the LOD from a GC-FID (gas chromatography–flame ionization detection) method is too high to allow measurement of analyte, success might be obtained by switching to a GC-EC (electron capture) or GC-MS (mass spectrometry) method. The *instrument detection limit* (IDL) is related, but is an operational parameter of the instrument chosen for final detection. It refers to the smallest signal from a pure standard of analyte that can be reliably detected by a given instrument above the instrument's inherent (primarily electronic) background level. IDLs are generally much lower than MDLs. IDL may be 1 pg/mL for GC-EC when a chlorinated hydrocarbon standard is injected in pure solvent on the fully tuned instrument. However, for a concentrated extract of animal tissue, it may take 100 pg/mL or more of the same analyte to produce a signal that stands out from tissue-derived interference, even after cleanup.

The limit of quantitation (LOQ) is different still (7). LOQ refers to the level above which quantitative results may be reported with a specified level of confidence. LOQ is generally several times higher than LOD reflecting the fact that most operators are not confident in reporting a residue whose signal is only twice the background. They might prefer one that is, at a minimum, 5 × or 10 × background as a reportable level that could be defended, if necessary, in a court of law.

The hypothetical chromatograms in Figure 1.2 illustrate the terms discussed above:

(*a*) *IDL.* Approximately 1-pg/μL in pure solvent shows a peak approximately twice the instrument electronic background noise.
(*b*) *Background.* The extract has so many peaks around the retention time of the analyte that a 1-pg/μL level of analyte would not be distinguishable above the background.
(*c*) *LOD.* At the 10-pg/μL level, the analyte peak can just be seen above the background, and could be measured and reported. But confidence in this result would not be great.
(*d*) *LOQ.* At the 50-pg/μL level, the analyte peak is clearly distinguishable above background, and much confidence could be placed in quantitation derived from it.

The treatment of LOD by Sutherland (5) emphasizes the importance of background, including its magnitude and its variability, in setting LOD

$$\text{Apparent LOD} = B + t\sqrt{\left(1 + \frac{1}{n}\right)s^2}$$

where B = mean of responding background at the same retention time, wavelength, and so on of the analyte of interest, determined from analysis of several replicates of the background matrix
t = student t test, from statistics table
n = number of determinations of background
s = standard deviation of background response

Several blanks (e.g., background, control, or untreated commodity) are individually run through a method (i.e., extracted, cleaned up, determined) in order to assess the background reading. These readings are averaged and the standard deviation of the averaged readings determined. Apparent LOD is calculated using a preset confidence limit (i.e., 95% or 99%). Sutherland's approach emphasized the importance of background variability in calculating the LOD; if the background from different samples of the untreated commodity is fairly uniform, a lower LOD may be achieved. If, on the other hand, the background from different samples is quite variable and random, then LOD will, correspondingly, be considerably higher.

It is clear that we can lower LOD by lowering both the magnitude and the variation in the background reading. "Background" is composed of several

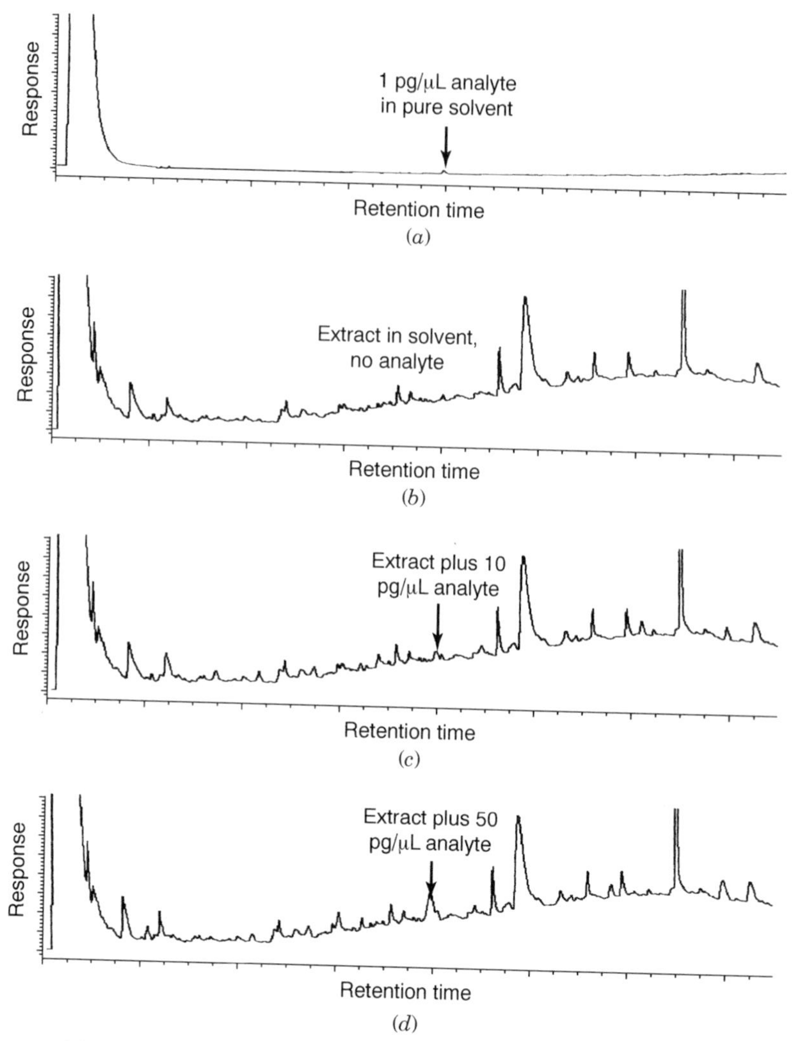

Figure 1.2. Hypothetical gas chromatograms run at same attenuation showing (*a*) instrument detection limit when the pure analyte is injected in pure solvent, (*b*) response due to matrix-derived background, (*c*) barely visible response due to analyte in the presence of matrix-derived background, (*d*) response due to analyte at a higher residue concentration. Chromatogram (*d*) would be suitable for quantification, with a high degree of confidence. If 1 mg equivalent of matrix were injected in (*d*), the response would correspond to 0.05 ppm.

components, each of which can be manipulated to a certain extent by the analyst:

1. *Background Due to Reagents and Glassware.* Can be lowered by buying the best grade of solvents (Nanograde, Resigrade, etc.), using pure reagents (recrystallized or redistilled if necessary), and rigorously cleaning all glassware (Chromerge soak, distilled water rinse, acetone rinse, high-temperature oven bakeout, etc.).
2. *Background Due to Matrix or Substrate.* Can be lowered by providing efficient cleanup (by solvent partitioning, Florisil column chromatography, etc.) in the method, or by using selective detectors that are "blind" to certain types of potential interference.
3. *Background Due to Instrument.* Each instruments' signal-to-noise ratio (S/N) must be optimized by proper tuning. Electronic noise filters, constant-current power sources, constant-temperature rooms, and similar devices may be helpful.

Background is clearly the "enemy" in pesticide residue analysis because measurements are typically made at low levels where background noise is a serious interference problem. Thus, much of the strategy in choosing approaches to pesticide residue analysis is based on methods that minimize background, by physically removing interferences or by electronically removing their signals.

Once an LOD is calculated from analysis of several portions of blank, untreated, or control material, recovery experiments are conducted to test the calculated LOD to see if in fact it is achievable under routine conditions.

A rigorous test protocol was published by Glaser (6) primarily for water analysis although it has more general applicability. Several portions of background matrix are spiked identically with analyte at a level chosen from the range 2–5 times the estimated LOD. Each of these replicates is run through the method, yielding an apparent residue value. The average of the apparent values is corrected by subtracting the background reading (from similar analyses of background samples or blanks). The percent recovery is calculated from

$$\% \text{ recovery} = \frac{\text{amount recovered}}{\text{amount spiked}} \times 100$$

The MDL is then calculated from:

$$\text{MDL} = t_{.99} \times s$$

where s is the standard deviation of readings from the identically spiked matrix portions and $t_{.99}$ is the confidence interval about the mean, as determined by the student t from statistics tables.

The results are considered satisfactory if the average of the apparent residues from the identically spiked portions is 2 times more than the MDL calculated from the standard deviation (above). If the average is less than 2 × MDL, the whole experiment must be repeated, at a higher spiking level, until satisfactory results are obtained, that is, greater than 2 × MDL.

Sutherland's method emphasizes the magnitude and the variability in responses of background interferences that are not the analyte but behave analytically as if they were, by producing a response mimicking that of analyte. Glaser's method looks at the variability in responses of analyte when spiked above, but fairly close to, the LOD. In fact, it is the variation in both the background response and the analyte response that must be considered in setting a limit for a method. By setting LOQs at several times higher than the LOD and MDL, analysts appear to have recognized this, although the question of how much confidence is needed remains a judgment call for the individual analyst or laboratory in the choice of factors (5 × , 10 × , . . .) for distinguishing an operating, defensible LOQ from the bare minimum LOD or MDL.

The preceding discussion emphasizes the irrelevance of reporting "zero" in residue reports (8). "Zero" literally means not a single molecule of contaminant is present, but no method has an LOD low enough to analyze at levels close to true zero. However, we will frequently find samples whose residue content is below LOQ and/or LOD. If the residue content is below LOQ but above LOD, the residue report should state the calculated value but modify it (with parenthesis or asterisk) to denote that less confidence can be placed in it than ones that are clearly above LOQ. Of course, the LOQ level adopted must be clearly stated in the report. If the residue value is below LOD or MDL, the term "not detected" (ND) is placed in the report with a footnote indicating the value for LOD or MDL. Thus, a good report will contain no zeroes and no fudge words such as "trace," but only (1) true residue values, (2) residue values that are above LOQ but less than LOD, and (3) NDs, with the LOQ and LOD clearly stated.

There is considerable current controversy over how best to prepare residue averages when the report has ND entries (9). Some authors include a zero whenever ND appears in the report. This biases the residue average to the low side of the true average because an ND can range anywhere from just below LOD to true zero. Other authors plug in the value for LOD whenever an ND is encountered, which of course prejudices the average to the high side of the true average. Still other authors use $\frac{1}{2}$ LOD for each ND value, which represents a middle ground perhaps more reflective of the true average. Table 1.1 provides some fictitious data that show the effect on the average of using zero, LOD, or $\frac{1}{2}$ LOD in preparing averages. Fortunately, in

Table 1.1. Consequences of Three Methods of Fictitious Residue Data

	Low-Residue Case			High-Residue Case		
	LOQ	0	0.5 of LOQ	LOQ	0	0.5 of LOQ
	1.5	1.5	1.5	5.5	5.5	5.5
	0.5	0.5	0.5	0.5	0.5	0.5
	0.1	0.1	0.1	0.1	0.1	0.1
	<0.1	0	0.05	<0.1	0	0.05
	<0.1	0	0.05	<0.1	0	0.05
	<0.1	0	0.05	<0.1	0	0.05
	<0.1	0	0.05	<0.1	0	0.05
	<0.1	0	0.05	<0.1	0	0.05
Average	0.32	0.26	0.29	0.83	0.76	0.79

[a] Sets (in ppm) in which nondetects [samples for which analyses show no residue detected at or above the LOQ (0.1)] are entered at LOQ, 0, and 0.5 of LOQ.

most cases the effects are not large ones, but it is still important to state in the final report how those nondetects were handled in averaging.

METHODS

Method Validation

Before actual samples can be run, data must be generated to show that the method works satisfactorily in the performing laboratory, at concentration levels and with the type of matrix expected in the actual samples. This validation involves running replicate spiked control samples, with fortifications at near the expected limit of quantitation (see preceding discussion) and at one or two higher levels that might reasonably be expected to bracket anticipated residue values. This is in addition to running background or control samples through the method in order to provide data for calculating the LOD and for correcting data by background subtraction.

Fortifications are generally done at the point of extraction. Many residue chemists have criticized this practice as not reflective of the extractability of incurred field residues that may have penetrated the matrix or become bound to it during weathering. However, fortifications at the point of extraction can still indicate whether subsequent steps in the method are operating with sufficient precision and accuracy. Once the replicated spiked samples are completed, the concentrations found may be corrected for background response (not generally encouraged, but done occasionally) and then the percent recovery is calculated for each spiked replicate. Two parameters are calculated:

1. *Relative error* (*RE*), a measure of method accuracy, from

$$\text{RE} = \frac{100 - \%\ \text{recovery}}{100} \times 100$$

Relative errors of 20% or less (i.e., recoveries of 80% or more) are considered satisfactory. When the best method available gives less than 80% recovery, it may still be used providing the percent recovery is reproducible.

2. *Relative standard deviation* (RSD), a measure of method precision, from

$$\text{RSD} = \frac{s}{\%\ \text{recovery (avg)}} \times 100$$

where s = standard deviation

Generalizing, RSD will increase with fortification level but RE is independent of fortification level, as shown in the detailed example in Table 1.2. The RE reflects losses from partitioning that are, in general, concentration-independent. Exceptions occur when adsorption to glassware, filters, and so on occurs, and when analyte volatilizes during an evaporation step. In these cases losses may increase (and thus RE increases) at lower concentration levels. It is useful to calculate a method's total error (10), from:

$$\text{Total error} = \text{RE} + 2\text{RSD}$$

Total errors tend to run high in trace analyses. A total error of <50% is considered good, 50–100% acceptable, and occasionally methods with >100% total error can still be usable if no better method exists.

Types of Method

There are two general analytical approaches to determining residues in food and environmental samples:

1. *Single-Residue Methods* (*SRM*). A residue method for quantitatively determining a single pesticide (and its toxicologically important

Table 1.2. Verification of Analytical Method by Fortification

Sample	Grams Analyzed (g)	Added ppm	Found Total ppm (Apparent)	Found Net ppm (Corrected)	% Recovered	Average (CI 95%)
Blank (control)	50	—	0.065	—	—	
		—	0.071	—	—	(0.071 ppm)
		—	0.077	—	—	
Blank spiked at 0.05 ppm	50	0.05	0.110	0.039	78	
		0.05	0.099	0.028	56	77% ± 26
		0.05	0.120	0.049	98	
Blank spiked at 0.10 ppm	50	0.10	0.164	0.093	93	
		0.10	0.152	0.081	81	91% ± 17
		0.10	0.165	0.094	94	
Blank spiked at 1.0 ppm	50	1.0	0.98	0.91	91	
		1.0	1.01	0.94	94	90% ± 10
		1.0	0.93	0.86	86	

conversion products) in samples of regulatory interest. This is generally the method submitted by the registrant to EPA, and eventually published in *PAM* II after registration is secured.

2. *Multiresidue Method* (*MRM*). A residue method capable of detecting and quantifying more than one pesticide in more than one food or environmental sample type. MRMs are commonly used by government agencies for surveillance and monitoring, to determine which pesticides are present in given samples and how much. FDA's MRMs for pesticides in foods are published in *PAM* I (*Pesticide Analytical Manual* I); the MRMs of state agencies, foreign governments, private industry, and academia are published in the open literature, or in special reports. MRMs may do one or both of the following:

   Screening—rapidly determine whether any pesticide is present, near or above tolerance or action level. This approach usually precedes a more detailed analysis. Cholinesterase enzyme inhibition tests screen for OPs and carbamate insecticides, and insect bioassays screen for any insecticide residue. Immunoassays are under development for screening targeted chemicals or classes of chemicals. Some of the new extraction methods can provide a basis for screening water and other liquids for a variety of pesticides (see Chapter 5).

Quantitation—detect and measure any and all pesticide residues that might be present in a given sample that the MRM is capable of including. Usually based on gas or liquid chromatography, some MRMs are more rapid than others (the term *screening* is often intended to denote rapidity in analysis), and some are more comprehensive than others.

The FDA (13) and other agencies [e.g., OTA (12)] often use simplified versions of MRMs in their surveillance program to determine whether violations exist in given samples, before proceeding to full quantitation with a more rigorous analytical method. Because all MRMs have limits in terms of numbers of chemicals accommodated, agencies use the more tedious SRMs only for targeted pesticides that are not included in the MRM, or in special circumstances such as when the public health is endangered by a single pesticide (as with illegal aldicarb residues in watermelons in California and Oregon in the late 1980s) or when a single pesticide comes under special review, and thus, special scrutiny for food residue content. SRMs are submitted to the EPA by pesticide registrants and published in *PAM* II or they are published in the open literature.

Factor in Selecting a Method

A number of considerations are taken into account when choosing an analytical method.

1. *SRM versus MRM.* SRMs are generally chosen when the sample is known or suspected to contain a residue of a chemical. MRMs are chosen when the residue history of the sample is unknown, and the question is "Are pesticides present and, if so, how much of each?" MRMs will provide information on a much broader range of pesticides than an SRM for a similar investment of time and energy.

2. *Breadth of Applicability.* The issue here is how many pesticides are included in the MRM. For primary FDA multiresidue methods, the number [as of May 1988 (13)] was 290 chemicals (pesticides and conversion products) out of 743 of interest. For pesticides with U.S. tolerances, the numbers were 163 out of 316. Typically *not* included are polar chemicals of high water solubility (e.g., paraquat, glyphosate), very volatile chemicals (fumigants such as methyl bromide, ethylene oxide, and phosphine), and compounds that are unstable to Florisil chromatography (some carbamates, some oxons). An aliquot of the sample (or its extract) must be split off and analyzed separately from the aliquot that goes to the MRM in order to expand the analytical scope to include such compounds in the report.

3. *Detection Limits.* All analytical methods have a limit below which the analyte could not be detected even if present. As noted previously, this limit is set by the noise level of extraneous, background material that is always present in the sample and the sensitivity of the instrumentation used to detect and quantify the analyte. To be of regulatory use, detection limits must be below established tolerances or action levels. To be of more general use, detection limits must be below the levels one might wish to measure for a given trial or sample type.

4. *Total Error.* The "total error" (accuracy plus precision) should not exceed 50–100%, and this must be assessed by the analyst running replicate spiked samples through the method. Greater total errors can only be accepted when no better method is available.

5. *Speed and Cost.* The time it takes to run a method (or the sample throughput per unit of time) is directly related to cost insofar as an analyst's time is involved. Thus, shorter methods are generally less expensive than longer, more tedious ones. Needless to say, the quickest, least expensive analytical method may not be "best" in terms of other criteria, so that choosing methods often represents a compromise of speed and quality.

6. *Instrumentation.* Most residue laboratories still rely on element-selective gas chromatography or selective HPLC determination. The more rigorous mass spectrometry-based methods—at least in the screening phase—may not be practical given the cost of mass spectrometry equipment. Thus, a factor in choosing a method is whether the equipment needed is available to the analyst. This is becoming less of a problem as analytical laboratories acquire more sophisticated instrumentation, and as technician time (rather than instrument capital costs) becomes a limiting budgetary component.

7. *Validation.* Methods that have been validated are generally more likely to provide reliable results than those that have not. The most rigorous level of validation is by a collaborative study of the method by several different laboratories. The Association of the Official Analytical Chemists (AOAC) conducts such validation studies and publishes the validated method in the *Official Methods of Analysis* of the AOAC (14). Because this is a time consuming process, most methods receive less rigorous validation involving perhaps one outside lab cross-check or even just an intralab cross-check. This is true for many SRMs in *PAM* II. Research methods published in peer-reviewed journals may not have been checked by anyone beside the developer of the method. If such a method is chosen for use, more time may be needed in modifying and validating it prior to its use for real samples. For methods from the open literature, a good check on utility is a determination of the number of times the method is cited by other literature sources.

Methods developed by registrants and submitted by them to EPA may receive only a paper-check by a chemist at EPA (or an EPA contractor), or increasingly in very recent times may be tried out using spiked samples. These methods can generally be considered quite reliable, because the registrant has used these methods extensively in obtaining data to fulfill EPA registration requirements.

Generalizing, one should choose a well-accepted method (AOAC, *PAM*, etc.) if one is available that can accommodate the analyte(s) of interest, at the levels and with the data quality of interest, and with the instrumentation available to run it. The development of a new method is justifiable only when it has been demonstrated that a suitable method fulfilling these criteria does not exist.

STEPS IN ANALYSIS

Pesticide residue methods may contain several discrete steps, as is true also for analytical methods for metals, drugs, and other agents of concern when present at relatively low levels. The steps are as follows:

Sample Preparation

- *Extraction*—to remove as much of the analyte from the matrix as practical, with a minimum extraction of extraneous materials that might interfere in the analysis.
- *Cleanup*—step(s) in which the analyte is concentrated and purified and the bulk of interfering coextractives are removed.
- *Derivitization*—conversion of the chemical of interest into a derivative, in order to enhance extractability, cleanup, or subsequent resolution and determination steps. This is an optional step, required for some chemicals and some methods, but not all.
- *Resolution*—step(s) in which the analyte is resolved from remaining coextractives, so that it may be subsequently measured without significant interference. This is usually done by some form of refined chromatography.

Sample Determination

- *Detection*—obtaining a response (usually an electronic signal) that is proportional to the amount of analyte present. *Selective detection* infers that the analyte will produce a signal several times higher (per unit of mass) than those originating from the background.

- *Determination*—calculating an amount of analyte present by reference to a standard, either external or internal.

Some authors would argue for sampling (15,16) and confirmation as integral parts of the analytical process, and a case can be made for GLP (good laboratory practice) requirements, such as freezer stability studies and recordkeeper as vital components as well. All of these aspects will be covered in this book, but in the following chapters, the five steps outlined above will provide the organizational framework.

CONCLUSIONS

The approach taken in analyzing for pesticide residues in foods, or for any chemical in any matrix at residue levels, involves an understanding of the asymptotic nature of decline or dissipation curves, and the relationship between residue decline and the occurrence of background interferences in the matrix of interest. These factors all influence the limit of detection—a fundamental concept in pesticide residue chemistry that typically works at or near this limit. There is still widespread misunderstanding of the concept of limit of detection, and this shows clearly in residue reports that, even today, may list "zero" as an entry, or may omit the methods limit of detection, or the steps taken to validate the method. This greatly limits the utility of what might otherwise be perfectly good data.

Because pesticide residue data are increasingly being used for risk assessment, or for making regulatory or economic decisions that can affect the availability of pest control chemicals to agriculture or the safety of the food supply, analytical chemists need to pay more attention to the quality and meaning of end data with less emphasis on simply running samples in order to work through the load that is present. The subjects of good laboratory practices and quality assurance/quality control are now much more familiar in the analytical laboratory than just 10 years ago, partly because of the need to impose a mentality that emphasizes quality and meaning in addition to speed and throughput.

REFERENCES

1. J. R. Coats, Pesticide degradation mechanisms and environmental activation, in *Pesticide Transformation Products: Fate and Significance in the Environment*, L.

Somasundaram, and J. R. Coats, eds., ACS Symposium Series 459, American Chemical Society, Washington, DC, 1991, pp. 10–31.

2. M. F. Wolfe, and J. N. Seiber, Environmental activation of pesticides, in *Occupational Medicine: State of the Art Reviews*, Hanley and Belfus, Philadelphia, Vol. 8 (July–Sept.), 1993, pp. 561–573.
3. B. W. Hermann, and J. N. Seiber, Sampling and determination of *S,S,S*-tributyl phosphorotrithioate, dibutyl disulfide and butyl mercaptan in field air, *Anal. Chem.* **53**, 1077–1082 (1981).
4. J. N. Seiber, General principles governing the fate of chemicals in the environment, in *Agricultural Chemicals of the Future* (J. L. Hilton, ed.), Rowman and Allenheld, Totawa, NJ, 1985, pp. 389–402.
5. G. A. Sutherland, Residue analytical limit of detectability, *Res. Rev.* **10**, 85–96 (1965).
6. J. A. Glaser, D. L. Foerst, G. D. McKee, S. A. Quave, and W. L. Budde, Trace analyses for wastewater. *Environ. Sci. Technol.* **14**, 1426 (1981).
7. L. H. Keith, W. Crummett, J. Deegan, Jr., R. A. Libby, J. K. Taylor, and G. Wentler, Principles of environmental analysis, *Anal. Chem.* **55**, 2210–2218 (1983).
8. G. Zweig, The vanishing zero: the evolution of pesticide analyses, in *Essays in Toxicology*, Academic Press, New York, 1970, Vol. 2.
9. NRC, *Pesticides in the Diets of Infants and Children*, National Research Council, National Academy of Press, Washington, DC, 1993, pp. 225, 274.
10. E. F. McFarren, R. J. Liskka and J. H. Parker, Criterion for judging acceptability of analytical methods. *Anal. Chem.* **42**, 358–365 (1970).
11. FDA, *Pesticide Analytical Manual*, 3rd ed., Food and Drug Administration, U.S. Department of Health and Human Services, Washington, DC, 1994.
12. OTA (Office of Technology Assessment), *Pesticide Residues in Food: Technologies for Detection*, U.S. Congress, Office of Technology Assessment, Washington, DC, 1988, p. 232.
13. B. McMahon, and J. Burke, Expanding and tracking the capabilities of pesticide residue methodology used in the Food and Drug Administration's pesticide monitoring programs, *Anal. Chem.* **20**, 1073 (1988).
14. AOAC, *Official Methods of Analysis*, 15th ed., Association of Official Analytical Chemists, Arlington, VA, 1990, Vols. 1, 2.
15. L. H. Keith, Principles of environmental sampling, *Environ. Sci. Technol.* **24**, 610–617 (1990).
16. L. H. Keith, *Environmental Sampling and Analysis: A Practical Guide*, Lewis Publishers, Boca Raton, FL, 1991.

CHAPTER

2

EXTRACTION, CLEANUP, AND FRACTIONATION METHODS

JAMES N. SEIBER

In the preparation of a sample for analysis, it is common practice to first extract the analyte away from the bulk of the matrix material and then to remove potentially interfering coextractives that will inevitably be present in the extract, by one or more cleanup steps. The strategy in choosing the proper extraction and cleanup conditions, and methods for subsequently fractionating samples into various subgroups of chemicals for separate determination, involve taking advantage of unique physical and chemical properties of the analyte that will allow it to stand out from the bulk of substances that occur in the matrix that could interfere in the determination step by responding to the detection system employed. The analyte physical properties of most utility are polarity, which governs the solubility and chromatographic behavior of the analyte; and volatility, which governs the vapor-condensed phase distribution of the analyte in such operations as codistillation, headspace transfer, and gas chromatography.

This chapter covers basic physicochemical properties that underlie analyte behavior during extraction, cleanup, and fractionation [taken in part from Biggar and Seiber (1)] and also discusses conventional experimental approaches to conducting these three operations. Newer and/or more advanced methods are covered in Chapters 5 and 6.

PHYSICOCHEMICAL PROPERTIES

Polarity and Water Solubility

A basic concept is that of *molecular polarity*. Its meaning is confounded by varying nuances in usage. A compound is referred to as "more polar" than

Pesticide Residues in Foods: Methods, Techniques, and Regulations, by W. G. Fong, H. A. Moye, J. N. Seiber, and J. P. Toth. Chemical Analysis Series, Vol. 151.
ISBN 0-471-57400-7 © 1999 John Wiley & Sons, Inc.

another, a designation often based on judgment rather than numerical data. In fact, no single measurement defines polarity as the term is commonly employed.

In essence, polarity is the unevenness of charge in a molecule. Water is polar because it is relatively negative in the region of the oxygen atom and relatively positive in the region of the hydrogen atoms in the nonlinear structure. The high dipole moment (μ) of 1.85 D (debyes) and the high dielectric constant (x) of 80 confirm this inherently polar character. Nitrobenzene is also a polar aromatic compound (dipole moment 4.21 D, dielectric constant 35.7). These values are reasonable on the basis of the polarizing effect of the nitro group.

Nitrobenzene, chlorobenzene ($\mu = 1.7$ D, $x = 5.7$), and toluene ($\mu = 0.37$ D, $x = 2.4$) are easily placed in a polarity series, on the basis of this part of the polarity concept. Their water solubilities also fall roughly in the order expected with respect to polarity and the ability of nitrobenzene to undergo dipole–dipole interactions with water. Some qualitative prediction is possible in structural series using the dipole moment contributions for various substituents (see Table 2.1). Symmetrical compounds (e.g., CCl_4) are less polar than nonsymmetrical compounds (e.g., $CHCl_3$), and symmetrically substituted compounds (e.g., *p*-dichlorobenzene) are less polar than their nonsymmetrically substituted isomers (e.g., *o*-dichlorobenzene). Some symmetrical compounds, such as benzene, can be polarized when placed in a polar milieu; this partly explains the water solubility of benzene

Table 2.1. Dipole Moments (in debyes) of Some Functional Groups

Functional Group	Dipole Moment (D)
Amine	0.8–1.4
Ether	1.2
Sulfide	1.4
Thiol	1.4
Carboxylic acid	1.7
Hydroxy	1.7
Halogen (F, Cl, Br, I)	1.6–1.8
Carboxylic ester	1.8
Aldehyde	2.5
Ketone	2.7
Nitro	3.2
Nitrile	3.5
Sulfoxide	3.5

(0.08%), which is higher than might be expected per other structural considerations.

The high water solubility of phenol (6.7%) when compared with nitrobenzene (0.19%) is difficult to explain, given that the dipole moment (1.45 D) and dielectric constant (9.8) of phenol are lower than corresponding values for nitrobenzene. The answer lies in the strong hydrogen-bonding ability of phenol and, specifically, the ability of phenol to act as a proton donor in water. Thus, the total interaction of solute with solvent (or with solid surface in a heterogeneous environment) involves several forces: polar interactions (dipole–dipole, dipole–induced dipole), hydrogen bonding, and dispersion interactions between adjacent molecules. These last, referred to as van der Waals forces, explain the ability of even nonpolar substances to associate in condensed phases. The preference of nonpolar DDT, for example, for a hydrocarbon oil over water involves both positive van der Waals attraction between DDT and the oil, and the fact that DDT is more energetically disposed to reside in oil than in water, where it would disrupt the large association energy between adjacent water molecules.

These considerations allow solvents to be ranked by polarity trending toward increased dielectric constants and dipole moments when moving from nonpolar hydrocarbon solvents to polar solvents such as methanol and water [Table 2.2 (2)]. A truly quantitative ranking based only on dielectric constant and dipole moment is not possible because hydrogen bonding may not be accounted for using these two parameters. Because of this, solvent polarity series are often constructed empirically on the basis of such factors as the solvent strength parameter observed in the ability of solvents to elute a series of test solutes from aluminum oxide absorbent (Table 2.2).

Molecular size must also be considered when estimating water solubility of solutes; the rule is that larger molecules are less soluble than smaller molecules when polarity is approximately constant. This explains the very low water solubility of high molecular weight aliphatic hydrocarbons, polynuclear aromatic hydrocarbons, chlorinated hydrocarbons, and polymers. Polarity and molecular size concepts provide a rationale for the solubilities of pesticides and other organic toxicants. For example, among organophosphates, paraoxon has a water solubility of 3640 ppm, compared with only 12.4 ppm for parathion (3), reflecting the much greater polarizing effect of P=O over P=S. Similarly, phorate sulfoxide (more than 8000 ppm) is much more soluble than phorate (17.9 ppm) because of the polarizing S=O group. The water solubility of methyl parathion (37.7 ppm) is greater than that of the ethyl analog (12.4 ppm), reflecting the addition of two hydrophobic methylene groups in the latter. A more complete compilation of water solubilities of pesticides may be found in Shiu et al. (4, 5) and various handbooks.

Table 2.2. Properties of Some Common Solvents

Solvent	Chemical formula	Water solubility in Solvent	Dielectric Constant (x)	Dipole Moment (u)	Solvent Strength Parameter (E) (Al_2O_3)
Pentane	C_5H_{12}	0.01	1.84	—[a]	0.00
Hexane	C_6H_{14}	0.01	1.88	—	0.00
Cyclohexane	C_6H_{12}	0.012	2.02	—	0.04
Carbon disulfide	CS_2	0.005	2.64	0	0.15
Carbon tetrachloride	CCl_4	0.008	2.24	0	0.18
Benzene	C_6H_6	0.058	2.30	0	0.32
Diethyl ether	$(C_2H_5)_2O$	1.3	4.3	1.15	0.38
Chloroform	$CHCl_3$	0.07	4.8	1.01	0.40
Methylene chloride	CH_2Cl_2	0.17	8.9	1.60	0.42
Tetrahydrofuran	C_4H_8O	Miscible	7.6	1.63	0.45
Acetone	CH_3COCH_3	Miscible	—	2.88	0.56
Dioxane	$C_4H_8O_2$	Miscible	2.2	0	0.56
Ethylene acetate	$CH_3COOC_2H_5$	9.8	6.0	1.78	0.58
1-Pentanol	$C_5H_{11}OH$	10.0	5.0	—	0.61
Acetonitrile	CH_3CN	Miscible	—	3.92	0.65
1-Propanol	C_3H_8OH	Miscible	20.3	1.68	0.82
Dimethylsulfoxide	$(CH_3)_2SO$	Miscible	4.7	3.96	0.75
Dimethylformamide	$HCON(CH_3)_2$	Miscible	36.7	3.82	—
Methanol	CH_3OH	Miscible	32.7	1.70	0.95
Water	H_2O	—	80.0	1.85	Large

Source: Adapted from Snyder and Kirkland (2).
[a]No data available for blank cells.

The extrinsic factors influencing water solubility have been summarized by Lyman et al. (6). They include the following:

Method of Measurement. The measurement of water solubility is straightforward for most compounds; it is the amount dissolved in water when an excess of the chemical is allowed to reach equilibrium with water at constant temperature. Centrifugation or filtration removes suspended material prior to measurement. Experimental variations on this basic method can produce rather large discrepancies (7). While precision appears to be lowest with hydrophobic compounds of very low solubility, a recent remeasurement revealed discrepancies with published values of up to a factor of 2 for several pesticides of moderate water solubility, and a factor of 100 for two pesticides (ronnel and bromophos) (3). A saturation column method generates precise and accurate solubility data much faster than the conventional method (8). This may stimulate an effort at remeasuring all pesticides under identical conditions.

Temperature. Water solubility is influenced by temperature, tending toward an increase in solubility with an increase in temperature. As a rule, solubility increases by a factor of 2 for a 10°C rise in temperature. There are, however, exceptions to this. Gases tend to be less soluble as temperature increases. Thiol carbamates show a reverse solubility–temperature relationship owing to the predominance of the less polar tautomer at higher temperature (9). Unfortunately, most published solubilities are given at only one temperature, so there is no basis for judging whether a regular or inverse relationship exists between water solubility and temperature for a given chemical. Furthermore, the temperature may not always be specified in the literature citation of solubility.

Salinity. Dissolved salts tend to lower the solubility of an organic solute, and this can be used to "salt out" organics to improve their recovery from aqueous solutions. For this reason, water of low, or at least known, ionic strength should be used when collecting solubility data and so specified in the report.

Dissolved Organic Matter. The presence of dissolved organic matter, such as the naturally occurring humic and fulvic acids (10), cosolvents such as acetone, or surface-active agents (surfactants), tend to increase the water solubility of organic compounds. Because many pesticides contain organic impurities or formulation additives, solubility may also be affected by these materials. For example, Bowman and Sans (3) determined a solubility of 56 ppm for solid technical methyl parathion, which contained xylene as the major impurity, and 37.7 ppm for recrystallized methyl parathion. The solubility of carbofuran, however, increased with increasing purity, suggesting that in this case the impurities may have been inorganic.

pH. The pH of water may dramatically affect the solubility of ionizable acidic or basic pesticides and may measurably affect the solubility even of "neutral" chemicals. Unfortunately, pH is rarely specified in literature reports of water solubility.

Physical State. The water solubility of the super-cooled liquid (S_{liquid}) exceeds that of the solid (S_{solid}) for a given chemical above its melting temperature (t_m). The following formula is one of several that approximate the relationship of the two solubilities (4, 5, 11):

$$\log S_{\text{solid}} = \log S_{\text{liquid}} - 0.0095(t_m - 25)$$

This may be an important correction in environmental fate calculations, because the supercooled liquid, rather than the solid (on which the solubility determinations are usually based), may be the state most relevant to environmental processes. The correction is greater for compounds of higher melting points.

Clearly, reported water solubilities, particularly those measured at single temperatures with no indications of replication or of purity, either of solute or water solvent, must be assigned a fairly large uncertainty factor (at least plus or minus 100%) when used to calculate distribution coefficients or other physical parameters. When water solubility is not reported in the literature, it may be estimated from the octanol–water partition coefficient (K_{ow}) or from structural parameters (6); in either case an even larger uncertainty factor is in order.

Octanol–Water Partition Coefficient

The partition coefficient is a useful distribution constant in pesticide chemistry, underlying calculations of bioconcentration, structure–activity relationships, and the choice of solvent conditions for extraction. Partition coefficient compilations for a variety of organic compounds are available in the literature (12). A solute will distribute itself between the two phases in accordance with $K = p/q$, where K is the partition coefficient, p is the fraction in the nonpolar phase, and q is the fraction in the polar phase. By knowing any one term, one can fix the other two, because $p + q = 1$. The solvent pair most frequently used comprises *n*-octanol and water, and the appropriate designation is K_{ow}. The laboratory measurement of K_{ow} is fairly straightforward, although error can creep in because of such things as a failure to use equilibrated solvents, inconstant temperatures, inaccurate measuring techniques, and other experimental variables, contributing to a fairly large uncertainty in literature values. A not atypical case is that of methyl parathion, for which the literature provides at least four values of log K_{ow} (6):

log K_{ow}	K_{ow}
2.04	109.6
2.99	977.2
1.91	81.3
3.22	1,659.6

Literature values of K_{ow} must be given a fairly broad latitude (plus or minus 1 order of magnitude for compounds of relatively high K_{ow}) unless they have been confirmed by more than one laboratory or by calculation.

The relatively new area of property estimation has perhaps been best developed for K_{ow}. Several physical properties can be calculated by knowing only molecular structure, obviating the need for a sample of the substance or any prior laboratory work with it. For K_{ow}, the calculation based on structure uses the fragment constant approach [Hansch and Leo (13) or early versions such as Leo et al. (12)]. Briefly, the method employs empirically derived atomic or group fragment constants (f) and structural factors (F):

$$\log K_{ow} = \text{sum of molecular fragments } (f) \text{ and bond factors } (F)$$

Several authors have compiled values for f and F (12, 13). The calculation for phenylacetic acid is as follows.

$C_6H_5CH_2COOH$		
$+f_{C_6H_5}$	=	1.90
$+f_C$	=	0.20
$+2f_H$	=	0.46
$+F^{IR}_{COOH}$	=	−1.03
$+(2-1)\ F_B$	=	−0.12
$\log K_{ow}$	=	1.41
(Observed	=	1.41, 1.51)

The calculation becomes more tedious (and the result more uncertain) for more complex structures.

Polarity and solubility considerations play an important role in the choice of extraction and cleanup conditions in the analysis of pesticides. Along with bioconcentration factors, they are also useful guides in designing sampling strategies. Apolar molecules, such as DDT, of low water solubility and high values for K_{ow} and BCF are much more likely to be associated with sediment and with biota than dissolved in water in the aquatic environment; sampling strategies for such compounds must focus on sediment and biota in such situations. Some examples of log K_{ow} values for organophosphates are listed in Table 2.3.

Volatility and Vapor Pressure

Where polarity-based properties govern the distribution, or partitioning, of solutes between condensed phases, volatility represents a parallel set of

Table 2.3. Some Physical Properties for Representative Organophosphorus Pesticides (GC Retentions and Vapor Pressure Show Approximate Inverse Relationship)

Chemical	Formula/ Molecular Weight	RTT[a]	$\log K_{ow}^{b}$	Vapor Pressure[c]	H_2O Solubility[d]	Henry's Constant[e]
Dichlorvos	$C_4H_9Cl_2O_4P$ 220.98	0.1	1.5	5.27×10^{-2} torr (25.0°C)	10,000 mg/L (20–25°C)	1.54×10^{-6}
Phorate	$C_7H_{17}O_2PS_3$ 260.40	0.37	3.8	5.54×10^{-4} (25.0°C)	85 (25°C)	2.15×10^{-6}
Diazinon	$C_{12}H_{21}N_2O_3PS$ 304.36	0.51	3.3	4.8×10^{-5} (25.3°C)	40 (20–25°C)	4.84×10^{-7}
Methyl Parathion	$C_8H_{10}NO_5PS$ 263.23	0.71	3.0	1.5×10^{-5} (25.4°C)	55 (25°C)	9.52×10^{-8}
Malathion	$C_{10}H_{19}O_6PS_2$ 330.36	0.91	2.8	7.95×10^{-6} (25.0°C)	75 (25°C)	4.34×10^{-8}
Ethyl Parathion	$C_{10}H_{14}NO_5PS$ 291.27	0.98	3.8	9.8×10^{-6} (25.3°C)	24 (25°C)	1.58×10^{-7}
Chlorpyrifos	$C_9H_{11}Cl_3NO_3PS$ 350.57	1.00	5.0	1.9×10^{-5} (25.0°C)	2.0 (25°C)	4.38×10^{-6}
Methidathion	$C_6H_{11}N_2O_4PS_3$ 302.31	5.20	2.7	1.4×10^{-6}	250 (20°C)	2.18×10^{-9}
Azinphosmethyl	$C_{10}H_{12}N_3O_3PS_2$ 317.34	—	2.4	2.2×10^{-7} (25.0°C)	29 (25°C)	3.18×10^{-9}

[a]Relative retention time on OV-101 (14).
[b]From compilation of Suntio et al. (15) in their Table 4.
[c]From gas saturation method, by Kim et al. (16) except for chlorpyrifos and azinphosmethyl, which are taken from Melnikov (17).
[d]From compilation of Shiu et al. (4, 5). Consensus values were selected when several values were listed.
[e]Calculated, from previous columns, in $(atm \cdot m^3)/mol$.

properties that govern distributions when one phase is a gas. It underpins such common analytical techniques as gas–liquid chromatography, head-space analysis, and purge-and-trap methods.

The physical properties that indicate the inherent volatility of a compound are the boiling point and vapor pressure. Because not all compounds are stable to temperatures required for boiling, even under reduced pressure, and because one normally needs to know the volatility at temperatures other than the boiling temperature, the vapor pressure tends to be the more useful property for analytical work and for environmental fate predictions.

An extensive compilation of pesticide Henry's law constants and vapor pressures is available (15), along with a review of comon methods for their measurement (18).

Because the vapor pressures of most pesticides are fairly low, manometric methods of measurement cannot be used, although the manometric unit, millimeters of mercury, mmHg (now torr), is still used to describe the property. The displacement distance for mercury in a manometer connected to a closed flask in which an equilibrium has been established between a chemical's condensed (solid or liquid) and vapor phase, measured in millimeters, is the origin of this unit. Thus, benzene vapors at 25°C exert a pressure that displaces mercury by 92 mm; water at 25°C, 24 mm; and so forth. Further, water at its boiling point (100°C) produces a vapor pressure of 760 mm: 1 atm. The important units are thus

mmHg (torr)
Atmospheres (760 torr = 1 atm)
Pascal (1 Pa = 133.2 × torr)

Temperature has a strong effect on vapor pressure and, in fact, the log P positively correlates linearly with T or, in the more common description, log P has an inverse linear correlation with $1/T$. For relatively volatile, common chemicals the correlations are tabulated in various handbooks so that one may select a vapor pressure for any temperature of interest. Unfortunately, the temperature dependence of vapor pressure for many pesticides is not known, and can be only roughly estimated by analogy with chemicals for which the dependence is known.

A common method for accurate measurement of vapor pressure is the gas saturation technique. An excess of the analyte coated on sand, glass beads, or some other inert surface is subjected to a constant, known flow of air or nitrogen in such a way that the gas becomes saturated with analyte. The saturated vapor phase is subsequently sampled and analyzed to determine the analyte's concentration, that is, the analyte's saturated vapor density at the temperature at which the experiment was run. The vapor density, that is, the number of moles of solute per unit volume of sweeping air or nitrogen, provides n/v in the gas-law equation:

$$PV = nRT \quad \text{or} \quad P = \frac{n}{v} \cdot RT$$

Because R and T, along with n/v, are then known, the saturated vapor pressure (P) may be calculated.

A generally quicker but somewhat less standard method is based on gas chromatographic retention data (16). The analyte of interest is run, along with a standard of similar polarity and known vapor presure, at several different temperatures on a GC phase of low polarity. The ratio of retention times for analyte and standard as a function of temperature, along with the known vapor pressures of the standard at the measurement temperatures and at the temperature at which the vapor pressure of the analyte is desired, provides the basic data for the calculation. The correlation between GC-estimated and gas-saturation-measured vapor pressures is quite good (16). This opens up the vast reservoir of GC retention data for pesticides (see, e.g., *Pesticide Analytical Manual* and Appendix in this volume) for at least a rough estimation of vapor pressure because, as a first approximation, vapor pressures decrease with increasing retention time on low polarity (i.e., "boiling point") GC phases such as the methyl silicones. For some common organophosphorus pesticides the decrease in vapor pressure through an increasing GC retention series is illustrated in Table 2.3.

In addition to its utility in correlating with GC retention data, vapor pressure determines the potential for loss of chemicals during evaporation steps in trace analysis. In the series of nine organophosphates (Table 2.3), losses during evaporation are more likely to occur for dichlorvos, phorate, and diazinon then for the other pesticides, particularly when the solvent is taken to complete dryness. By the same token, extraction methods that rely on forced volatilization, such as sweep co-distillation, will be more successful for dichlorvos, phorate, and diazinon than, say, methidathion.

Henry's Law Constant

Vapor pressure influences the tendency of a compound to volatilize from an aqueous solution, but this is only part of the picture. Water solubility (or, more generally, polarity) also exerts an influence. In fact, the tendency of a solute to volatilize from water is a function of both the vapor pressure and water solubility:

$$\text{Volatilization tendency} = f\left(\frac{P}{S}\right)$$

Henry's law describes the equilibrium concentrations for a solute distributed between water and air in an enclosed system of equal phase volumes:

$$H' = \frac{\text{concentration in air}}{\text{concentration in water}}$$

H′ is the dimensionless Henry's constant (provided identical concentration units are used for both numerator and denominator). Commonly, however, Henry's law constant is expressed in units derived from the equation $H = P/S$. If P is in atmospheres, and S in mol/m^3, the units for H are $(\text{atm} \cdot \text{m}^3)/\text{mole}$ and $H' = H/RT$ where R is 0.082 $(\text{m}^3 \cdot \text{atm})/(\text{deg} \cdot \text{mol})$.

The Henry constant (either H or H′) may thus be calculated if both the water solubility and vapor pressure are known at the temperature of interest. The H constant may also be measured experimentally by any of several techniques (15), most commonly by measuring the rate of loss of a solute from an aqueous solution subjected to sparging with an air or nitrogen stream (19). A less common, but simpler method is by direct headspace analysis of the vapor above an equilibrated vapor-solution chamber (20).

Both vapor pressure and water solubility are important in determining the value of the Henry law constant. For example, chlorpyrifos has a lower vapor pressure than diazinon (Table 2.3), and it has a higher Henry law constant because of its much lower water solubility (2 vs. 40 mg/L). If both chemicals were present in water in an open beaker (or pond), chlorpyrifos would have approximately a 300% greater rate of volatilization than diazinon. Alternatively, when rotary-evaporating a water solution of the two, chlorpyrifos loss from water could be greater than for diazinon. For some chemicals of moderate vapor pressures and very low water solubilities, loss by evaporation from water during transfer operations in the laboratory can be very rapid. Cases in point are α- and β-BHC, and hexachlorobenzene, which are difficult to contain in water unless sealed and chilled.

EXTRACTION

It is somewhat arbitrary to separate "extraction" from "cleanup" because many techniques do part or all of both virtually simultaneously. Supercritical fluid extraction (SFE), for example, can extract the analyte; it can crudely separate analytes into fractions by pressure, modifier, or flow programming, and some SFE instruments allow for selective trapping and elution of the SFE extractives using a solid-phase mini-column at the exit end of the SFE. But traditionally extraction has been done primarily with organic solvents that, once collected and concentrated, yield an extract which is cleaned up separately by one or more partition, chromatography, codistillation, or chemical treatment techniques. We will follow the traditional approach in discussing these two topics, but with frequent cross-over in the discussion.

Physical properties really come into play in choosing the strategy for extraction. For extractions using solvents or supercritical fluids, conditions are chosen so that the polarity of the extractant matches the polarity of the analyte ("like dissolves like") so that the analyte enters solution but the bulk of the matrix remains undissolved. For solventless extractions that volatilize the analyte, conditions should be sufficient to promote efficient removal of analyte but not so harsh that the matrix decomposes and/or volatilizes as well. Strong solvents, acids or bases, high temperatures, and other extremes should be avoided, because they may add significantly to the difficulty in the remainder of the procedure. Also, many pesticides are unstable in acid or base solution. There are exceptions, when the entire matrix must be dissolved in order to get at the analyte, such as in determining mercury in fish or ores when acid digestion is used.

Water

Extracting water is a fairly simple operation because water samples are relatively uniform in contrast with biological tissue or soil. Many methods have been worked out for water in considerable detail. Nonpolar analytes, such as DDT, PCBs, and HCB, whose K_{ow} values are in the range of 10^4–10^6, are readily extractable using a nonpolar solvent such as hexane, isooctane, benzene, or ethyl acetate. The phase ratio can be adjusted in the analyst's favor by using, for example, 10 mL of hexane to extract 100 mL of water, resulting in a 10 : 1 concentration without solvent evaporation. A volumetric flask and magnetic stirring setup (Fig. 2.1*a*) is frequently all that is needed, with the hexane recovered by pipetting from the top phase into a suitable receiver, such as a calibrated hematocrit, which can be subsequently concentrated further using a nitrogen blowdown, distillation through a Snyder column, or rotary evaporation (Fig. 3.1*b*–*d*). A separatory funnel can replace the volumetric flask for initial extraction. In the example at hand, if 100 mL of hexane were used as extractant, and the hexane reduced to 1 mL by evaporation (caution must be exercised here, so that analyte is not lost by volatilization during the evaporation step), a 100 : 1 concentration would

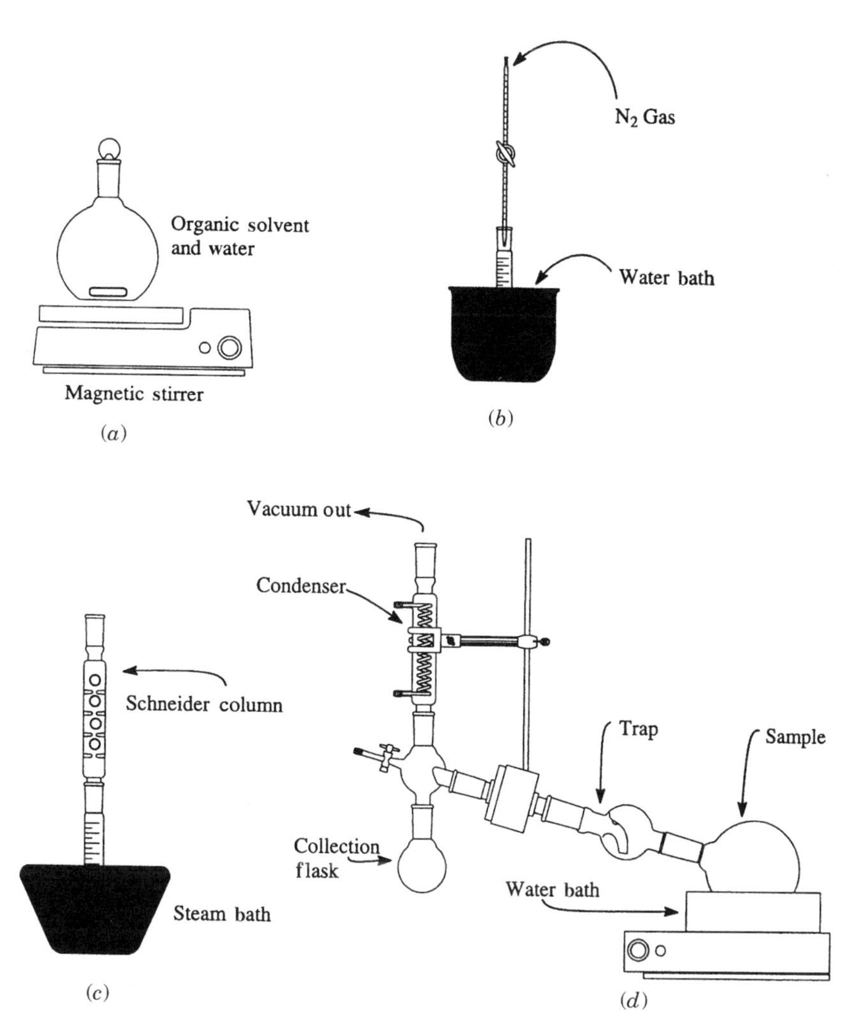

Figure 2.1. Common techniques for extracting (*a*) and concentrating extracts (*b–d*) from aqueous media such as water.

have resulted. If the analyte level were 1 μg/L (1 ppb) in water, it would be 0.1 μg/mL (or 100 ng/mL) in the final extractant. Injection of 1 μL on the GC column would deliver 100 pg to the column—well within the sensitivity range of electron capture, NP-TSD, GC-MS, and several other selective detectors. Details for analysis of organochlorine compounds in water are given in the first AOAC supplement (21). One liter of water was extracted with 300 mL methylene chloride. Solvent was concentrated using a K-D concentrator, exchanged to methyl *t*-butyl ether, concentrated further, then analyzed by EC-GC on a DB-5 (primary) and DB1701 (secondary) column. MDLs ranged from 0.0015 for α- and γ- benzene hexachloride to 5 μg/L for chlorobenzilate among 29 OCs analyzed.

For analytes of K_{ow} less than $\sim 10^3 - 10^4$, it may be necessary to adopt correspondingly more stringent extraction conditions:

1. Increase the volume of hexane, or use multiple extractions with hexane.
2. Choose a more polar solvent than hexane, from within the polarity series hexane < benzene < methylene chloride < ethyl acetate, and so on (see Table 2.2).
3. Add salt (prepurified sodium chloride or ammonium acetate) to the water, to "salt out" (i.e., reduce the solubility of) the analyte.
4. Adjust pH: If the analyte is acidic, adjustment to pH ~ 3 with acid will protonate the analyte and reduce its water solubility; if the analyte is an amine, adjustment to pH ~ 10 with sodium carbonate or ammonia will deprotonate and reduce its water solubility.

A continuous extractor can be used for recovering analytes of unfavorable partition characteristics, or when large volumes of water must be handled. When dealing with biological fluids (urine, plasma, milk, etc.), some analytes may be present as a conjugate and require treatment with an enzyme or strong acid to break the conjugate prior to extraction. Morphine, for example, exists in urine as the glucuronide (85%) and as free morphine (15%). Treatment with acid breaks the conjugate, liberating the less soluble free morphine, which can then be recovered efficiently. Pesticide-derived phenols and some amines behave similarly.

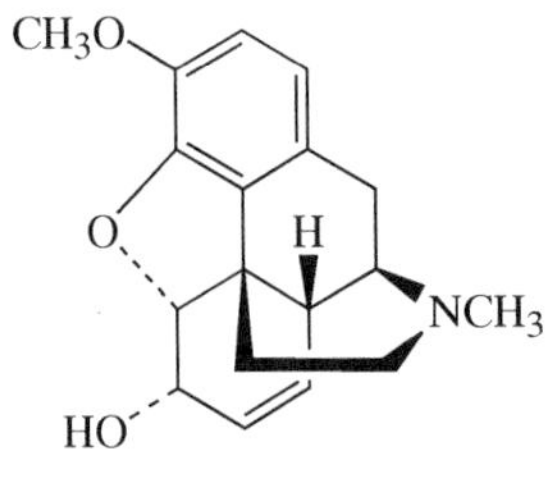

Morphine

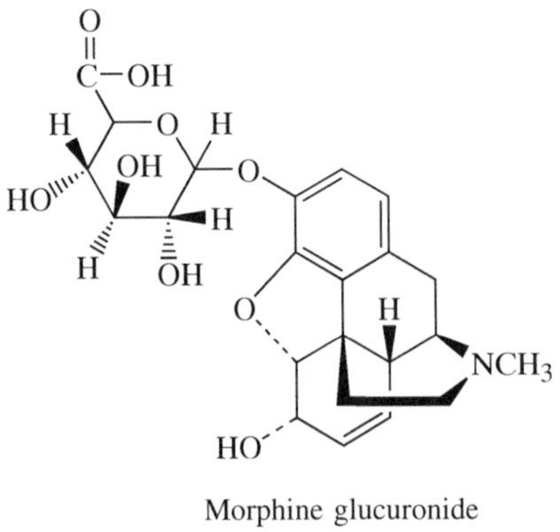

Morphine glucuronide

While extraction of water is fairly straightforward, milk, blood, fruit juices, urine, and other biological fluids can pose special problems. Direct solvent extraction of milk, for example, may give low analyte recovery even when the partition coefficient is favorable, because the analyte may be stabilized in a fatty micelle or colloid that is surrounded by a hydrophilic phospholipid membrane. The solution is to disrupt the membrane, by adding ethanol and salt to the milk, prior to extraction with the organic solvent. Data for aldrin and ethyl parathion show this, and also point out that the problem is more acute for nonpolar organochlorines rather than more polar organophosphates:

	Direct Extraction (%)	**Extraction after Adding EtOH (%)**
Aldrin	8.4	91.5
Parathion	63.0	90.0

For some fluids, such as blood, simply diluting the fluid (whole blood or plasma) with water ruptures cells and allows recovery of analyte. For juices, the principal extraction problem is one of emulsions, for which salt addition and/or contrifugation may be needed for phase separation.

Solid-phase extraction (SPE) and solid-phase microextraction (SPME) devices offer excellent alternatives to organic solvent extraction for water, juices, biological fluids, and other largely aquatic media, and even some nonaqueous liquids. A summary of extraction techniques that do not involve organic solvents, or use minimal organic solvents, is in Table 2.4. Details are given in Boyd-Boland et al. (22)

When very large volumes of water are to be sampled, or when conducting air sampling, porous macroreticular organic polymeric resins, principally XAD-4 (crosslinked polystyrene) and polyurethane are frequently used. These materials combine high adsorptive capacity (particularly for low- to medium-polarity organics) with high flow-through characteristics. The structural grid of XAD-4 (and the related XAD-2) suggests that they might sorb particularly well any material extractable with benzene or toluene and, indeed, both have been used extensively to sample pesticides from water (23, 24).

Polyurethane foam—the same basic material as used in upholstery cushions—finds similar applications to water (25) and air (26). Both materials have the advantage over charcoal of releasing sorbed analytes more efficiently where charcoal tends to irreversibly bind many analytes, particularly the more polar ones.

Table 2.4. Modern Techniques for Extracting Water and Other Solutions that Do Not Involve, or Minimize, the Use of Organic Solvents

Technique	Abbreviation	Definition
Solid-phase extraction	SPE	Solution containing analyte is filtered through a column or filter pad containing sorbent
Solid-phase microextraction	SPME	A droplet of extractant or, more commonly, a fiber coated with extractant, is suspended in solution to be extracted, then transferred to an analytical device such as GC
Accelerated solvent extraction	ASE	Extracting solvent, under super-ambient temperature and/or pressure, is flushed through the matrix, removing analyte in a smaller volume of solvent than normally employed; solvent may be water
Microwave-assisted solvent extraction	MASE	Extracting solvent, plus microwave energy, are combined to remove analyte from matrix; solvent may be water
Supercritical fluid extraction	SFE (or SCFE)	A fluid such as carbon dioxide is pumped under supercritical temperature–pressure conditions through the contaminated matrix
Passive sampling device	PSD	General term for any sampling device that can be left in the environmental medium (water, air, soil) of interest to sorb analyte, then extracted separately by a variety of techniques
Semipermeable membrane device	SPMD	A type of PSD in which triolein or some other lipophilic liquid deployed on the inside walls of a polymer membrane tube is suspended in the environmental medium (water, air) of interest

$$\left[\begin{array}{c} -CH_2-CH(C_6H_5)-CH_2-CH(C_6H_4)-CH_2-CH(C_6H_5)- \\ | \\ -CH_2-CH(C_6H_5)-CH_2-CH-CH_2-CH(C_6H_5)- \end{array}\right]_n$$

XAD-4 polymer

Organic solvent, solid-phase, and macroreticular resin extractions are based on the affinity of the medium for the analyte by, primarily, polarity-based partitioning. For the more volatile analytes, particularly those which have relatively high vapor pressures and/or Henry law constants, it may be more convenient and efficient to "extract" analytes by volatilizing them away from the matix bulk. Clearly, less volatile potential interferences would be left behind and also no organic solvent would necessarily be involved, thus preserving many of the advantages noted previously for SPEs. Purge-and-trap and headspace extraction systems make use of volatilization for analyte recovery.

Purge-and-trap systems were developed primarily for analyzing volatile organic compounds (VOCs) in polluted water. The VOCs of interest are those that appear on the EPA Priority Pollutant List (27) as purgeable organics: 31 solvents (benzene, toluene, carbon tetrachloride, trichloroethylene, etc.), monomers (vinyl chloride, bischloromethyl ether, acrylonitrile, etc.), chlorofluorocarbons (trichlorofluoromethane, dichlorodifluoromethane), and miscellaneous compounds (chlorobenzene, ethylbenzene, acrolein, etc.), including at least one pesticide (methyl bromide). Purge-and-trap analysis usually entails use of a commercial instrument that is attached to a flame ionization gas chromatograph, automated in its handling of a carousel of water samples. There are four major steps in P–T analysis (Fig. 2.2):

1. *Purge*—nitrogen is bubbled through the aqueous sample, volatilizing in the process the "purgeables"— that is, those compounds with Henry's law constant of $\sim 1 \times 10^{-3}(\text{atm} \cdot \text{m}^3)/\text{mol}$ or greater.
2. *Trap*—purged VOCs are trapped from the exit nitrogen-purging gas by sorption on a column of Tenax.
3. *Desorb*—Tenax-sorbed VOCs are thermally desorbed with carrier gas from the flash-heated Tenax, and flushed directly into a GC, or

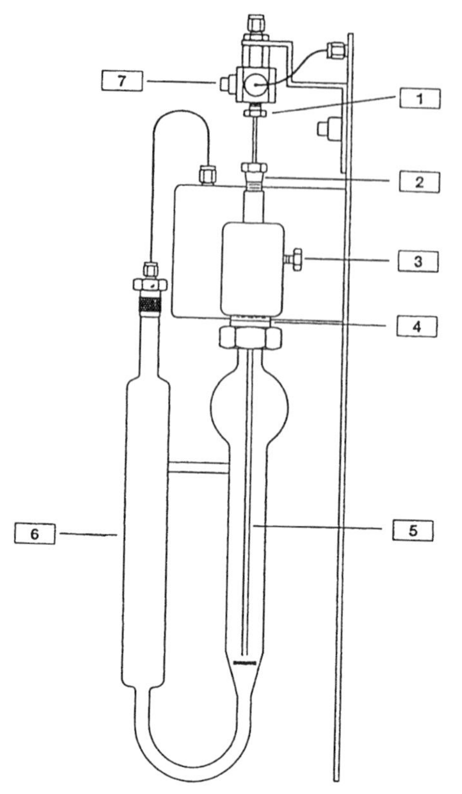

Figure 2.2. Purge-and-trap apparatus for analysis of trace volatile contaminants in water: (5) denotes sample needle; (6) denotes glass sparger. Other numbered components are ferrules and connector. (Perkin-Elmer Corporation.)

indirectly after first condensing the liberated VOCs in a cryogenic trap or on the cooled front end of the GC column.

4. *Gas chromatography*—discrete VOCs are resolved on a GC column, which in the past was a packed column but more recently has involved a capillary column.

Virtually all of the $P-T$ steps, and the criteria by which data quality are assessed, are mandated in EPA methods for VOCs (28). $P-T$ may also be used for analyzing pesticide and pesticide metabolites that are sufficiently volatile to be purged from water; methyl bromide, methyl isothiocyanate (MITC), ethylene oxide, and propylene oxide are examples. Beverages, juices, and water suspensions may also be analyzed by *P–T*, for volatile contaminants.

Headspace sampling is a related technique, but it may accommodate aqueous, other liquid, and solid samples. Early applications were to volatile flavor and aroma components in fruits and vegetables but, increasingly, applications are to contaminants and pollutants in a variety of media (29). Most modern headspace sampling is done with an automated attachment for use with GC—a technique termed *headspace gas chromatography* (HSGC).

Figure 2.3 shows the primary operations in HSGC, which entail (1) equilibrating the sample in a septum-sealed vial in which headspace–condensed-phase equilibration occurs with heating, (2) pressurizing the vial by forcing carrier gas through a syringe inserted through the septum into the headspace, and (3) sampling of the pressurized headspace through the same syringe, but with the released vapor injected on to the GC column.

Virtually any analyte capable of measurable existence in the equilibrated headspace above a condensed-phase matrix (water, biological fluid or tissue, soil, etc.) may be analyzed by HSGC, with examples provided by ethanol in blood, methyl bromide in charcoal-accumulated air samples (23), various pesticides in contaminated soil and water (30), and ethylene oxide in spices (31). HSGC may be regarded as still another solvent-less extraction method, for which more applications will undoubtedly be forthcoming in the future. In the ethylene oxide procedures, small samples of spices such as black pepper, oregano, and ginger are placed, along with 1-octanol as a desorbing solvent, in sealed headspace vials. After 20 min of equilibration at 60°, an aliquot of headspace was analyzed by HSGC on a porous layer open-tube (PLOT) column using a flame-ionization detector (FID). Representative chromatograms are presented in Figure 2.4.

Crop and Animal Tissue

Literally dozens of single solvent and solvent mixtures have been employed for extracting pesticides from crop tissue, extracts, derived foodstuffs, and other substances; the more common ones may be found in standard references [14, 32]. One common technique involves blending the matrix with solvent in a Waring blender or a Polytron or Virtis homogenizer, and then filtering the mixture to recover the solvent for subsequent cleanup. Alternatively, a heated Soxhlet extractor (Fig. 2.5) may be used for more exhaustive extractions, or when the analyte is incorporated in or bound to the matrix in such a way as to be poorly recovered by blending at or near room temperature.

The use of water-miscible solvents, such as acetonitrile, acetone, and methanol, provides recovery of a broad range of pesticides without

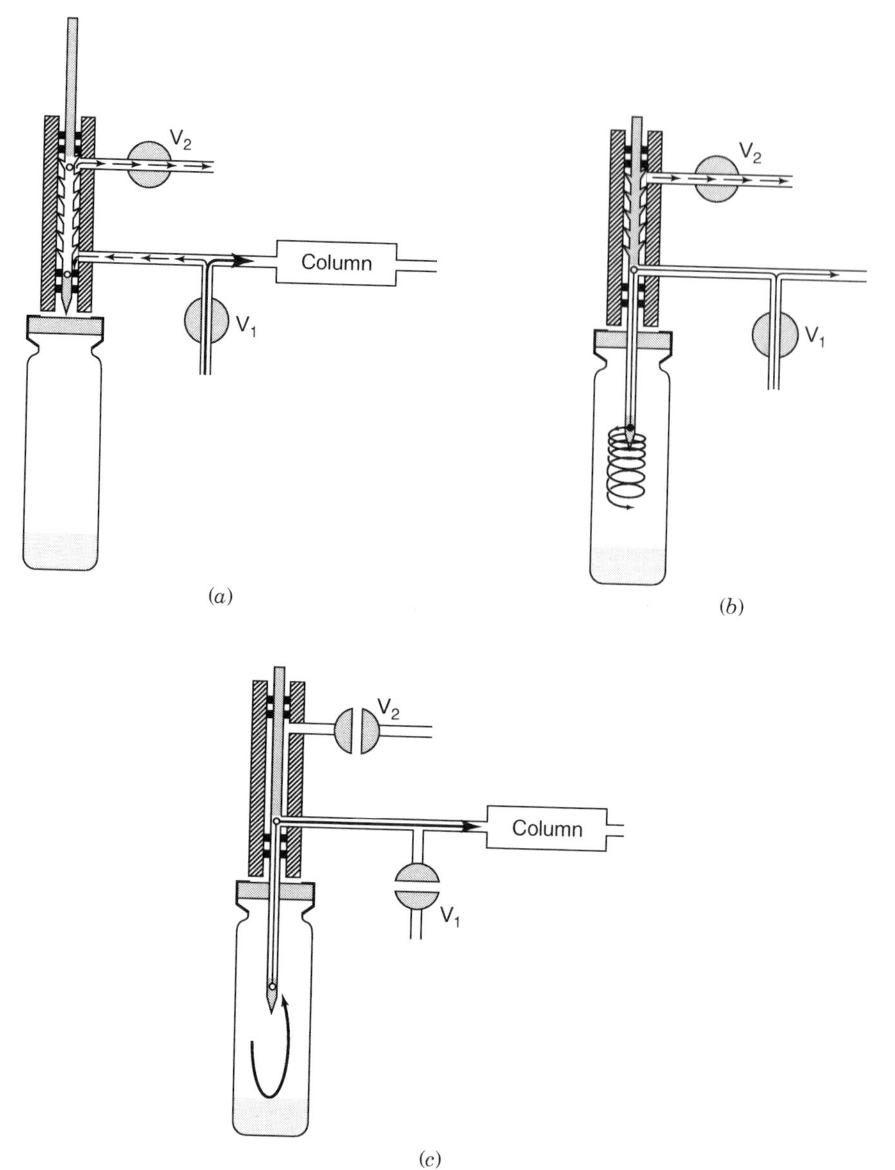

Figure 2.3. Sampling sequence for high pressure sampling in headspace gas chromatography. (*a*) Thermostatting phase. During the thermostatting phase (standby) the sampling needle is in the upper position. The carrier gas flows through solenoid value V_1 to the column; at the same time the needle cylinder is purged by a small cross flow vented through solenoid value V_2. The cross flow

(*contd*)

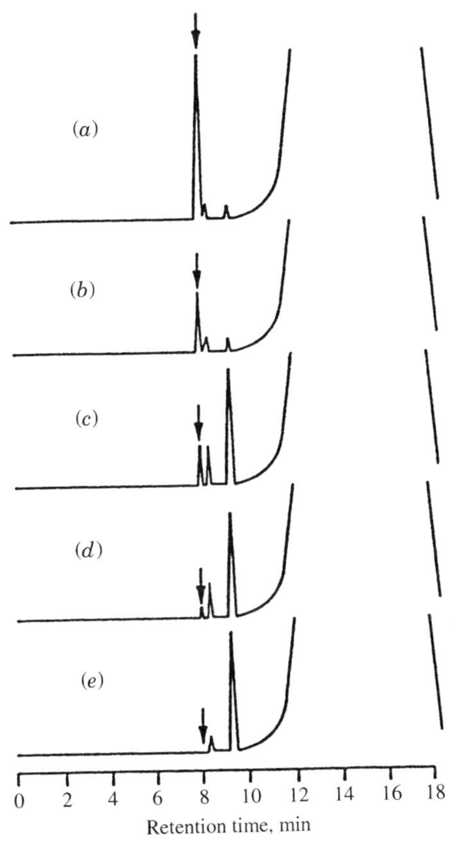

Figure 2.4. Headspace gas chromatograms of ethylene oxide; 3 μg standard (*a*); 1 μg standard (*b*); 3 μg in spice matrix (*c*); 1 μg in spice matrix (*d*); and spice matrix alone (*e*) (31).

Figure 2.3. (*contd*) prevents the buildup of sample residues. (*b*) Pressurization phase. After completion of the thermostatting phase, the sampling needle moves to the lower position, piercing the sample vial septum. The carrier gas flows into the vial headspace, pressurizing it to equal the column head pressure. (*c*) Injection phase. After the pressurization phase, the solenoid valves V_1V_2 are closed, stopping the carrier gas flow. The compressed gas in the vial flows onto the column (sampling). After the preselected injection time the solenoid values V_1V_2 are again opened, completing the sampling phase. The carrier gas now flows directly onto the column and branches to the sample vial stopping the sample vapor from reaching the column (Perkin–Elmer Corporation).

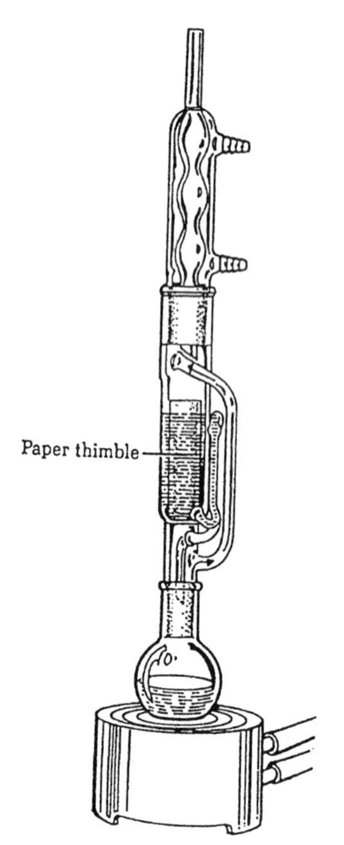

Figure 2.5. Soxhlet extractor. For exhaustive extraction the solid is packed into a paper thimble. Solvent vapor rises in the side-arm tube on the right, and condensed solvent drops onto the solid in the paper thimble, leaches out soluble material, and, after an automatic siphon, carries it to the boiling flask where extracted material accumulates.

extracting the large lipid volumes that later pose separation problems. In the acetone-based multiresidue extraction method of Luke (33), pesticides are recovered from the extract by partitioning with petroleum ether–dichloromethane and then may, after concentration, be clean enough for direct GC analysis without further cleanup. In the California Department of Food and Agriculture (CDFA) multiresidue method, designed for fruits and vegetables of high moisture content, acetonitrile is used to extract the sample and no additional solvent is used for partitioning. Again, the samples may be clean enough (following removal of saltwater and exchange to another solvent) for direct GC or HPLC analysis (34). An excellent review of extraction methodology for pesticides—including the so-called universal extraction solvents—is by Steinwandter (35). As pointed out in this review, the trend is toward smaller sample sizes (miniaturization), resulting in savings in solvent costs, lessening the time and potential losses associated with evaporation steps, and decreasing the size of glassware. Several chapters in the Office of Technology Assessment report (36) address extraction methods for pesticides in foods.

A problem with all extraction systems is how to assess whether the pesticide of interest is efficiently removed from the matrix. Usual spiking procedures involve adding a known amount of analyte to a blank or control matrix portion at the point of extraction, and then determining the percent recovery in the prepared extract. There are two problems with this procedure: First, a field-treated matrix may have incorporated or bound the analyte in such a way that simple extraction can not get it out efficiently. The spike at extraction is not similarly incorporated or bound, and thus provides no good measure of extractability. Second, there is a potential for loss of analyte from the field sample during its preservation in the freezer while awaiting analysis—another loss not assessed by spiking at the point of extraction. Dealing with the first problem is difficult, and approaches involve comparing recovery from a field-treated sample using an exhaustive extraction method (e.g., Soxhlet extraction) with that from a solvent blending operation, or specially preparing a field sample treated with ^{14}C-labelled analyte for subsequently assessing radiolabel recovery during a variety of extraction steps (37). The second problem may be adequately handled by spiking the blank or control matrix at the time it is placed in the freezer ("freezer spike") and comparing recovery with a similar portion spiked at the point of extraction. If there is no difference in recovery, it is assumed that the analyte is stable during the period of freezer preservation.

Soil

Soil is one of the hardest sample types to extract because of its unique binding potential for a variety of pesticides. A common procedure involves (1) saturating the soil with water to its "field capacity" (5% or so) to release adsorbed residues and (2) blending or, better, Soxhlet-extracting the moist soil with a strong solvent (e.g., acetone) or solvent blend (e.g., 2 : 1 isopropanol-benzene).

Paraquat and other bipyridilium herbicides represent special problems because they are bound so tightly to the mineral fraction of soil, as well as to crops and other solid matrices. Strong mineral acid, such as 18N sulfuric acid, is required for extraction of these chemicals from soil, and 5N sulfuric is used for extraction from plants (38). Seiber and Woodrow (39) used 5N hydrochloric acid to extract paraquat from airborne particulate matter prior to determination by GC-NP-TSD as a reduced tertiary amine derivative. Tightly bound residues, in soil and crop tissue, may be extracted more completely using supercritical fluid extraction (40) and other special techniques.

CLEANUP AND FRACTIONATION

Cleanup refers to a step or series of steps in the analytical procedure in which the bulk of the potentially interfering coextractives are removed, primarily by physical and/or chemical methods. For the analysis of one, or a few closely related analytes [as in single-residue methods (SRMs)], the cleanup can be fairly drastic and tailored to the physicochemical properties of the analyte(s) in question. For the analysis of multiple analyte residues [multiresidue methods (MRMs)], the cleanup step is generally softer and more universal, fractionating residues with somewhat differing physicochemical properties into polarity- or volatility-based classes for separate determination. All cleanup methods will represent a compromise between time and cost, on one hand, and the detection limit and/or breadth of analytes accommodated on the other hand (41, 42).

Nature of Potentially Interfering Coextractives

A recognition of the nature of potential coextractives that may be present in food and other biological matrices is important in selecting the most appropriate cleanup procedures. Some major classes for foodstuffs are included in Table 2.5:

Table 2.5. Common Interferences in Pesticide Residue Analysis

Class	Types
Lipids	Waxes, fats, oils
Pigments	Chlorophylls, xanthophylls, anthocyanins
Amino acid derivatives	Proteins, peptides, alkaloids, amino acids
Carbohydrates	Sugars, starches, alcohols
Lignin	Phenols and phenolic derivatives
Terpenes	Monoterpenes, sesquiterpenes, diterpenes, etc.
Miscellaneous environmental contaminants	Most classes of organic compounds, minerals, sulfur, PCBs, phthalate esters, hydrocarbons

Lipids

The class of chemicals designated as lipids includes the waxes, fats, and oils. These are either hydrocarbons or esters containing a fatty acid and an alcohol portion. Esters in which the alcohol is glycerol constitute the classes of fats and oils. Waxes may include fatty esters, and also long-chain hydrocarbons such as exist in the surface cuticle of most leaves and fruits. Lipids occur in large amounts in crop and animal products. Butter, for example, is 81% fat, and the remainder is water. Cheeses range to above 30% fat and chicken eggs, 61%. Even some plant-derived foods, such as peanuts (49%) and avocados (17%) can contain high lipid levels. A complete listing of the fat, water, and sugar content of foods is in the *Pepticide Analytical Manual*, Section 201 (16). Because lipids are soluble in many common organic solvents, they may be present in the crude extract. Lipid esters are saponifiable with acids and bases, and degraded further with oxidizing acids. They tend not to be volatile, but their sheer bulk can still be problematic in gas chromatography because they tend to coat the column and injector ports, changing analyte chromatographic characteristics, and they may slowly degrade thermally into volatile products that survive to the detector.

Pigments

This group includes a number of structurally unrelated colored compounds, such as chlorophylls, xanthophylls, and anthocyanins. Pigments are soluble in some organic solvents, particularly the more polar ones, are destroyed by corrosive reagents, and are principal interferences in colorimetric and spectrophotometric determination steps.

Other Agents

Peptides and amino acids contain nitrogen and often sulfur, and thus are potential interferences in determinations using *N*- or *S*-selective detection systems. Carbohydrates are colorless, nonvolatile, and only sparingly soluble in organic solvents and thus present problems only with pesticides that possess low volatility and high water-solubility characteristics. Lignins are somewhat similar, but with the caveat that they may degrade, hydrolytically or thermally, to phenols, which may interfere with analysis of phenolic metabolites of pesticides such as the carbamates and phenoxy acids. Certain vitamins, such as vitamin K, have physicochemical properties similar to those of many pesticides and drugs, and may thus interfere.

Nonpesticidal environmental chemicals often cause serious interference problems in pesticide analysis. Sulfur has extensive pesticidal uses on crops (it is the leading agrochemical pest control agent in California, for example), and its allotropic S_4, S_6, and S_8 forms are extractable and gas chromatographable, responding to electron capture, electrolytic conductivity, and flame photometric detectors. In one notable legal case in the 1970s involving the United Farm Workers and grape growers in California, sulfur was misidentified as aldrin since the two chemicals have nearly identical retention times and EC response characteristics (43). It was alleged that aldrin was present at levels that may have affected farmworker health where, in fact, sulfur was the chemical being detected. PCB and phthalate ester plasticizers may also be present in environmental samples, but of perhaps even greater concern, may contaminate samples in the laboratory because of their presence in some technical solvents, plastic tubing, bottle caps, and other equipment. The misidentification of organochlorine pesticides in samples contaminated in the field or lab with plasticizers has occurred in several cases, particularly in the earlier days of residue chemistry before mass spectrometric confirmation was widely practiced.

Specifically, then, it is the purpose of the cleanup to reduce the amount of these potential interferences to a sufficiently low level to permit determination of the pesticides of interest. The general strategy is to take advantage of physicochemical property differences between the analyte(s) and the background matrix as a basis for cleanup. Both physical and chemical cleanup procedures exist:

Physical	**Chemical**
Partition (polarity)	Partition (acid–base)
Precipitation (solubility)	Concentrated acids, alkalis (degradation)
Volatilization (volatility)	Oxidation (degradation)
Chromatography (polarity)	Derivatization (modification of analyte)

Methods based on partitioning between immiscible solvents, and on chromatography (primarily using open-column adsorption chromatography) tend to be the most heavily used for general pesticide cleanup, with the other methods serving for specialty cleanup for analytes of somewhat unusual properties, or for "difficult" matrices.

Multiresidue Cleanup Methods Based Primarily on Partitioning and Chromatography

The U.S. Food and Drug Administration (FDA) has developed extensive methodology to underlie its need to analyze large numbers of foodstuffs for both registered and unregistered pesticides, and their toxicologically important metabolites, in the U.S. food supply. The evolution of FDA's multiresidue methods has been reviewed by Sawyer (44, 45) and Luke (46) and those in present-day use by FDA are presented in detail in the *Pesticide Analytical Manual* (14). All three of the FDAs general MRMs — Luke, Mills (47) (non-fatty foods) and Mills (48) (fatty foods) — as well as those used by other agencies use solvent partition and/or column chromatography as the "heart" of the cleanup (Table 2.6) (33, 34, 47–49). The Mills procedure (Fig. 2.6), termed PAM methods 303 and 304, was originally developed for analyzing organochlorine pesticides in human and animal tissue samples, as well as certain dairy and crop items. It includes a petroleum ether/acetonitrile partitioning step in order to separate the bulk of the lipids (petroleum ether phase) from the pesticides (acetonitrile phase). Partition p values somewhat analogous to octanol–water partition coefficients have been determined (50) in order to define the partition behavior of pesticides between petroleum ether (or hexane, which is equivalent) and acetronitrile. Some representative p values (defined as the fraction in the hexane phase when equal volumes of preequilibrated hexane and acetonitrile are used) are in Table 2.7. Thus aldrin and DDT will be inefficiently removed from lipid, and parathion and carbofuran will be efficiently removed, in a single partition between hexane (lipid phase) and acetonitrile (pesticide extract phase). The Mills pocedure uses three partitionings with acetonitrile and unequal phase volumes to improve the separation for chemicals such as aldrin and DDT, but even then there will be some loss of these materials in the hexane or petroleum ether (lipid) layer. This is an example of the type of compromise which may be necessary in cleanup procedures.

The pesticides recovered in the acetonitrile phase of the Mills procedure and its later modifications [the general methodology is presented in the *Pesticide Analytical Manual* (14)] are then back-extracted by adding saltwater to the acetonitrile and partitioning this solution against hexane or petroleum ether. Most of the common nonionic pesticides are recovered, and are then ready for separation into polarity groupings by column chromatography (see Fig 2.6). Very polar and/or ionic compounds (paraquat, glyphosate, some phenoxy acids) will either not be extracted in the first extraction step, or will be lost in the saltwater–acetonitrile solution, which is discarded; as noted, the very nonpolar compounds, represented by aldrin (the least polar of all common pesticides) will be lost, at least partially, in the extracted hexane or petroleum ether phase.

In the remainder of the Mills procedure, pesticides in the defatted hexane or petroleum ether phase are subjected to column chromatography on Florisil—a synthetic magnesium silicate material originally supplied by the Floridin Corporation. Specifications for this series of operations, from activation of the Florisil by heating, to packing the column, to the elution of

Table 2.6. Multiresidue Methods (MRM) in General Use for Pesticides in Foods with Major Variables Listed[a]

Name/PAM No.	Variables to Fit Matrix and Compound Class				References
	Extraction Solvent	Cleanup	Column	Detector	
Luke/302 (nonfatty foods)	Acetone Aqueous acetone	Florisil/$MeCl_2$ Charcoal/MgO Charcoal/silanized celite C-18 cartridge Florisil/mixed ether	OV-101 OV-17 OV-225 DEGS C-8 (HPLC)	EC FPD-P FPD-S ELCD-N ELCD-X Postcolumn fluorescence (HPLC)	33, 49
Mills/303 (nonfatty foods)	Acetonitrile Water/acetonitrile	Petroleum ether/acetone partition Florisil/mixed ether Florisil/$MeCl_2$	OV-101	EC FPD-P ELCD-X ELCD-N NP-TSD	47
Mills/304	Petroleum ether Other solvents	Petroleum ether/acetonitrile partition Florisil/mixed ether Florisil/$MeCl_2$ Gel permeation Gel permeation plus florisil Alumina plus florisil	OV-101 OV-17	EC FPD-P ELCD-X ELCD-N NP-TSD	48
CDFA	Acetonitrile	RP-SPE cartridge Acetonitrile–saltwater partition	DB-5 C-18 (HPLC)	FPD-P ELCD-X Postcolumn fluorescence (HPLC)	34

[a]See *Pesticide Analytical Manual* (14) and literature references for details.

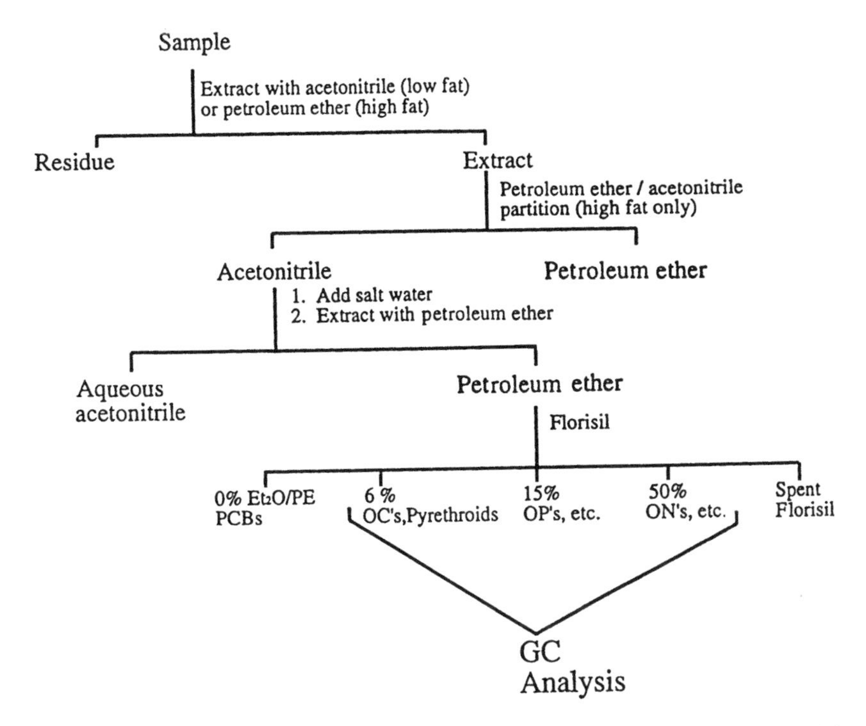

Figure 2.6. General schematic of FDA Mills multiresidue procedure for pesticides in foods (PAM methods 303 and 304). OC = organochlorines; OP = organophosphates; ON = organonitrogen compounds.

compounds in the three primary eluting solvent mixtures, may be found in detail in the *Pesticide Analytical Manual* (14). In the more general procedure, elution with 6% diethyl ether in petroleum ether recovers the less polar, primarily organochlorine pesticides, elution with 15% diethyl ether in petroleum ether will recover moderately polar OCs and OPs, and the 50% elution fraction will recover the more polar OPs such as malathion and some additional chemicals (Table 2.8). This elution scheme is not perfect; some compounds may be split between more than one fraction, and the more polar groupings, including the *N*-methylcarbamates, may not be eluted or may degrade on the Florisil. It is possible to remove PCB interferences by a Florisil column elution with only petroleum ether. The *Pesticide Analytical Manual* (14) tabulates the recovery of individual chemicals through the various multiresidue methods, and general recovery information may be found in the Appendix (at the end of this volume), also taken from *PAM*. Table 2.8 shows Florisil elution patterns for some representative organochlorine and organophosphorus pesticides.

Table 2.7. Beroza *p* Values (Hexane–Acetonitrile) for Representative Pesticides (50)

Pesticide	*p* value
Aldrin	0.73
Heptachlor	0.55
PCNB	0.41
p,p′-DDT	0.38
Dieldrin	0.33
Heptachlor epoxide	0.29
Diazinon	0.28
Trifluralin	0.23
Dicofol	0.15
Tetradifon	0.10
Ethyl parathion	0.044
Malathion	0.042
Methyl parathion	0.022
Azinphosmethyl	0.008

The Mills-type methodology has limitations, including those noted above and the fact that it uses fairly large volumes of expensive solvents, but it has stood the test of time and thus represents a good yardstick for comparing other methods. The Luke method (*Pesticide Analytical Manual*, method 302) represents a major departure, in that a water miscible extracting solvent (acetone) is used, the acetone is either dried or partitioned against a petroleum ether–dichloromethane solution, and Florisil cleanup is one of several options for cleanup (Fig. 2.7). A "dilute-and-shoot" GC analysis is possible without the more rigorous cleanup and fractionation provided by column chromatography. The Luke method still provides opportunity for loss of some pesticides, but it is a rapid and reproducible method whose broad utility has led to its dominant use for the majority of FDA enforcement analyses in the 1980s and 1990s. In the "Luke II" version, SPE cleanup was used in place of some of the solvent–solvent partition steps (46). A broad polarity range of pesticides is included, and extracts are generally cleaner and amenable to such determinations as GC-MS (51).

Another variation is provided by the multiresidue method used by the California Department of Food and Agriculture, which annually analyzes nearly as many samples of fruits and vegetables as FDA. In the current version (Fig. 2.8) (34), acetonitrile is used as extracting solvent, and the aqueous acetonitrile is cleaned up via reversed phase SPE. The acetonitrile is recovered by a salting out process, and then exchanged by evaporation to hexane, for analysis of OCs by electrolytic conductivity GC, or acetone for

Table 2.8. Elution Patterns and Recovery Data for Representative Organochlorine and Organophosphate Pesticides on 130°C Activated Florisil[a]

	Elution Increments (mL) (Et_2O in hexane)						
	6% Fraction		15% Fraction		50% Fraction		
Compound	0–100	100–200	200–300	300–400	400–500	500–600	Recovery, %
α-BHC	100						97
β-BHC	100						95
Lindane	100						96
Heptachlor	100						91
Aldrin	100						100
Heptachlor epoxide	78	22					105
Dieldrin			85	15			96
Endrin			89	11			99.6
p,p′-DDE	100						97
o,p′-DDT	100						99.6
p,p′-DDT	100						90
Ronnel	100						93
Methyl parathion			47	53			103
Malathion					100		99
Ethyl parathion			78	22			96
Diazinon			100				83
Trithion	100						43

Source: *Pesticide Analytical Manual*, 1972 ed.

[a]Numerical values represent the relative percentage of each compound eluting in each cut. Recoveries in right hand column are absolute recoveries.

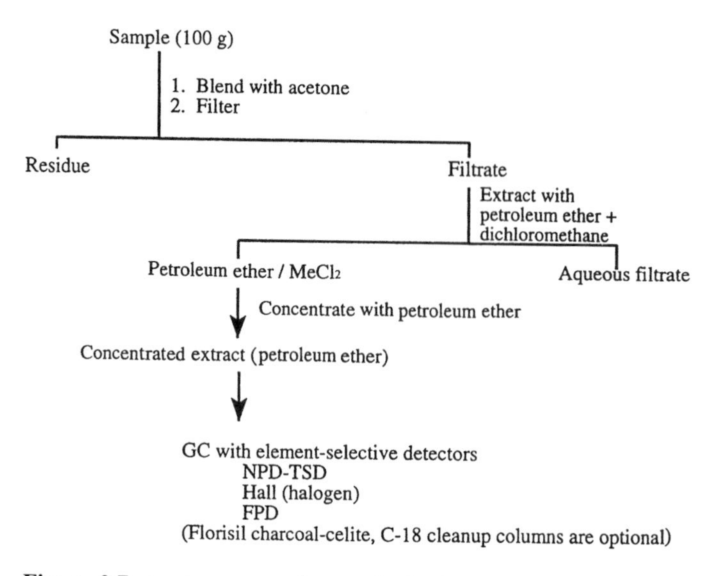

Figure 2.7. Acetone extraction method of Luke (PAM method 302).

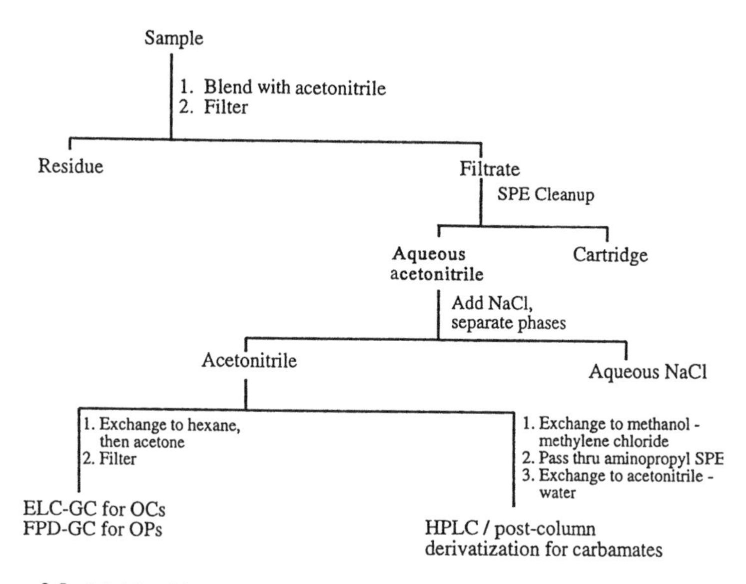

Figure 2.8. Multiresidue method used by California Department of Food and Agriculture for fruit and vegetable samples (34).

analysis of OPs and some organonitrogen compounds by FPD or NP-TSD GC. A separate aliquot is exchanged to methanol–methylene chloride, cleaned up on a solid-phase extraction column, and then exchanged to acetonitrile–water for HPLC with postcolumn derivatization for analysis of carbamates.

Krause (52) reported a multiresidue method for measuring *N*-methylcarbamate insecticides in crops, including some metabolites such as the sulfoxide and sulfone of aldicarb, using HPLC. The method was developed by fortifying nine crops with the pesticides and their metabolites at 0.05 and 1.0 ppm. Methanol was used to extract the crop, followed by solvent partitioning and use of a charcoal-silanized Celite column, which removed most of the coextractives. The residues were analyzed on a reverse-phase HPLC column using a gradient of acetonitrile and water. Eluted compounds were detected using a postcolumn derivatization technique based upon the methyl amine hydrolysis product. Recovery was quite good, averaging 99% for both levels; an exception was provided by aldicarb sulfoxide, which gave 56% because of loss in the partitioning step. The carbamate method has been incorporated in the Luke method (302, *Pesticide Analytical Manual*) and the CDFA method (34).

There are literally dozens of variations, of either or both the partition and column chromatography cleanup steps, in the literature. For example, Laws and Webley (53) used a partition between petroleum ether and aqueous methanol to separate OPs into polarity groups after extraction from vegetables with dichloromethane. Chromatography was done on alumina for the less polar OPs (petroleum ether phase), or on carbon for the more polar compounds (aqueous methanol phase). In addition to Florisil, alumina, and carbon, silica gel is used fairly often as an all-purpose chromatographic adsorbent for cleanup-fractionation. Some compounds, such as *N*-methylcarbamates, which are not stable on Florisil or alumina, may be recovered intact from silica gel, particularly when this adsorbent has been suitably deactivated with water. Sherma and Shafik (54) fractionated wide ranges of pesticide types recovered from air samples on a minicolumn of silica gel, deactivated with 20% by weight of water, eluted with hexane, 60% benzene in hexane, and 5% actetonitrile in benzene. A variation of this theme was presented by Seiber et al. (55), who developed an HPLC-based fractionation scheme using a Partisil silica column eluted with a gradient ranging from hexane to methyl *t*-butyl ether. The method was developed for air and water samples, which have very low contents of matrix-derived interference bulk, but it was extended to derivatized glyphosate from fruit/vegetable matrices as a "finish" cleanup to remove stubborn interferences (56). Much work is currently being done using mini-columns, both conventional and SPE or HPLC types, for cleanup in order to reduce the solvent volumes required, reduce the time, encompass broader groups of chemicals, or automate the cleanup (35). Because conventional cleanup is so labor-intensive, there is much driving force to develop alternatives amenable to automation, or to simply bypass cleanup altogether, leaving separations to the realm of the GC or mass spectrometry systems used in the determination step.

Advances in GC and HPLC detection may necessitate development of new/modified cleanup procedures. Accordingly, new methods are under active development for use with ion-selective GC-MS, atomic emission detection, and HPLC-mass spectrometry. By way of example, Cairns et al. (51) reported a new cleanup procedure for pesticides in a range of fruits and vegetables, specifically for use with ion-trap GC-MS. It is a variation on the Luke MRM, in which three solid-phase extractions (C_{18}, anion exchange, and aminopropyl) were used to clean hydrocarbons, colored compounds, flavors, and sugars from the aqueous acetone extract. Satisfactory recoveries were obtained for 24 pesticides (OC, OP, carbamate, and pyrethroid) in strawberry, carrot, tomato, and lettuce matrices. Lehotay and Eller (57) and Lehotay et al. (58) reported development of a supercritical fluid extraction and GC/MS method applicable to 46 pesticides in fruits and vegetables. Their results compared favorably with those obtained using traditional GC detectors and solvent-based extractions.

There has been and continues to be much activity in the realm of automating cleanup methodologies. The largest inroads have been made with gel permeation chromatography in which separations involve a combination of size-based separations (lipids from organochlorine and organophosphate compounds, proteins from OP and organonitrogen compounds) and polarity-based separations. For example, an automated gel permeation chromatography system described by Johnson et al. (59), based on the system of Stallings et al. (60), uses Bio-Beads SX-3 column and a toluene-ethyl acetate mobile phase to clean up fish and other high-lipid extracts for recovery of organochlorine compounds. The system will run 24 h per day, with reuseable columns, and it also has the capability to "cut" classes of compounds into discrete fractions for further, independent analyses. A commercial version has received wide acceptability from laboratories faced with large sample loads of high-lipid matrices. An example of application of GPC to organophosphorus pesticides is in Holstege et al. (61). Robotic automation of cleanup, and all other steps in residue analysis, is of much current interest (62).

Although it has found only limited use in multiresidue cleanup methods, extraction of acidic or basic pesticides (phenoxy acids, anilines, sulfonic acids, etc.) from the bulk of the coextacted matrix material can be done relatively quickly and conveniently by simply partitioning the extract, usually held in an all-purpose, water-immiscible solvent such as dichloromethane, against aqueous acid or aqueous base. This methodology is used frequently in multiresidue analysis for drugs, most of which tend to be either acidic, basic, or amphoteric, and in the standard scheme for priority pollutants (27), which includes a fairly sizeable number of acidic (phenolic) compounds, some of which are hydrolysis products of pesticides. A relatively straightforward example in residue analysis is the recovery of phenoxy acids from organic extracts of water or crops by an extraction with dilute base, with subsequent recovery of the acids in an organic solvent after neutralizing the basic extract, methylation of the phenoxy acids with diazomethane, and subsequent GC determination of the methyl esters [*Pesticide Analytical Manual* (14)].

Codistillation/Forced Volatilization Methods

Purge-and-trap and headspace methods for "extracting" volatile analytes from water and less volatile matrices were described previously in this chapter. Related techniques of codistillation and forced volatilization have been used to clean up samples that contain less volatile impurities and require additional cleanup prior to determination.

Sweep codistillation is a cleanup technique that relies on the volatilization of pesticide away from lipids and other coextractives under the influence of a stream of inert gas and, optionally, solvent vapors. The volatilized pesticides are recovered by condensation from the sweeping stream in a cold trap or on a column of adsorbent. The technique was first described by Storherr and Watts (63), primarily for cleanup of organophosphate pesticides. Sample extracts were introduced in a suitable extracting solvent, such as ethyl acetate, onto glass wool or glass beads, at a temperature sufficiently high to promote the desired volatilization, but not too high so as to volatilize or thermally degrade the matrix material. Continuous sweep of the glass wool–glass bead zone is carried out until a reasonable recovery of desired analyte(s) has been achieved.

Over 40 papers were published and a commercial "sweep codistiller" was available between 1965 and 1973, but the technique fell into neglect because of unreliable recovery and unpredictability. It was also somewhat of a burden to clean out the device of its baked-on residual contents between runs. Recently, the technique has made a modest comeback (64). A new commercial system is available with improved control over primary variables, such as codistillation temperature, gas flows, and condensed phase recovery. Figure 2.9 shows the newer UNITREX design. Recovery of a variety of OC, OP, triazine, and other pesticide groups from animal and plant samples, and of PCBs from human breast milk, have been demonstrated with

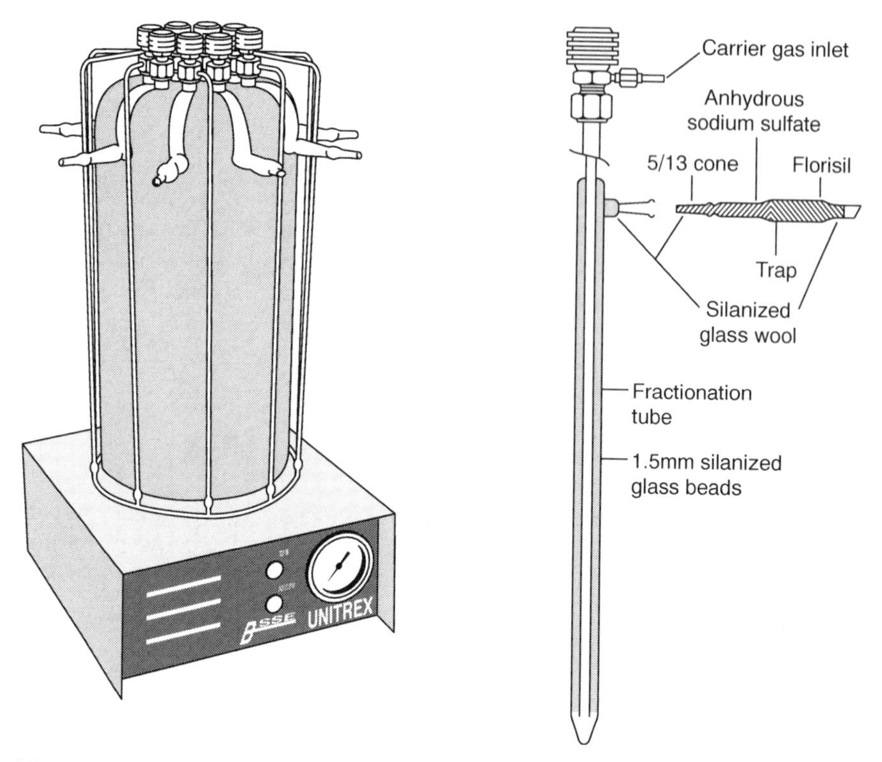

Figure 2.9. UNITREX sweep codistiller. Assembled equipment (left) and distillation tube and trap design (right) (64).

the newer device (64). Literally any pesticide with appreciable volatility, and thermal stability, is a candidate for cleanup from a relatively dirty extract by this technique.

A number of single residue methods (SRMs) make use of simple codistillation for recovery of volatile analyte (or a hydrolysis product) from aqueous solution. For example, ethylene dibromide (EDB) can be recovered from an aqueous slurry of flour, juices, and other substances by steam codistillation. A method for analysis of metaldehyde utilized steam codistillation of acetaldehyde generated from the cyclic tetramer by digestion in aqueous sulfuric acid (65). And older methods for substituted urea herbicides made use of a base digestion to generate the parent substituted aniline, followed by steam distillation of the aniline away from interferences subsequent to determination by a colorimetric diazo coupling reaction or by GC. These methods, which may be found in detail in early versions of the *Pesticide Analytical Manual*, and the series edited by Zweig

(Ref. 41 and other volumes), are no longer used as much because of the advent of HPLC and capillary GC, which allow sensitive determination of the parent compounds without hydrolysis and subsequent derivatization. It is desireable to separately analyze the parent and breakdown product, which was difficult by the older methodology.

Precipitation Cleanup Methods

Fats and waxes can be made to crystallize out at low temperatures, and thus be removed from more soluble analytes. In a typical procedure, an acetone extract of plant or animal tissue is cooled to $-80°C$, precipitating the lipids and some pigments. The acetone supernatant is then further cleaned up by column chromatography. In this regard, the precipitation process can take the place of liquid–liquid partition, as used in the Mills and related multiresidue methods described above (66).

Chemical Cleanup Methods

Methods which destroy, or otherwise "digest away" the matrix, leaving the pesticide or a predictable breakdown product for further analysis, find favor in some procedures. These methods were, for the most part, developed for the stable organochlorine pesticides. For example, fuming sulfuric acid has been used to saponify, and subsequently remove, lipids and pigments from extracts prior to analysis of such stable OCs as toxaphene, chlordane, aldrin, heptachlor, and the chlorinated dibenzodioxins. The procedure can be carried out in a separatory funnel although, for safety reasons, a Celite column coated with the acid is preferable.

Similarly, concentrated alkalis have been used, again primarily to saponify and remove lipids. For example, refluxing a crude extract with KOH in ethanol, followed by dilution with water, and extraction in hexane leads to removal of lipids to the water layer (as salts of the fatty acids and glycerol), and recovery of pesticides such as aldrin, and PCBs in the hexane phase. Many pesticides undergo specific transformations during this step which may be beneficial from an analytical viewpoint. DDT forms DDE, so that GC analysis before and after base treatment can help confirm the presence of DDT (50). Lindane forms 1,2,4-trimethylbenzene, and toxaphene forms dehydrochlorinated toxaphene (67); both transformations are useful for confirmation and, in the case of toxaphene, the transformed material is more readily quantified because it contains fewer peaks, of shorter retention time, than the parent.

Digestion of extracts with oxidizing acids, such as nitric–sulfuric mixture, perchloric acid, or permanganate is used to oxidize all materials to inorganic ions. For methyl mercury, the product is mercuric ion, which may then be determined by reduction to atomic mercury and flameless atomic absorption spectrophotometry. Under mild oxidation with chromic acid, sulfur contaminants were removed (as SO_2) from an aldrin extract of crop; aldrin was epoxidized to dieldrin in the process, and quantitated as such by GC. This reaction set was used to solve the aldrin-in-grapes controversy alluded to previously.

The use of enzymes should be mentioned at least in passing, although their primary use is in extraction of conjugates rather than in cleanup. Papain or pepsin could break down and remove troublesome peptides, although relatively few examples have been forthcoming. Pyrolysis has also found specialty use, to free up bound materials from matrix. Treatment of rice plants with radiolabelled propanil resulted in recovery of only 70% label during conventional extraction. The remaining 30% of "bound residue" was recovered, as dichloroaniline, after pyrolysis treatment and subsequent counting of the appropriate region on the chromatogram (37).

Ion Exchange

This cleanup technique occupies a niche between physical and chemical methods because it relies on a chemical ion-exchange process to effect separation but the analyte, or matrix, is not chemically modified in the overall process. Ion-exchange cleanup has been useful for a number of acidic and basic pesticides and transformation products, but it is essential for the cleanup of such materials as paraquat, a quaternary dication, and glyphosate, an amphoteric aminophosphonic acid. A method for paraquat involves extraction in 5N sulfuric acid, followed by neutralization and ion-exchange chromatography on a strong cation-exchange column eluted with saturated ammonium chloride. Paraquat is subsequently determined by absorption spectrophotometry as the intensely blue-colored radical cation reduction product (38). Modern developments include the use of ion-exchange HPLC using straight UV absorption for detection, or postcolumn reduction to the cation radical for determination at 396 or 600 nm (56).

The lengthy PAM method for gyphosate, which involves the use of two ion-exchange cleanup columns, was considerably simplified by Moye and St. John (68) with the introduction of an HPLC system consisting of an Aminex A27 column operated at 62°C, to which the extract was applied and subsequently eluted with a mobile phase of 0.1M phosphoric acid. Postcolumn detection, of both glyphosate and its resolved aminomethylphosphonic acid metabolite, was provided by postcolumn derivatization with orthophthalaldehyde solution to a fluorescent derivative. Analysis time was considerably shorter than the *Pesticide Analytical Manual* (14) method and, predictably, this is now the standard method for analysis of glyphosate.

Pardue (69) described a multiresidue method for triazine herbicides and their metabolites using solvent partition and cation-exchange solid-phase extraction cleanup steps. Nineteen herbicides and four metabolites were determined in six agricultural products, with recoveries ranging within 81–106% for the parents and 59–87% for the metabolites.

Combination Methods

The toxicological importance of some compounds dictates the development of heroic methods to detect residues at extremely low levels. Because interferences generally increase as the detection limit decreases, multiple cleanup methods may be required. Such is the case for 2,3,7,8-tetrachlorodibenzodioxin (TCDD) (Fig. 2.10) and some of the other dibenzodioxins and dibenzofurans. For TCDD, the method of Langhorst and Shadoff (70) (Fig. 2.8) employed both acid- and base-impregnated

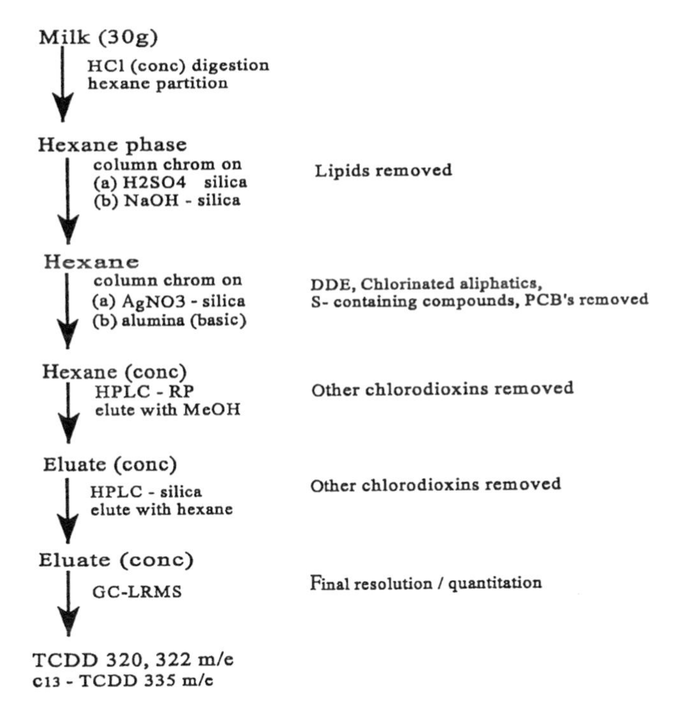

Figure 2.10. Schematic of analytical method for 2,3,7,8-tetrachlorodibenzodioxin, and other dioxins, in human milk (70).

cleanup columns to remove lipids; column chromatography on a silver nitrate-impregnated column followed by a basic alumina column to separate out DDE, chlorinated aliphatics, sulfur compounds and PCBs; and both reverse-phase and normal-phase preparative HPLC to separate the 2,3,7,8 isomer from other chlorodibenzodioxins. Coupled with GC-MS determination, the resulting method had a limit of detection of 1 ppt in human milk (but a very low sample throughput). Subsequent modifications, including the use of a novel multiadsorbent cleanup column, have been made that reduce the number of cleanup steps, increase the GC resolution and reliability of the determination step, and result in an overall increase in the sample throughput (71).

GENERAL SOURCE OF INFORMATION

More information on cleanup/fractionation, as well as other steps of residue analysis, can be found in the following sources:

- "*Official Methods of Analysis*", 15th ed., Association of Official Analytical Chemists, 1990 (21)
- *Pesticide Analytical Manual*, 1994 (14)
- G. Zweig and J. Sherma, eds., *Analytical Methods for Pesticides and Plant Growth Regulators*, Academic Press, 1972 (41). (a series of 20 + volumes).
- G. Ware, ed., *Reviews of Environmental Contamination and Toxicology* (a series of approximately 150 volumes dating from 1965, previously titled *Residue Reviews*, F. A. Gunther and J. D. Gunther, eds. Springer-Verlag, New York.
- J. Sherma, Pesticides, *Anal. Chem.*, **67**, IR–20R (1995).

REFERENCES

1. J. W. Biggar, and J. N. Seiber, eds., *Fate of Pesticides in the Environment*, Chapter 5, Publication 3320, Agricultural Experiment Station, Division of Agriculture and Natural Resources, Univ. California, Oakland, CA, 1987.
2. L. R. Snyder, and J. J. Kirkland, *Introduction to Modern Liquid Chromatography*, 2nd ed., Wiley, New York, 1979, pp. 248–251.
3. B. T. Bowman, and W. K. Sans, The aqueous solubility of twenty-seven insecticides and related compounds, *J. Environ. Sci. Health*, **6**, 625–634 (1979).
4. W. Y. Shiu, K. C. Ma, D. Mackay, J. N. Seiber, and R. D. Wauchope, Solubilities of pesticide chemicals in water. Part I. Environmental physical chemistry, *Rev. Environ. Contamin. Toxicol.* **116**, 1–13 (1990).

5. W. Y. Shiu, K. C. Ma, D. Mackay, J. N. Seiber, and R. D. Wauchope, Solubilities of pesticide chemicals in water. Part II. Data compilation. *Rev. Environ. Contamin. Toxicol.* **116**, 15–187 (1990).
6. W. J. Lyman, W. F. Reehl, and D. H. Rosenblatt, *Handbook of Chemical Property Estimation Methods*, McGraw-Hill, New York, 1982.
7. F. A. Gunther, W. E. Westlake, and P. S. Jaglan, Reported solubilities of 738 pesticide chemicals in water, *Residue Rev.* **20**, 1–148 (1968).
8. S. P. Wasik, M. M. Miller, Y. B. Tewari, W. E. May, W. J. Sonnefeld, H. DeVoe, and W. H. Zoller, Determination of the vapor pressure, aqueous solubility, and octanol-water partition coefficient of hydrophobic substances by coupled generator column/liquid chromatographic methods, *Res. Rev.* **85**, 29–42 (1983).
9. F. N. A. Rummens, and F. J. A. Louman, Proton magnetic resonance spectroscopy of thiocarbamate herbicides, *J. Agric. Food Chem.* **18**, 1161–1164 (1970).
10. C. T. Chiou, R. L. Malcolm, T. I. Brinton, and D. E. Kile, Water solubility enhancement of some organic pollutants and pesticides by dissolved humic and fulvic acids, *Environ. Sci. Technol.* **20**, 501–508 (1986).
11. R. P. Schwarzenbach, P. M. Gschwend, and D. M. Imboden, *Environmental Organic Chemistry*, Wiley, New York, 1993, Chapter 7.
12. A. Leo, C. Hansch, and D. Elkins, Partition coefficients and their uses, *Chem. Rev.* **71**, 525–621 (1971).
13. C. Hansch, and A. J. Leo, *Substituent Constants for Correlation Analysis in Chemistry and Biology*, Wiley, New York, (1979).
14. USDHHS, *Pesticide Analytical Manual*, 1994 and prior versions, U.S. Department of Health and Human Services, Washington, DC.
15. L. R. Suntio, W. Y. Shiu, D. Mackay, J. N. Seiber, and D. E. Glotfelty, Critical review of Henry's law constants for pesticides, *Rev. Environ. Contamin. Toxicol.* **103**, 1–59 (1988).
16. Y. H. Kim, J. E. Woodrow, and J. N. Seiber, Evaluation of a gas chromatographic method for calculating vapor pressures with organophosphorus pesticides, *J. Chromatogr.* **314**, 37–53 (1984).
17. N. N. Melnikov, Chemistry of pesticides, *Res. Rev.* **36**, 1–427 (1971).
18. W. F. Spencer, and M. M. Cliath, Measurement of pesticide vapor pressures, *Res. Rev.* **85**, 57–71 (1983).
19. D. Mackay, and A. W. Wolkoff, Rate of evaporation of low-solubility contaminants from water bodies to atmosphere, *Environ. Sci. Technol.* **7**, 611–614 (1972).
20. J. C. Sagebiel, J. N. Seiber, and J. E. Woodrow, Comparison of headspace and gas-stripping methods for determining the Henry's Law constant (H) for organic compounds of low to intermediate H, *Chemosphere* **25**, 1736–1768 (1992).
21. Association of Official Analytical Chemists, *Official Methods of Analysis*, 1st suppl. 15th ed. Association of Official Analytical Chemists Inc., Arlington, VA, 1990, pp. 8–12.

22. A. A. Boyd-Boland, M. Chai, Y. Luo, Z. Zhang, M. Yang, J. Rawliszyn, and T. Goreck, New solvent-free sample preparation techniques, *Environ. Sci. Technol.* **28**, 569A (1994).
23. J. E. Woodrow, M. S. Majewski, and J. N. Seiber, Accumulative sampling of trace pesticides and other organics in surface water using XAD-4 resin, *J. Environ. Sci. Health* **21** (2), 143–164 (1986).
24. J. N. Seiber, J. E. Woodrow, Methods for studying pesticide atomospheric dispersal and fate at treated areas, *Res. Rev.* **85**, 217–229 (1983).
25. R. L. Jolley, and I. H. Suffet, Concentration techniques for isolating organic constituents in environmental water samples, in I. H. Suffet, and Malaiyandi, eds., *Organic Pollutants in Water*, ACS Symposium Series, American Chemical Society, Washington, DC, (1987).
26. B. C. Turner, and D. E. Glotfelty, Field sampling of pesticide vapors with polyurethane foam, *Anal. Chem.* **49**, 7 (1977).
27. L. H. Keith, and W.A. Telliard, Priority pollutants: A perspective view. *Environ. Sci. Technol.* **13**, 416–417 (1979).
28. L. H. Keith, ed., *Compilation of EPA's Sampling and Analysis Methods*, Lewis Publishers, Chelsea, MI, 1992.
29. B. Kolb, *Applied Headspace Gas Chromatography*, Heyden, London, 1982.
30. J. E. Woodrow, and J. N. Seiber, Two chamber methods for the determination of pesticide flux from contaminated soil and water *Chemosphere* **23**, 291–304 (1991).
31. J. E. Woodrow, M. M. McChesney, and J. N. Seiber, Determination of ethylene oxide in spices using headspace gas chromatography *J. Agric. Food. Chem.* **43**, 2126–2129 (1995).
32. Environmental Protection Agency, *Manual of Methods for Pesticides in Human and Environmental Samples*, Washington, DC (1980).
33. M. A. Luke, J. E. Froberg, and H. T. Matsumoto, Extraction and cleanup of organochlorine organophosphate, organonitrogen, and hydrocarbon pesticides in produce for determination by gas-liquid chromatography, *J. Assoc. Off. Anal. Chem. Intl.* **58**, 1020–1026 (1975).
34. S. M. Lee, M. L. Papathakis, H. C. fong, G. F. Hunter, and J. E. Carr, Multipesticide residue method for fruits and vegetables: California Department of Food and Agriculture, *Fresinius J. Anal. Chem.* **399**, 376–383 (1991).
35. H. Steinwandter, Universal extraction and cleanup methods, in *Analytical Methods for Pesticides and Plant Growth Regulators*, J. Sherma ed., Academic Press, San Diego, 1989, Chapter 2, pp. 35–73.
36. Office of Technology Assessment, *Pesticide Residues in Food. Technologies for Detection.* U.S. Congress, OTA-F-398, Washington, DC, Oct. 1988.
37. G. G. Still, F. A. Norris, and J. Iwan, Solubilization of bound residues from 3.4-dichloroaniline-^{14}C and propanil-phenyl-^{14}C treated rice root tissue, in *Bound and Conjugated Pesticide Residues*, D. D. Kaufman, G. G. Still, G. D. Paulson,

and S. K. Brandal, eds., ACS Symposium Series 29, American Chemical Society, Washington, DC, 1976.

38. D. E. Pack, Paraquat, in *Analytical Methods for Pesticides, Plant Growth Regulators, and Food Additives*, G. Zweig, ed., Academic Press, New York, 1967 Chapter 32, pp. 473–481.
39. J. N. Seiber, and J. E. Woodrow, Sampling and analysis of airborne residues of paraquat in treated cotton field environments, *Arch. Environ. Contam. Toxicol.* **10** 133–149 (1981).
40. S. U. Khan, Supercritical fluid extraction of bound pesticide residues from soil and food commodoties, *J. Agric. Food. Chem.* **43**, 1718 (1995).
41. G. Zweig, and J. Sherma, eds., Sample preparation, in *Analytical Methods for Pesticides and Plant Growth Regulators*, Vol. VI, Academic Press, New York, 1972, pp. 1–38.
42. J. N. Seiber, Conventional pesticide analytical methods: Can they be improved? in *Office of Technology Assessment,* Pesticide Residues in Food Technologies for Detection. U.S. Congress, OTA-F-398, Washington, DC, Oct. 1988, pp. 142–152.
43. J. R. Pearson, F. D. Aldrich, and A. W. Stone, Identification of the aldrin artifact. *J. Agric. Food Chem.*, **15**, 938–939.
44. L. D. Sawyer, The development of analytical methods for pesticide residues, in *Pesticide Residues in Foods. Technologies for Detection*, Office of Technology Assessment, U.S. Congress, OTA-F-358, Washington, DC, Oct. 1988, pp. 112–122.
45. L. D. Sawyer, B. M. McMahon, W. H. Newsome, and G. A. Parker, Pesticide and industrial chemical residues, in *Official Methods of Analysis of the Association of Official Analytical Chemists*, Association of Official Analytical Chemists, 15th ed., Arlington, VA, 1990, Chapter 10.
46. M. A. Luke, The evolution of a multiresidue pesticide method, in *8th Intl. Cong. of Pesticide Chemistry: Options 2000*, N. N. Ragsdale, P. C. Kearney, and J. R. Plimmer, eds., American Chemical Society, Washington, DC, 1995, 174–182.
47. P. A. Mills, et al., Rapid method for chlorinated pesticide residues in non-fatty foods, *J. Assoc. Off. Anal. Chem. Intl.* **46**, 186–191 (1963).
48. P. A. Mills, Detection and semiquantitative estimation of chlorinated organic pesticide residues in foods by paper chromatography, *J. Assoc. Off. Agric. Chem.* **42**, 738–740 (1959).
49. M. A. Luke, J. E. Froberg, G. M. Doose, and H. T. Masumoto, Improved multiresidue gas chromatographic determination of organophosphorus, organonitrogen, and organohalogen pesticides in produce using flame photometric and electrolytic conductivity detectors, *J. Assoc. Off. Anal. Chem. Intl.* **64**, 1187–1195 (1981).
50. M. Beroza, M. N. Inscoe, and M. C. Bowman, Distribution of pesticides in immiscible binary solvent systems for cleanup and identification and its application in the extraction of pesticides from milk, *Res. Rev.* **30**, 1–62.

51. T. Cairns, M. A. Luke, K. S. Chiu, D. Navarro, and E. Siegmund, Multiresidue pesticide analysis by ion-trap technology: a clean-up approach for mass spectral analysis, *Rapid Commun. Mass Spec.* **7**, 1070–1076 (1993).
52. R. T. Krause, Multiresidue method for determining N-methylcarbamate insecticides in crops using high performance liquid chromatography. *J. Assoc. Off. Anal. Chem.* **63**, 1114–1124 (1980).
53. E. Q. Laws, and D. J. Webley, The determination of organophosphorous insecticides in vegetables, *Analyst*, **86**, 249–255 (1961).
54. J. Sherma, and T. M. Shafik, A multiclass, multiresidue method for pesticides in air. *Arch. Env. Contamin. Toxicol.* **3**, 55 (1975).
55. J. N. Seiber, D. E. Glotfelty, A. D. Lucas, M. M. McChesney, J. C. Sagebiel and T. A. Wehner, A multiresidue method by high performance liquid chromatography-based fractionation and gas chromatographic determination of trace levels of pesticides in air and water, *Arch. Environ. Contamin. Toxicol.* **19**, 583–592 (1990).
56. J. N. Seiber, M. M. McChesney, R. Kon, and R. A. Leavitt, Analysis of glyphosate residues in kiwi fruit and asparagus using high performance liquid chromatography of derivatized glyphosate as a cleanup step, *J. Agric. Food Chem.* **32**, 678–681 (1984).
57. S. J. Lehotay, K. I. Eller, Development of a method of analysis for 46 pesticides in fruits and vegetables, by supercritical fluid extraction and gas chromatography / ion trap spectrometry, *J. Assoc. Off. Anal. Chem. Intl.* **78**, 821–830 (1995).
58. S. J. Lehotay, N. Aharonson, E. Pfeil, and M. A. Ibrahim, Development of a sample preparation technique for supercritical fluid extraction for multiresidue analysis of pesticides in produce, *J. Assoc. Off. Anal. Chem. Intl.* **78**, 831–840 (1995).
59. J. J. Johnson, E. E. Sturino, and S. Bourne, *An Automated Gas Chromatographic System for Pesticide Residue Analysis*, Report, 905/4-77-001 EPA, (1977).
60. D. L. Stallings, R. C. Tindle, and J. L. Johnson, Cleanup of pesticide and polychlorinated biphenyl residues in fish extracts by gel permeation chromatography. *J. Assoc. Off. Anal. Chem.* **55**, 32–38 (1972).
61. D. M. Holstege, D. L. Scharberg, E. R. Richardson, and G. Moller, Multiresidue screen for organophosphorus insecticides using gel permeation chromatography—silica gel cleanup, *J. Assoc. Off. Anal. Chem.* **74**, 394–399 (1991).
62. B. E. Kropscott, C. N. Peck, and H. G. Lento, The role of robotic automation in the laboratory, in *Office of Technology Assessment, Pesticide Residues in Foods: Technologies for Detection* U.S. Congress, OTA-F-398, Washington, DC, Oct. 1988. pp. 163–170.
63. R. W. Storherr, and R. R. Watts, A sweep co-distillation cleanup method for organophosphate pesticides I. Recoveries from fortified crops, *J. Assoc. Off. Anal. Chem.* **48**, 1154–1158 (1965).

64. B. G. Luke, Sweep codistillation: recent developments and applications, in *Analytical Methods for Pesticides and Plant Growth Regulators*, J. Sherma. ed., Academic Press, San Diego, 1989, Vol. XVII, Chapter 3, pp. 75–117.
65. Y. Kimura, V. L. Miller, A modified analytical method for microgram amounts of metaldehyde in plant material, *J. Agric. Food Chem.* **12**, 249 (1964).
66. H. A. McLeod, Low temperature apparatus for cleanup of biological sample extracts, *Anal. Chem.* **44**, 1328 (1972).
67. T. E. Archer, and D. G. Crosby, Dehydrohalogenation of toxaphene, *Bull. Environ. Contamin. Toxicol.* **1**, 70–75 (1966).
68. H. A. Moye, and P. A. St. John, A critical comparison of pre-column and post-column fluorogenic labeling for the HPLC analysis of pesticide residues, in *Pesticide Analytical Methodology*, J. Harvey, Jr., J and G. Zweig, eds., ACS Symposium Series 136, American Chemical Society, Washington, DC (1980) Chapter 6, pp. 89–102.
69. J. R. Pardue, Multiresidue method for the chromatographic determination of triazine herbicides and their metabolites in raw agricultural products, *J. Assoc. Off. Anal. Chem. Intl.* **78**, 856–862 (1995).
70. M. L. Langhorst, and L. A. Shadoff, Determination of parts-per-trillion concentrations of tetra-, hexa-, hepta-, and octachlorodibenzo-*p*-dioxins in human milk samples, *Anal. Chem.* **52**, 2037–2044 (1980).
71. R. E. Clement, Utratrace dioxin and dibenzofuran analysis: 30 years of advances, *Anal. Chem.* **63**, 1130A–1139A (1991).

CHAPTER

3

DETERMINATION METHODS

JAMES N. SEIBER

Once a residue-containing extract is provided, with or without cleanup, the steps of resolution, detection, measurement, quantitation, and confirmation are performed, sometimes after derivatization. Advances in the technologies available in these five areas have been responsible for the improvements in both selectivity and detection limits over the past half-century [Fig. 3.1 (1)]. In the 1940s and 1950s, gravimetric and bioassay techniques were the mainstays in "trace" analysis, extending detection limits to the then-remarkable levels of about 1 ppm. The introduction of colorimetric and spectrophotometric methods in the 1950s and early 1960s provided more selectivity over bioassay and gravimetric methods, which responded to whole classes of compounds rather than to individual chemical species, and also improved detection limits at least an order of magnitude. The colorimetric and spectrophotometric methods generally involved derivatization in order to convert colorless or nonabsorbing pesticides into colored or UV-absorbing derivatives for detection and measurement. The Schechter–Haller method for DDT and Averill-Norris method for parathion were examples of these rather complex, wet chemistry–based methods (2).

The inroads of chromatography began in the 1950s with paper and thin-layer chromatography, adding a new dimension in terms of the analyst's ability to resolve individual chemicals contained in fairly complex mixtures. Paper and thin-layer chromatography were primarily qualitative techniques best suited to "screening" samples for the presence or absence of specific pesticides. They could, also be used for quantitation by comparing spot intensities, either visually or through densitometry, with standards. Presently, and beginning in about the 1960s, gas and high-performance liquid chromatography (HPLC) have largely supplanted (but not completely eliminated) PC and TLC. These are resolution techniques par excellence,

Pesticide Residues in Foods: Methods, Techniques, and Regulations, by W. G. Fong, H. A. Moye, J. N. Seiber, and J. P. Toth. Chemical Analysis Series, Vol. 151
ISBN 0-471-57400-7

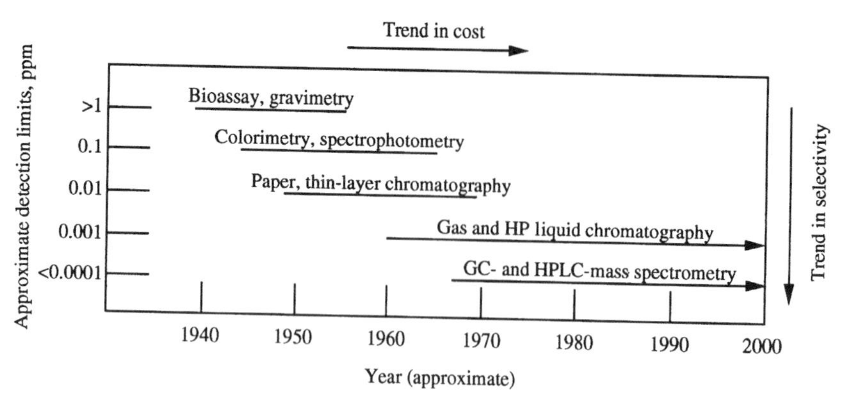

Figure 3.1. The evolution of analytical methodology for organic toxicants in environmental samples (Adapted from Ref. 1.).

with the added dimension of quite precise quantitation made possible by very sensitive and selective detectors. Detection limits of low ppb and even ppt are attainable, particularly for rather clean substrates such as water and air, and occasionally for soil and biological matrices following cleanup. The use of mass spectrometry in connection with GC or HPLC has added still another order-of-magnitude increase in detectability and selectivity, although the costs have, until very recently, relegated these promising techniques to occasional confirmation of analytes measured primarily by GC or HPLC. Spectrophotometry, including IR, UV, and visible absorption and fluorescence, has enjoyed a resurgence during the past 20 years or so as the basis for detection systems for HPLC and, to a lesser extent, GC.

These represent the array of techniques used for 99% + of all pesticide residue analyses, at least through the mid 1990s. Immunoassay, which has the potential for cutting into this chromatography-based bastion of pesticide residue methodology, is currently under rapid development and is beginning to be accepted in "official" methods of analysis. Immunoassay, and some of the newer GC- and HPLC-MS techniques, will be discussed in detail elsewhere in this volume. This chapter deals primarily with GC and HPLC.

GENERAL ASPECTS OF CHROMATOGRAPHY

Chromatography includes a family of techniques ranging from classical gravity-fed columns of the type used by Michael Tswett in 1906, who conducted the earliest chromatographic separations (of plant pigments on a chalk-based adsorbent), to sophisticated high-performance GC and HPLC as

Table 3.1. Classification of Chromatography According to Modes of Separation

Mode	Stationary Phase	Moving Phase	Technique
Adsorption	Solid adsorbent	Liquid Gas Supercritical fluid	Liquid–solid (LSC) Gas–solid (GSC) Supercritical fluid–solid (SCFC)
Partition	Liquid	Liquid Gas Supercritical fluid	Liquid–liquid (LLC) Gas–liquid (GLC) Supercritical fluid–liquid (SCFLC)
Ion exchange	Ion exchanger (resin or glass)	Liquid (usually aqueous buffer)	Ion exchange (IEC) Cation exchange (CEC) Anion exchange (AEC)
Permeation	Porous solid (resin or glass)	Liquid	Gel permeation (GPC)

practiced today. In general, *chromatography* may be defined as a technique for the separation of components of a mixture the basis of their differential movement through a stationary phase under the influence of a moving (mobile) phase. This family of chromatographic techniques may be classified, according to the physical interactions involved between the solute and the stationary phase, into adsorption, partition, ion-exchange, and gel-permeation chromatography (Table 3.1). Useful subdivisions are based on whether the chromatography experiment is done in a column or on a plate, paper, or film (column chromatography vs. planar chromatography), or whether the moving phase is a gas or a liquid or a supercritical fluid (i.e., gas chromatography, liquid chromatography, supercritical fluid chromatography). Although some books deal with all the modes of chromatography [e.g., Willard et al. (4)], it is more common to find books that deal just with GC, or just with HPLC or with TLC, and so on, which, unfortunately, reinforces the impression that chromatography is a series of disjointed techniques rather than emphasizing the commonality of the various forms of chromatography.

In adsorption chromatography, movement of solutes results from the sum of countless adsorption–desorption processes at the molecular level as the solutes percolate through a bed of stationary phase. It depends on the existence of weaker van der Waals forces and/or hydrógen bonding. The greater the affinity a solute has for the adsorbent stationary phase, the longer it will take to traverse through the bed. Adsorption chromatography may be divided into gas–solid or liquid–solid chromatography in terms of the physical state of the moving phase. LSC is the more common of these two, and may be carried out on a plate (TLC) or in a column. LSC, particularly when carried out in a column, is an excellent cleanup–fractionation technique in common use in pesticide residue analysis, as noted in Chapter 2. It can also be used for preparative scale work, to isolate compounds in milligram–gram quantities for further examination. And it is used in HPLC to both fractionate and quantitate various analytes.

In partition chromatography, separation is based on a selective partitioning of solutes between a moving phase (either a gas or liquid) and a stationary liquid phase. The factor that influences the movement of a solute in true partition chromatography is the relative solubility of the solute in the stationary and moving phases, which is a function of the chemical's gas–liquid or liquid–liquid partition coefficients. It can be divided into GLC and LLC, and LLC can be further divided into column, paper, or thin-layer chromatography depending on the means of supporting the liquid stationary phase. LLC can also be subdivided according to whether the stationary liquid is more polar than (normal phase LLC) or less polar than the moving-phase (reverse-phase) LLC.

Ion-exchange chromatography is a form of adsorption chromatography in which electrovalent or ionic linkages are set up between the solute and the stationary phase, contrasted with the van der Waals and hydrogen-bonding forces in adsorption chromatography. Separation is based on the affinity of solute ions for the stationary ion-exchange phase, particularly the ease with which they are displaced from ion-exchange sites by other ions in the moving phase. Ion-exchange is almost always carried out in a column, but further subdivision is possible depending on whether a cation exchange or anion exchange stationary phase is being employed. The moving phase is generally an aqueous buffer. Affinity chromatography has some analogy with ion-exchange chromatography. It designates a form of chromatography in which the stationary phase contains sites identical or similar to enzyme receptors that can "bind" specific substrate solutes. Ion-exchange and, increasingly, affinity chromatography are used to clean up samples prior to determination, or in HPLC columns where resolution and detection are performed in a single instrument.

Gel chromatography employs a porous polymer or glass as stationary phase. Separation of solutes is based on differential diffusion into the pores of the stationary phase. In gel-exclusion chromatography, the most common form of gel chromatography, larger molecules are excluded from entering the pores because of their size, and elute before smaller molecules that can enter the pores and be detained. Partition and ion exchange often occur simultaneously with permeation, so that gel chromatography often represents a blend of several chromatographic separation modes. Gel chromatography has found much use as a cleanup tool for fatty samples, in which the generally larger lipids are resolved from the generally smaller pesticides (5). It is seldom used in the determination step, however.

Many specialty chromatographic techniques are not currently used extensively in pesticide analysis, but may be in the future [Table 3.2 (6–12)]. These include electrophoresis and countercurrent, centrifuge, radial, and field-flow chromatography. Many of these are described, at least briefly, in Willard et al. (4) or in feature section articles in the journal *Analytical Chemistry*.

Table 3.2. Special Techniques of Chromatography and Chromatography-like Separations

Technique	Brief Description	References
Ion chromatography	Separates cations or anions by ion-exchange chromatography with (generally) conductivity detection	6
Electrophoresis	Separates molecules such as proteins, and nucleotides, based on migration rates under the influence of an electrical field; includes modes of moving boundary, isotachophoresis, zone, and isoelective focusing	7
Countercurrent chromatography	Support-free liquid–liquid chromatography with two immiscible liquids	8
Affinity chromotography	Immobilized biochemicals, or affinity ligands, are the stationary phase for bioselective binding of analytes	9
Chiral chromatography	HPLC or GC separation of enantiomers using, usually, a chiral stationary phase	10
Supercritical fluid chromatography	Analogous to HPLC, except that moving phase is a supercritical fluid and stationary phase is an adsorbent or partitioning agent	11
Fluid-flow fractionation	A family of high-resolution techniques, for separating high-molecular-weight and colloidal materials using an external field or gradient	12

CHROMATOGRAPHIC TERMS

Resolution

Three basic terms are used to express the key parameters in all chromatography experiments: retention, resolution, and efficiency. These terms are described briefly here, and in more detail in such references as Snyder and Kirkland (13), Willard et al. (4), and Jennings (14).

The chromatogram reflects the equilibrium of solute distributed between the moving and stationary phase. Each chromatographic peak may be regarded as a Gaussian distribution of solute molecules eluting during chromatography (Fig. 3.2). The retention time, or time for elution of the peak maximum, is independent of solute amount since true chromatography is an equilibrium condition, but it is dependent on a number of factors expressed in the equation

$$t_r = \frac{L}{v}\left(1 + \frac{KV_s}{V_m}\right)$$

where t_r = retention time
L = column length
v = mobile-phase velocity (directly related to mobile-phase flow rate)
K = equilibrium distribution coefficient (concentration of solute in stationary phase/concentration of solute in mobile phase)

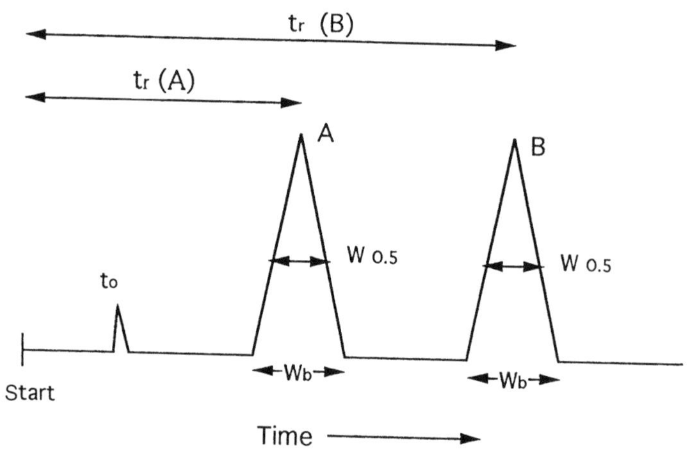

Figure 3.2. Generalized chromatogram showing the resolution of solutes A and B.

V_s = "volume" of the stationary phase (liquid-phase on column in LLC or GLC, or the stationary phase's surface area in LSC)
V_m = volume occupied by the mobile phase in the column (sometimes called the *void volume*).

A useful definition is $k' = KV_s/V_m$, where k' is the capacity factor. From this it follows that

$$k' = \frac{t_r - t_0}{t_0}$$

where t_0, the time it takes for a totally nonretained solute to traverse the length of the column, is directly related to V_m. The capacity factor can be used in place of t_r to express chromatographic retention. But because absolute retention times, and capacity factors, are difficult to reproduce from day to day in the same lab, or from lab to lab, relative retentions are generally preferred, and used in compilations of retention data. In GC of pesticides, for example, aldrin is often used as a reference chemical, and retentions are expressed relative to aldrin:

$$\text{Relative retention time} = \frac{t_r \text{ (pesticide)}}{t_r \text{ (aldrin)}}$$

In capillary GC, t_r-t_0 (the adjusted retention time) should be used in place of t_r owing to the large void volume of capillary columns. The Kovats retention index is a more universal system for expressing GC retention. It uses *n*-alkanes as reference compounds. The retention index of an unknown may be expressed relative to the *n*-alkanes (4). The efficiency of a chromatographic column is expressed as its theoretical plates—hypothetical units within the column where the solute reaches equilibrium between the stationary and moving phases. The number of theoretical plates (*n*) is calculated from

$$N = 16\left(\frac{t_r}{w_b}\right)^2 = 5.54\left(\frac{t_r}{w_{0.5}}\right)^2$$

where w_b is the width of the peak at its base (see Fig. 3.1) and $w_{0.5}$ is the width of a peak at its half-height. The sharper the peak, the greater the *N*, and the more efficient the column toward that solute. Chromatographers must master the variables of their chromatographic system that affect efficiency in order to reliably apply chromatography to an analytical situation. These variables include the moving-phase flow rate, the moving-phase composition, the nature and quality of the stationary phase, and the length and diameter of the column. Generalizing, those chromatographic systems that promote rapid equilibration of the solute between the stationary and moving phases, and minimize dead zones where the solute may be trapped in the column, provide the greatest efficiency. For gas chromatography, these conditions are best fulfilled by capillary columns in which the liquid phase is bonded to the walls of the column, and in liquid chromatography by fairly short columns packed with microparticles of either bonded liquid-to-solid stationary phase or adsorbent stationary phase.

The resolution of a column refers to its ability to separate or resolve two solutes. It is calculated from (Fig. 3.1):

$$R_S = \frac{2\Delta t_r}{w_1 + w_2}$$

A resolution of about 1.5 corresponds to complete (100%) resolution of two components with no overlap between them. In order to achieve this baseline resolution between two closely spaced peaks, the chromatographer may need to adjust one or more of several parameters, such as the column length, mobile-phase flow rate, column temperature (particularly in GC), and any of the parameters described above that affect efficiency (because of the occurrence of the peak width terms in the denominator of the resolution equation). If overlap still occurs after all the efficiency variables have been optimized, a different mobile-phase/stationary-phase combination must be selected, that is, a combination that gives a greater difference in K values for the solutes in question. For HPLC, one may vary either or both the moving phase and the stationary phase, while for GLC only the stationary phase will materially alter the K values.

Detectors

Characterization of detectors is generally based on several parameters. One of these is their sensitivity, which designates the slope of the response curve in a plot of response (peak area or peak height) versus amount (mass or concentration) injected. If two detectors were compared for their sensitivity toward a test solute, and one gave a larger increment of response per increment of injected test solute, it would be judged the more sensitive of the two detectors toward that solute. Sensitivity says nothing about the limit of detection, although the two terms are often used (erroneously) interchangeably. The limit of detection of a detector is set by the noise level, just as the limit of detection of a method is set by the background from the sample matrix. By convention, the limit of detection of a detector is defined as the

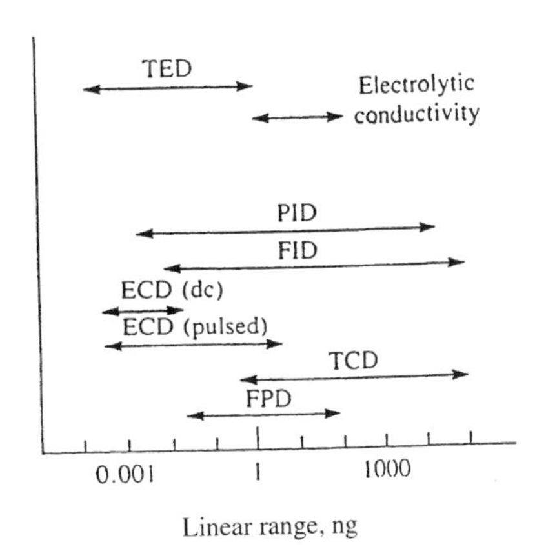

Figure 3.3. Linear dynamic ranges of gas chromatographic detectors: TED, thermionic emission; PID, photoionization; FID, flame ionization; ECD, electron capture; TCD, thermal conductivity; and FPD, flame photometric (4).

quantity of test solute that gives a peak twice the detector's background noise level when the detector's attenuator is set as low as practical to get a steady background signal. Some find it more meaningful to define detectability as the limit of detection (in units of solute weight) divided by the peak width (in seconds). Thus a peak corresponding to 0.1 ng (twice the noise level in this example) with a width of 10 s yields a detectability of 10×10^{-12} g/s.

The detector's linear range is the range of solute amounts over which a plot of response versus amount gives a straight line. For the purpose of characterizing detectors, the lower limit of the range generally corresponds to the limit of detection. A comparison of common detectors is given in Figure 3.3, from which it can be seen that the flame-ionization and photoionization detectors have very broad linear ranges, over about 7 orders of magnitude, while the electron capture detector has only a 100 or, at best, 1000-fold linear range.

The *selectivity* of a detector refers to the relative response of the detector toward equal amounts of two test solutes. For solutes A and B, if equal amounts result in a peak area for A which is 10 times that of B, then

$$\frac{\text{Area A}}{\text{Area B}} = \frac{\text{response factor A}}{\text{response factor B}} = \frac{10}{1}$$

The relative response factor, normalizing on component B, is 10, and the detector is said to be selective toward solute A (and, generalizing, solutes of the class represented by A) by a factor of 10 relative to solute B (and other members of solute B's class). For the "universal" detectors, such as flame ionization and thermal conductivity, F values are close to unity because virtually all solutes produce about the same response. For FID, for example, F ranges from about 0.2 (methyl acetate) to 1.1 (benzene) for combustible organics. For the so-called selective detectors, which include virtually all the detectors commonly used for pesticide analysis, F values may be in the 1000s and higher. The number of selective detectors is less for HPLC than for GC, which has always been a limitation of HPLC for trace residue work. However, the HPLC fluorescence detector is highly selective toward compounds that fluoresce, and the HPLC UV detector may be quite selective at a specific wavelength or operation for a given solute, or solute class.

GAS CHROMATOGRAPHY

A general methodology for pesticide residue analysis evolved that was tailored toward pesticides of relatively high stability and low polarity, and that contained a heteroatom(s) because these predominated in the first synthetic organic pesticide classes of high usage. Practically all the organochlorine pesticides (DDT-type, cyclodienes, benzene hexachlorides, etc.) and organophosphate esters were, in fact, relatively nonpolar so that they could be extracted with an organic solvent, of relatively high stability so that they could be cleaned and/or fractionated by column adsorption chromatography on Florisil or silica gel, and also stable to gas chromatographic temperatures ($\geqslant 100°C$) in the resolution step. And because they contained halogens and/or phosphorus, they were detectable at very low levels and with high selectivity over the matrix background constituents using the early selective GC detectors such as electron-capture (halogens), microcoulometric (chlorine, bromine), and alkali-flame ionization (phosphorus). In fact, the co-occurrence of the first reliable GC instruments (late 1950s–early 1960s) and the electron-capture detector (1959–1960) combined to make GC the method of choice for pesticide residue analysis from about 1960 to the present. The principle was to focus on a unique structural feature of the analyte (presence of a heteroatom), which would make the analyte stand out even when other chemicals were present in the extract at higher concentrations. Background from the interferences that lacked the heteroatom was thus suppressed and the analyte signal was enhanced, resulting in a substantially increased signal-to-noise ratio and a lowered limit of detection. Other operations that suppressed background (cleanup, use of ultrapure solvents, etc.) or enhanced the analyte signal (e.g., derivatization) were included to support the lead role of the selective detector. With this technology, detection limits of 0.01 ppm were readily attainable.

Gas chromatography GC columns for pesticide residue analysis were almost exclusively packed columns until about 1980, when capillary columns began to appear in pesticide labs. By 1990, more capillary columns were in use than packed columns. The standard packed columns and associated instrumental parameters for pesticide residue analysis are described in Zweig and Sherma's *Analytical Methods for Pesticides and Plant Growth Regulators*, Volume 6 (15) and the earlier versions of the FDA's *Pesticide Analytical Manual* (16). The primary variables for the standard 2-m glass column were the choice of solid support, liquid stationary phase, and percent of liquid phase coating on the solid support. The carrier-gas mobile phase was generally not a major variable, and governed more by availability in pure form and compatibility with various detectors. Theoretical considerations from the van Deemter equation indicated that helium would give superior efficiency, all other things being equal (17). In fact, nitrogen was used most commonly because of availability and cost.

The solid support, whose purpose was to hold the liquid stationary phase in place, was chosen to have a large surface area ($0.5–5\,m^2/g$), to be chemically inert and of low sorptive activity toward common pesticide analytes, and to have good mechanical strength to prevent fracture of the coated particles during loading into the column and subsequent handling. Diatomaceous earth, composed of hydrous silica plus its impurities, was used, with Chromosorbs as the principal marketing brand name. Chromosorb G offered the best qualities, and could be purchased deactivated (by acid washing and silanization) in the mesh sizes (60/80 or 80/100) appropriate for the 2-m × 4-mm-diameter glass column. Poropak (crosslinked polystyrene) had specialty use, without need for a liquid phase, for analysis of volatile gases, such as methyl bromide and ethylene oxide. The GC liquid phases were primarily silicone-based oils of high temperature stability. A range of polarities was available, extending from methyl silicone (OV-1, OV-101, SE-30, DC-200) to methyl phenyl silicone (OV-17, SE-52) to methyl trifluoropropyl silicone (OV-210, QF-1) to methyl cyanoethyl silicone (OV-225, AN-600) (15) (Table 3.3). Virtually any pesticide determination amenable to GC could be done with one of these four phases, or occasionally with a phase blend. Nonsilicone liquid phases such as the polyether alcohol Carbowax 20 M or the polyester diethylene glycol succinate (DEGS) were available for specialty use, particularly with the more polar solutes. A typical retention-time table (Table 3.4) provided relative retention time and detector response factors for the common pesticides, from which it was possible to choose the correct GC conditions for a given separation problem. For example, if one wished to slow the elution of the more polar OPs such as

Table 3.3. Common Stationary Phases in Gas Chromatography

Liquid Phase (Type; Similar Phases)	McReynolds constants[a]					
	x'	y'	z'	u'	s'	z'
For Boiling-Point Separation of Broad Molecular-Weight Range of Compounds						
Squalene (2,6,10,15,19,23-hexamethyltetracosane)	0	0	0	0	0	0
OV-101 (polydimethylsiloxane: SE-30, DC-200)	17	57	45	67	43	229
SE 54 (polydiphenylvinyldimethylsiloxane, 5%/1%/94%)	33	72	66	99	67	337
OV-3 (polydiphenyldimethylsiloxane, 10%/90%)	44	86	81	124	88	423
Dexsil 300 (polycarboranemethylsiloxane)	47	80	103	148	96	474
For Unsaturated Hydrocarbons and Other Semipolar Compounds						
OV-17 (polydiphenyldimethylsiloxane, 50%/50%)	119	158	162	243	202	884
Dexsil 410 (polycarboranemethylcyano-ethylsilicone)	72	286	174	249	171	952
OV-225 (polycyanopropylphenylmethylsiloxane; XE 60)	228	369	338	492	386	1813
For Nitrogen Compounds						
Poly-A 103 (polyamide)	115	331	149	263	214	1072
XF-1150 (polycyanoethylmethylsilicone, 50%/50%)	308	520	470	669	528	2495
Retards Compounds with Keto Groups; for Halogen Compounds						
OV-210 (polytrifluoropropylsiloxane; QF-1, SP-2401)	146	238	358	468	310	1520
For Alcohols, Esters, Ketones, and Acetates						
SP-2300 (polyethylene glycol; Carbowax 20 M, FFAP)	316	495	446	637	530	2424
SP-2340 (tetracyanoethylated pentaerythritol)	523	757	659	942	801	3682
For Fatty-Acid Methyl Esters						
Diethylene glycol adipate (SP-2330, HI-EFF-1AP, LAC-1-R-296)	378	603	460	665	658	2764
Diethylene glycol succinate (HI-EFF-1BP, LAC-3-R-728)	499	751	593	840	860	3543
OV-275 (polydicyanoallylsilicone)	781	1006	885	1177	1089	4938
Absolute Index Values on Squalene for Reference Compounds						
	653	590	627	652	699	

Source: Adapted from Willard et al. (4).

[a] *Key*: $x' = \Delta I$ for benzene (intermolecular forces typical of aromatic and olefin compounds); $y' = \Delta I$ for 1-butanol (an electron attractor typical of alcohols, nitriles, acids, nitro compounds, and alkyl mono-, di-, and trichlorides); $z' = \Delta I$ for 2-pentanone (an electron repeller typical of ketones, ethers, aldehydes, esters, epoxides, and dimethylamino derivatives); $u' = \Delta I$ for 1-nitropropane (typical of nitro and nitrile compounds); $s' = \Delta I$ for pyridine.

Table 3.4. Retention Times and Response Values of 53 Pesticides and Related Compounds[a]

	3% OV-1: 180°C, 70 mL / min		5% QF-1: 180°C, 45 mL / min		3% DEGS: 195°C, 70 mL / min	
Pesticide	RRT[b]	RPH[c]	RRT[b]	RPH[c]	RRT[b]	RPH[c]
CDAA	0.10	0.02	0.36	0.13	0.39	0.072
Phosdrin	0.16	0.034	0.69	0.016	*0.74*, 0.88	0.02
CDEC	0.36	1.22	0.68	0.64	0.86	0.52
Phorate	0.37	0.039	0.65	0.023	0.71	0.017
α-BHC	0.35	3.4	0.64	1.8	1.24	0.91
2,4-D (isopropyl ester)	0.40	0.11	0.93	0.12	0.96	0.10
Simazine	0.40	0.002	0.92	0.001	3.8	0.0003
Atrazine	0.41	0.004	0.92	0.001	2.86	0.006
β-BHC	0.40	0.74	0.96	0.34	5.7	0.082
Dimethoate	0.46	0.032	2.36	0.01	5.9	0.011
Lindane	0.44	2.6	0.82	1.8	1.89	0.73
δ-BHC	0.47	2.0	1.12	1.1	5.2	0.093
Diazinon	0.52	0.04	0.77	0.026	0.72	0.024
Daconil 2787	0.51	1.1	2.45	0.32	3.3	0.13
2,4-D (butyl ester)	0.060, *0.69*	0.057	1.29, *1.45*	0.037	1.33, *1.61*	0.031
2,4,5-T (isopropyl ester)	0.67	0.49	1.30	0.32	1.41	0.23
Methyl parathion	0.70	0.22	2.55	0.082	5.1	0.036
Heptachlor	0.78	1.6	0.86	1.2	1.00	1.1
Malathion	0.94	0.054	2.96	0.022	3.3	0.017
Aldrin (reference)[d]	**1.00**	**1.00**	**1.00**	**1.00**	**1.00**	**1.00**
Ethyl parathion	1.00	0.12	3.92	0.047	4.3	0.04
Dyrene	1.27	0.21	1.97	0.081	7.8	0.005
Heptachlor epoxide	1.28	0.66	1.88	0.61	2.68	0.40
Chlorbenside	1.41	0.33	1.86	0.30	3.8	0.29
2,4-D (butoxyethanol ester)	1.59, *1.89*	0.026	2.91, *3.41*	0.018	3.6, *4.7*	0.015
2,4-D (P,G,B, ether ester)	1.61	0.036	2.78	0.028	2.61	0.028
o,p′-DDE	1.60	0.30	1.56	0.34	2.64	0.24
Thiodan	*1.62*, 2.25	0.23	*2.51*, 4.64	0.183	*3.1*, 10.0	0.16
DDA (methyl ester)	1.72	0.005	2.83	0.003	5.7	0.001
Dieldrin	1.93	0.45	2.95	0.37	4.0	0.26
p,p′-DDE	2.00	0.71	2.04	0.48	3.4	0.33
o,p′-DDD	2.04	0.31	2.49	0.20	5.9	0.086
2,4-D (isooctyl ester)	2.20, *2.64*	0.009	3.61, *4.34*	0.011	2.89, 3.3, *3.7*	0.008
Endrin	*2.18*, 2.50	0.21	3.56	0.19	4.4	0.22
Perthane	2.48	0.003	1.90	0.003	3.9	0.002
2,4,5-T (butoxyethanol ester)	2.61, *3.11*, 3.34	0.057	4.33, *5.05*	0.052	5.7, *7.25*	0.022
p,p′-DDD	2.61	0.48	3.64	0.23	10.2	0.079
o,p′-DDT	2.69	0.31	2.60	0.23	4.6	0.15

(continued)

Table 3.4. *(continued)*

	3% OV-1: 180°C, 70 mL / min		5% QF-1: 180°C, 45 mL / min		3% DEGS: 195°C, 70 mL / min	
Pesticide	RRT[b]	RPH[c]	RRT[b]	RPH[c]	RRT[b]	RPH[c]
Ethion	2.84	0.055	5.43	0.034	6.0	0.03
Kepone	2.91	0.079	2.65	0.076	—	—
Dilan	3.16, *3.94*	0.051	10.4, *11.7*	0.02	—	—
Carbophenothion	3.22	0.094	4.80	0.09	8.1	0.057
2,4,5-T (isooctyl ester)	3.56, *4.31*	0.025	5.21, *6.23*	0.013	0.80, 4.3, *5.5*	*0.031*
p,p'-DDT	3.50	0.25	3.93	0.20	8.3	0.079
Methoxychlor	5.70	0.046	6.45	0.054	—	—
Guthion	6.25	0.005	—	—	—	—
Tetradifon	6.05	0.054	18.5	0.10	—	—
Mirex	6.59	0.07	2.65, *3.77*	0.08	—	—
			Multipeak Compounds			
Fenthion	*0.98*, 2.12, 3.10	0.0002	—	—	3.7, 7.7	0.00004
Chlorobenzilate	1.58, 1.98, *2.56*	0.004	1.58, 2.08, *4.19*	0.003	—	—
Dicofol	0.83, *1.00*, 2.06	0.081	1.67, *1.96*	0.093	—	—
Chlordane	0.69, *0.79*, 1.17 1.47, 1.67, 2.67	0.13	*0.87*, 1.29, 1.80, 1.98, 3.27,	0.16	—	—
Strobane	1.79, *2.37*, 2.81, 3.21, 3.82	0.007	1.57, 2.62, 2.90, 3.08, *4.84*, 6.5	0.034	—	—
Toxaphene	1.82, 2.38, 2.85, *3.32*, 3.88, 5.64	0.009	2.72, 2.98, 3.2, 5.0, 6.1, 6.7, 7.8,	0.007	—	—

Source: Adapted from Zweig and Sherma (15).

[a] Order is based on relative retention time on OV-1 (= OV-101, SE-30, DC-200).

[b] Computed as retention relative to aldrin, measurements from injection point. For multipeak compounds, the major peak is indicated in italic. GLC columns (Pyrex 6 f × $\frac{1}{6}$ in. OD − $\frac{5}{32}$ in. ID); tritium source, parallel-plate electron-capture detector.

[c] Computed as peak height response relative to aldrin. Where more than one peak appeared, only the major peak was computed.

[d] Retentions are relative to aldrin.

methyl and ethyl parathion, or malathion, relative to less polar organochlorine compounds, OV-210 or DEGS would be a good choice (Table 3.4). Or if one desired to have pesticides elute in the order of decreasing volatility, a "boiling point" column such as methylsilicone oil (≡OV-1, OV-101, SE-

30, DC-200), would be the logical choice. Use of a higher liquid phase load (10% or so) decreased adsorptive interactions between the solute and the solid support, and also increased the column capacity, preventing overloading from dirty samples, but also requiring use of higher column temperatures for elution.

Capillary column technology has represented a major step forward, because of greatly enhanced column efficiencies (theoretical plate values of 10,000 and higher versus 1000 and less for packed columns) and less tailing of the more adsorptive, more polar pesticides. Capillary, or open-tube GC is a technique in which the column packing material is eliminated, and the liquid stationary phase is coated on, or bonded to the column inside walls. In a typical wall-coated open-tube column (WCOT), a ~0.25-mm-ID (inner diameter) capillary tubing contains a deposit of ~0.25 μm thickness of liquid stationary phase. Higher efficiency results from the elimination of the multipath effect (van Deemter *A* term), which limits the efficiency of packed columns, because the capillary is open to laminar gas flow (14). Also, the open capillary tubing presents no noticeable pressure drop between the column inlet and exit, so that columns could be much longer (⩾10–100 m) compared to 6 ft for the packed column. This factor alone makes possible the much larger numbers of theoretical plates in capillary columns. Finally, the WCOT column has no solid support whatsoever to promote the adsorptive tailing and occasionally decomposition of the more polar, less stable pesticides. Thus compounds such as aldicarb and other carbamates, and the more polar organophosphates and phenols, can often be chromatographed without the need for derivatization. For multicomponent pesticides, such as toxaphene, capillary GC provides much more information on residue behavior, including the fate of individual component. Figure 3.4 compares capillary and packed-column chromatograms for toxaphene residues on cotton leaves showing the loss of the more volatile, early-eluting components on field weathering (18). Drawbacks to capillary column technology include the need for an inlet splitter (19) to prevent overloading when a relatively fine-bore (most efficient) capillary column is used, the need for addition of detector makeup gas to compensate for the lower carrier-gas flow rates in capillary (~1-mL/min) vs packed (~30-mL/min) columns, and the requirement for helium as a carrier gas to promote efficient mass transfer from the carrier gas and the liquid stationary phase, and vice versa, in capillary chromatography. Because of the first drawback, and to alleviate the second, the megabore capillary column was introduced in the 1980s as a compromise between the fine-bore capillary and the packed column. The megabore column is about 0.5 mm ID, and uses 10–20 mL/min carrier-gas flow, usually helium. It does not require an inlet splitter, and a detector makeup gas adapter is optional. The megabore retains some of the higher

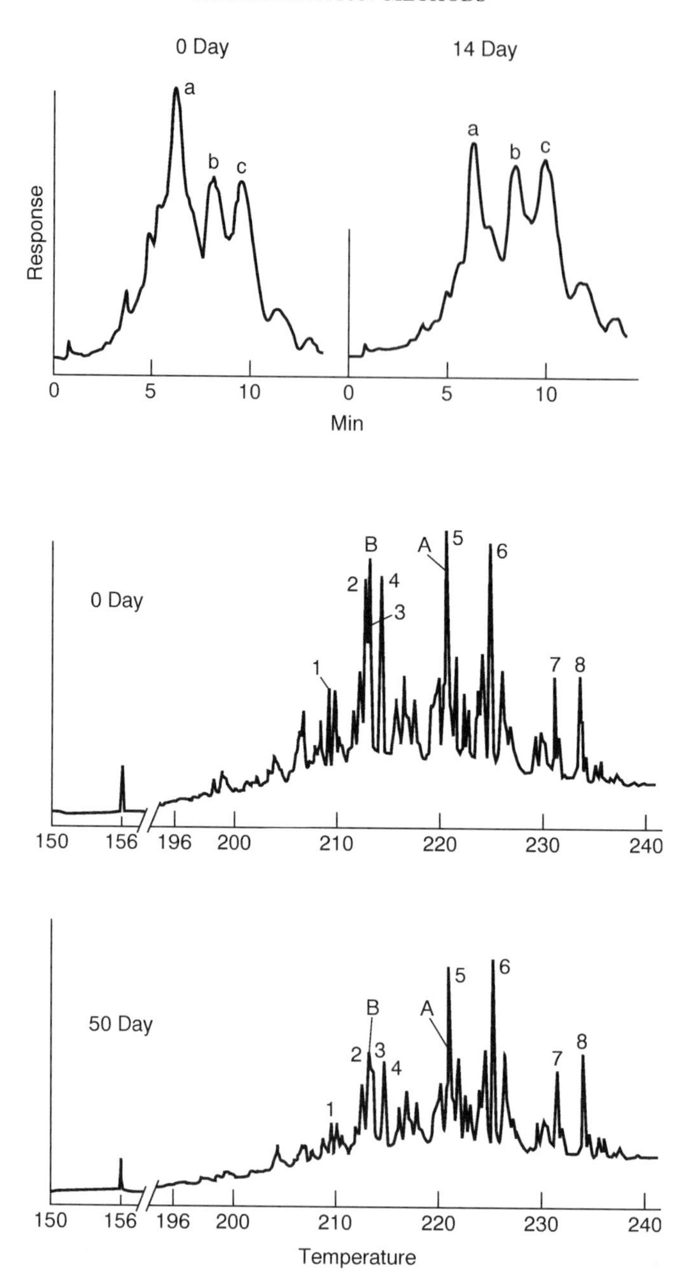

Figure 3.4. Packed (top) and capillary column gas chromatograms of toxaphene residues on treated cotton leaves (18).

Table 3.5. Retention Data for N- and P-Containing Pesticides and Related Compounds on Open-Tube Column (DB-1, NPD, 190°C Isothermal, 1–124 Split injection) and Comparison with Packed Column Retention Data

Compound	t(min)[a]	t'(min)	k'	t'/t' par[b]	Lit[c]
Solvent	1.45	—	—	—	—
p-Chlorotoluidine	1.93	0.47	0.32	0.08	—
Mevinphos	2.11	0.66	0.46	0.12	0.13
Butylate	2.21	0.76	0.52	0.14	—
p-Nitrophenol	2.41	0.96	0.66	0.18	—
Oxamyl oxime	2.48	1.03	0.71	0.19	—
Methomyl	2.53	1.08	0.74	0.20	—
Molinate	2.67	1.22	0.84	0.22	—
Propoxur	2.82	1.37	0.94	0.25	—
Ethoprop	3.01	1.56	1.08	0.28	0.33
Azobenzene	3.06	1.61	1.11	0.29	—
CIPC	3.08	1.63	1.12	0.30	—
2,3,5-Landrin	3.10	1.65	1.14	0.30	—
Chlordimeform	3.22	1.76	1.21	0.32	—
Trifluralin	3.29	1.84	1.27	0.34	—
Phorate	3.43	1.98	1.36	0.36	0.38
Dimethoate	3.53	2.08	1.43	0.38	0.41
Carbofuran	3.56	2.11	1.46	0.38	—
Aminocarb	3.88	2.43	1.68	0.44	—
Diazinon	4.25	2.80	1.93	0.51	0.53
3-Ketocarbofuran	4.33	2.88	1.99	0.52	—
3-Hydroxycarbofuran	5.10	3.65	2.52	0.67	—
Methyl parathion	5.27	3.82	2.63	0.70	0.70
Carbaryl	5.35	3.90	2.69	0.71	—
Terbutol	5.46	4.01	2.76	0.73	—
Paraoxon	5.61	4.16	2.87	0.75	0.78
Ronnel	5.92	4.47	3.08	0.81	0.82
Methiocarb	6.06	4.61	3.18	0.84	—
Malathion	6.46	5.04	3.48	0.92	0.90
Parathion	6.94	5.49	3.79	1.00	1.00
Chlorpyrifos	7.05	5.60	3.86	1.02	1.02
Methidaoxon	7.08	5.63	3.88	1.03	—
Cruformate	7.28	5.83	4.02	1.06	1.10
Methidathion	9.26	7.81	5.39	1.42	1.43
Def	12.41	10.96	7.56	2.00	1.92
Methyl trithion	14.08	12.63	8.71	2.30	2.20
Azinphosmethyl	32.20	30.85	21.28	5.62	5.20

Source: Adapted from Wehner and Seiber (21).

[a] t = retention time, $t' = t$-t(solvent), $k' = t'/t$(solvent).

[b] t'/t' par = retention time relative to parathion.

[c] Lit = relative retention time (to parathion) in *Pesticide Analytical Manual* (16) for 10% DC-200 (SE-30) at 200°C.

efficiency advantages of the fine capillary, and also provides more rapid elution of solutes at generally lower column temperatures. In general, it provides better separation than packed columns is one-third the time. It is the column of choice in most applications of GC in the mid-1990s.

For both the fine-capillary and the megabore capillary, fused silica is the column construction material of choice (20). Fused silica is much more inert than glass, and it can be flexed normally without breaking. Its susceptibility to air oxidation is overcome by coating the column exterior with a polyimide polymer. The liquid phase is bonded to the fused-silica inside wall, solving a major problem with earlier capillary coating techniques in which the liquid phase could move within the column to create puddles, or even volatilize out of the column in high-temperature operation. A bonded-phase column can even be cleaned by washing with solvent without losing its liquid phase. A bonded-phase fused-silica open-tube (FSOT) column is the most popular type in the 1990s.

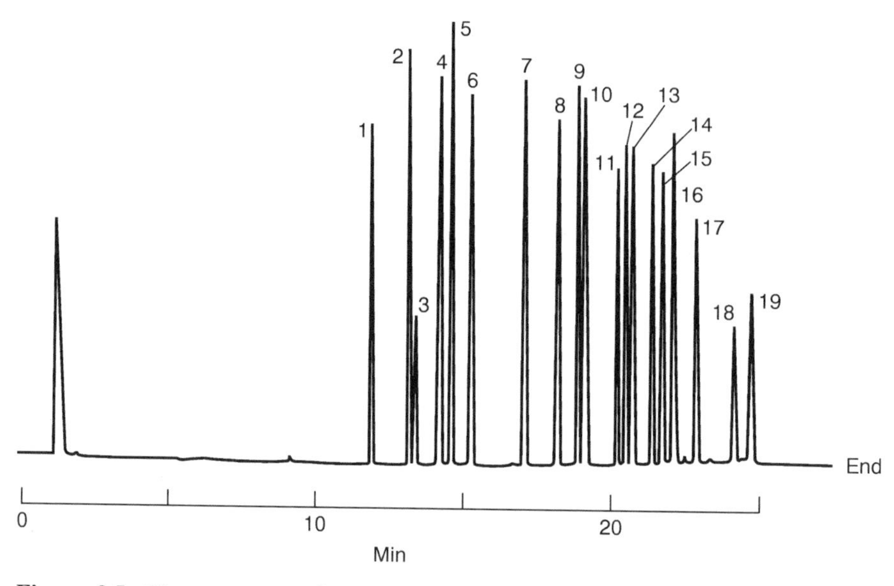

Figure 3.5. Chromatogram of a pesticide mixture on a 30-m × 0.53-mm experimental "DB-608" column. Helium carrier at 6 mL/min; 140°C for 0.5 min, 6°C/min to 275°C; ECD with nitrogen makeup at 30 mL/min. Solutes (200 pg per component): (1) α-BHC, (2) β-BHC, (3) γ-BHC, (4) heptachlor, (5) δ-BHC, (6) aldrin, (7) heptachlor epoxide, (8) endosulfan I, (9) 4,4′-DDE, (10) dieldrin, (11) endrin, (12) 4,4′-DDD, (13) endosulfan II, (14) 4,4′-DDT, (15) endrin aldehyde, (16) endosulfan sulfate, (17) dibutyl chlorendate, (18) 4,4′-methoxychlor, (19) hexabromobenzene (surrogate). [See Jennings (14) for further details.]

The same types of liquid stationary phases are available in capillary technology as for packed columns so that, fortunately, relative retention tables developed for packed columns can still be used as a guide in selecting capillary columns. For example, J & W Company's DB-l is the same composition as OV-1 (≡OV-101, SE-30, DC-200), that is, methyl silicone. DB-5 is the same composition as OV-5, and so on. This relationship can be seen in Table 3.5 (21), generated from various N- and P-containing pesticides on DB-1, compared with literature packed column OV-1 (SE-30) retention data. Figure 3.5 is of a complex mixture of pesticide standards, emphasizing the outstanding efficiencies and resolution afforded by the capillary column.

Temperature programming is widely practiced with capillary GC, in order to allow early-eluting interferences to be removed at the lower temperatures, to promote more rapid elution of medium-to-high-boiling solutes, and to clean the column of impurities prior to a subsequent run. Bonded-phase columns can withstand these wide swings in column temperature without losing stationary phase. Specialty-phase WCOT columns are available for difficult separation problems. An example provided by Onuska et al. (22) is for a cyanopropyltolylalkyl siloxane phase that provided nearly complete resolution for all 22 tetrachlorodibenzodioxin (TCDD) isomers.

DETECTORS

Practically all of the GC detectors commonly used for pesticide residue analysis are of the selective category (Table 3.6). The reason is that, even with very efficient cleanup, a food or soil extract may still contain hundreds of potential interferences, many at much higher concentrations than the analyte(s) of interest. Because one does not want to see the response of these other components, a detector is chosen that is essentially "blind" toward them, that is, not responsive to them, but still responds to the analyte of interest.

Many of the selective GC detectors are variants of the flame-ionization detector, which is well described in virtually all books on gas chromatography or instrumental analysis [see, e.g., Willard et al. (4) and Grob (17)]. The thermionic, or alkali flame-ionization detector (AFID) is essentially an FID with the addition of an alkali salt (potassium bromide, cesium bromide, rubidium sulfate, etc.) to the flame base. It was discovered by chance, by FDA pesticide residue chemist Laura Giuffrida, who inadvertently left a salt deposit on the detector base during a cleaning operation (23). The contaminated detector showed a much enhanced response, and extreme selectivity, toward organophosphorus compounds—which continues as the principal application of the AFID. The detector may be prepared by a

Table 3.6. Element-Selective GC Detectors

Detector	Basis for Selectivity	Year First Reported or Commercialized (Approx.)
Electron capture (EC)	Halogen	1959
Microcoulometric (MC)	Cl, Br, N, S	1961
Alkali flame ionization (thermoionic) (AFID)	P, N	1964
NP-thermionic selective detector (ND-TSD)	P, N	1974
Electrolytic conductivity		
Coulson (CECD)	Cl, Br, N, S	1965
Hall (HECD)	Cl, Br, N, S	1974
Flame photometric (FPD)	P, S	1966
Thermal energy analyzer (TEA)	NO	1975
Photoionization (PID)	Halogen, S, aromatic	1978
GC-MS (bench top)		
Ion-trap detector (ITD)	Diagnostic ions	1983
Mass selective detector (MSD)	Diagnostic ions	1984
Atomic emission detector (AED)	Several elements	1988

relatively straightforward modification of the FID, by inserting a salt pellet on top of the flame base or by placing a metal cup containing the compressed salt as a substitute for the plain metal or quartz base. More commonly, the AFID may be purchased as a unit, with the advantage that the electrode and gas flow geometries are pre-adapted to the many idiosyncrasies of the AFID mode of operation. The gas flows are critical to successful operation, such that accurate flow controllers and needle valves are used with the carrier gas, hydrogen, and air. Although the exact mechanism of the enhanced response of the AFID to P-containing compounds is unknown, it is believed to involve an electron transfer between the alkali metal—formed in the reducing edge of the relatively cool AFID flame—and various P-containing radicals produced from partial combustion of the OP (Fig 3.6). The vapor-phase redox reaction results in a flux of alkali metal cations and P-containing anions that lower the resistance between the electrodes, and set up a current flow in the external circuit that is amplified in the electrometer to provide the recorder signal. A somewhat lower, but still useful, enhancement of response to organic nitrogen compounds in the AFID similarly involves a vapor-phase redox chemistry, between the alkali metal reducing agent and cyano radicals,

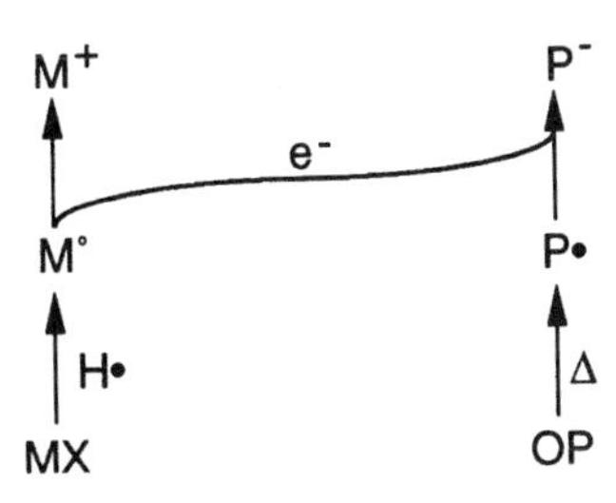

Figure 3.6. Mechanism for enhanced response of alkali-flame-ionization detector to organophosphorus compounds. M° = Alkali metal (Rb, Cs, K); X = Anion (Cl^-, Br^-, SO_4^{2-}; P• = Phosphorous-containing radical (PH_2, PO, PO_2); H• = Hydrogen atom from hydrogen-rich flame.

producing alkali metal cations and cyanide ions. Response factors, and mass detection limits, for typical organophosphorus and organonitrogen compounds are as follows:

	Response Factor (*F*) Relative to Hydrocarbons	**Detection Limit**
Organophosphorus	5000	1–10 pg
Organonitrogen	50	1–10 ng

Because of difficulty in reproducing the response characteristics of the AFID during routine operation, Kolb and Bischoff (24) introduced a major modification that is now the standard for commercial thermionic detectors. The modified detector, termed the *nitrogen–phosphorus thermionic selective detector* (NP-TSD), replaced the flame with a hydrogen–air plasma created around the surface of an alkali-salt bead. The latter is heated by means of resistance wire. Because the extent of heating of the salt bead could be more carefully controlled electrically, the NP-TSD is much more stable and reproducible than the AFID. The lower temperature of the plasma also appears to increase the yield of cyano radicals from organonitrogen compounds, so that the NP-TSD is now a viable detector for trace analysis of organonitrogen pesticides, particularly those, like the triazines, that have several nitrogen atoms. The response factors for the NP-TSD are of the same magnitude as those listed above for AFID, except that the *F* for organonitrogen to hydrocarbon is more favorable, on the order of 500 or more. The NP-TSD is a fairly inexpensive detector with good detection limits for trace analysis of OP, carbamate, urea, triazine, and other classes of pesticides in relatively clean substrates such as water and air. Application to soil and biological matrices may also be made, but efficient cleanup is usually needed and confirmation is essential—because of possible interference from high levels of non-P- and N-containing compounds which are also detected by NP-TSD.

The flame photometric detector (FPD) is another type of flame detector, but is based on the principle that visible light of distinctive wavelengths is emitted when compounds containing specific elements are combusted. The same physical phenomenon that produces distinctive colors such as when lithium (red), sodium (yellow), potassium (violet), calcium (red), and barium (green), are placed in a flame—namely, excitation of outer electrons to higher energy levels, with reversion to the ground state accompanied by emission of light—is involved in FPD. For OP compounds, combustion produces electronically excited HPO radicals, which emit as an envelope of wavelengths centered on 526 nm in returning to the ground state. For organosulfur compounds, excited S_2 is produced that emits at 394 nm in returning to the ground state. The emitted light is focused through an optical interference filter selected to transmit either 526-nm (P-mode) or 394-nm (S mode) light, to a photomultiplier tube that converts the incident light to an electric current. The detector is highly selective, with F values of close to 10,000 for P- and S-containing compounds relative to hydrocarbons. The usable limits of detection are roughly 10–100 pg for OPs and 0.1–10 ng for organosulfur compounds. In the sulfur mode the FPD response is exponential rather than linear owing to the requisite formation of S_2 by a bimolecular collision in the flame. Figure 3.7 shows a chromatogram of four OP insecticides and their oxon metabolites clearly resolved and readily measurable at the 50-pg level.

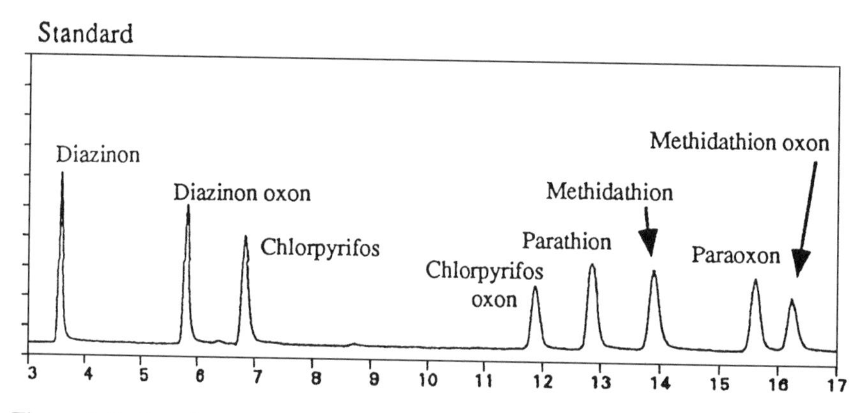

Figure 3.7. Gas chromatogram of 50 pg each of standards of four OP insecticides and their oxon metabolites resolved on a 30-m DB-210 megabore column with FPD detection in the phosphorus mode. Column program was 175–195°C at 2°C/min and then to 230°C at 20°C/min. Units are in minutes.

The expense of the FPD is well worth the investment. Much time is saved by bypassing the need for the extensive cleanup required for soil and biological extracts when using the less selective AFID or electron-capture detectors, and in assigning identities to peaks. In fact, a direct sample introduction system for blended fruit and vegetable samples, coupled with GC using a pulsed-flame photometric detector was described recently (25). Low-ppb detectability was demonstrated for several insecticides and fungicides with much more rapid throughput than is possible using other analytical methods. Also, the response of FPD is very stable, making it the detector of choice for monitoring OP pesticides and S-containing air pollutants. The linear range (10^5 for OPs) and baseline stabilities are other attributes of FPD. Identification of OP pesticides can be materially aided by observing response characteristics on a dual P- and S-mode FPD; response on both modes, for example, indicates presence of both P and S, as might be expected for a thion pesticide, while response only to the P- mode might indicate the presence of an oxon (or nonpesticidal organophosphorus plasticizer such as tricresyl phosphate) (26).

The photoionization detector (PID) owes its response to ions produced by irradiating column effluent peaks with high-energy ultraviolet light, entirely replacing the need for a flame. By varying the energy of the light, the PID can be used as a nonselective, universal detector (11.7 eV) or a selective detector responding primarily to relatively easily ionized organosulfur or polynuclear aromatic hydrocarbons (9.5 eV). Since the energy from the light source may be better regulated than that in a flame, PID gives much less baseline noise, and thus a better limit of detection than FID. Its linear range (10^8; see Fig. 3.3) is the broadest of any of the commercial detectors. PID is perhaps the detector of choice from among readily affordable ones for aromatic compounds that cannot be detected by element-selective detectors, and for sulfur compounds. Detection limits, 1–10 pg for these two classes, are substantially lower than those offered by FPD (sulfur) and FID, which represent alternatives among existing, non-mass-spectrometry-based detectors.

The electron-capture (EC) detector was one of the first of the selective detectors to be introduced for GC, and is still widely used for analysis of polyhalogenated compounds to which it is particularly sensitive. The detector owes its response to the loss of electrical signal when electrons, produced by radioactive beta emitters, are captured by organic analytes as they elute from the GC column. The relatively high standing current produced by beta particles and their descendent thermal electrons, produced following ionization of nitrogen or another carrier gas by the beta particles, is reduced when the electron capture takes place. Operating variable include type of beta source, with Nickel-63 now preferred over tritium (as titanium tritide) because of its higher temperature limit and lower tendency to "bleed" radioactivity from the detector; and mode of signal production, formerly constant potential with varying current as the resistance in the electrode gap increased following electron capture, but now pulse-modulated constant-current mode with output response due to the increase in pulse frequency required to maintain a constant current in the external circuit. This latter mode of operation provides a somewhat increased linear range (up to 10^4) over the very restricted linear range of the older tritium-based constant potential EC detectors (see Fig. 3.3), and also allows for the use of nitrogen as carrier and makeup gas.

The response of EC is very structure-dependent. Contrasts in F values (relative to chlorobenzene) [(see Table 3.7) (27)] are as follows:

Carbon tetrachloride	7000	versus	ethane	0.01
Quinone	5000	versus	hydroquinone	0.1
Nitrobenzene	390	versus	benzene	0.01

In general, polyhalogenated compounds, compounds with electron-withdrawing groups (e.g., nitro), those with conjugated carbonyl systems, and those with sulfur, respond well to EC. In some cases, derivatization is employed to convert non-EC-responsive compounds to EC responders; an example is provided by the conversion of phenols to dinitrophenyl ethers by reaction with dinitrophenyl flouride [several examples of derivatization to enhance EC response are provided in the review by Seiber, (28)]. A major weakness of EC is the wide variety of naturally occurring compounds that possess one or more of the above structural features, presenting an array of potential interferences that can totally obliterate whole regions of the chromatogram unless removed by fairly efficient cleanup. Also, a number of common lab contaminants, such as phthalate ester plasticizers, chlorinated solvents and reagents, and elemental sulfur, can enter valuable samples during storage and in-lab processing, complicating the analysis considerably. Nevertheless, EC is still the detector of choice for monitoring various organochlorine compounds in relatively clean substrates such as water and air, and can be used for more complex matrices provided efficient cleanup and rigorous confirmation are included in the analytical protocol.

The electrolytic conductivity detector is selective for compounds that can be pyrolyzed to gases that, when absorbed in water, increase the conductivity of the resulting solution. For a chlorinated compound, for example, the steps are as follows:

1. Compound is pyrolyzed to release HCl.
2. HCl is absorbed in water.

Table 3.7. Electron Affinities of Different Compounds [a]

Compound with High Electron Affinity	Electrophore [b]	Affinity	Compound with Low Electron Affinity	Affinity
Ethyl pyruvate	–CO · CO–	1700	Ethyl acetate	<0.01
Diacetyl	–CO · CO–	2000	Acetylacetone	1.2
Dimethyloxal acetate	–CO · CO–	1300	Ethyl acrylate	0.1
Diethyl fumarate	–CO · CH:CH · CO–	1500	Diethyl malate	0.01
Diethyl maleate	–CO · CH:CH · CO–	1700	Diethyl succinate	<0.01
Quinone	–CO · CH:CH · CO–	5000	Hydroquinone	0.1
Dimethylnaphthyl quinone	–CO · CH:CH · CO–	4100	Benzyl alcohol	0.05
Benzaldehyde	–CO · Ph	48	Styrene	<0.01
Cinnamaldehyde	–CO · CH:CH · Ph	310	Methyl amine	0.01
cis-Stilbene	Ph · CH:CH · Ph	1	Aniline	0.1
trans-Stilbene	Ph · CH:CH · Ph	4	Pyridine	0.02
Acetamide	–CO · NH ·	12	Phenanthrene	0.05
Azobenzene	Ph · N:N · Ph	9	Naphthalene	<0.01
Diacetyldihydro pyridine	–CO–Py	4000	Cyclopentadiene	<0.01
Anthracene	Aromatic ring	12	Ethane	<0.01
Azulene	Aromatic ring	340	Benzene	<0.01
Cyclooctatetraene	–CH:CH–CH:CH–	210		
Carbon tetrachloride	–Cl	7000		
Chloroform	–Cl	800		
Hexachlorobenzene	–Cl	1100		
Chlorobenzene	–Cl	1		
Bromobenzene	–Br	6		
Iodobenzene	–I	370		
Nitrobenzene	$–NO_2$	390		
Dinitrophenol	$–NO_2$	1450		

Source: McNair and Bonelli (27).

[a] Relative to chlorobenzene = 1.

[b] P, phenyl radical; C, quinone structure.

3. Resulting solution is passed between two electrodes in a flowthrough cell, lowering resistance and thus increasing current in the external circuit.
4. Aqueous solution is deionized and then recirculated to absorber.

The detector may be operated in an oxidative mode (air or oxygen is added to carrier gas as it enters the pyrolysis furnace), in which case sulfur produces SO_3 and, in water, H_2SO_4, and chlorine produces HCl, or in a reductive mode (hydrogen is added rather than air or oxygen), in which case chlorine produces HCl and nitrogen produces ammonia. With the proper selection of combustion gas, "scrubbers" are used to remove unwanted combustion product gases prior to absorbing the desired product gas in water [e.g., $Sr(OH)_2$ is used to scrub HCl when nitrogen is to be detected]. The detector can be selective to compounds containing sulfur, chlorine (and bromine), and nitrogen. The older Coulsen electrolytic conductivity detector (CECD) (29) responded to about 0.1 ng as a lower practical limit of detection depending, in part, on the number of the heteroatoms per molecule. The Coulsen version was, however, fairly clumsy because of its size and use of fragile glass components in the absorber–conductivity cell unit and proneness to contamination requiring frequent servicing. In the Hall electrolytic conductivity detector (HECD) (30), Hall and co-workers redesigned the pyrolysis chamber and gas–liquid absorber, resulting in a reduction in size, and substituted concentric electrodes and an alcohol-based absorbing solution, enhancing the signal-to-noise ratio and overall detectability. Limits of detection of 10 pg and less were attainable with lindane and other polychlorinated hydrocarbons, with F values (relative to hydrocarbons) of 10^5, and a linear range of 10^5. The HECD, and more recent OI version, are detectors of choice for monitoring organohalogen compounds and some organonitrogen and organosulfur compounds in environmental and biological matrices, even though they may require more operator skill in maintaining at high performance levels than is needed for EC and NP-TSD.

The atomic emission detector (AED) (Table 3.8) is the most recent of the non-mass-spectrometry-based selective detectors, introduced in 1988 by Hewlett-Packard Corporation. A high-energy plasma is used to degrade compounds to their basic atomic elements (C, H, N, O, Cl, Br, S, etc.) in their respective electronically excited states. A characteristic wavelength of radiative emission (C, 495.7; H, 486.1; Cl, 479.5, etc.) is monitored to provide element selectivity (Table 3.6). The AED is very sensitive (limit of detection in the low-picogram range) and extremely selective and can also accommodate over 15 different heteroatoms (including tin, for organotin pesticides such as cyhexatin, and mercury), obviating the need for an array of different detectors, each accommodating one or a few of the heteroatoms of potential interest. Its ability to isolate on carbon or hydrogen extends it to general use for organic compounds of literally any type. Its downside is its expense—even greater than that of some of the benchtop GC-MS detectors.

Summaries of applications of the various GC detectors to pesticide residue analyses can be found in review articles/chapters by Beaver (31), Farwell et al. (32), Holland and Greenhalgh, (33) and several others. Specific applications can be found in the *Pesticide Analytical Manual* (16), various

Table 3.8. Atomic Emission Detector Wavelengths for Various Elements

Group	Element	Symbol	Wavelength (nm)
1	Nitrogen	N	174
	Sulfur	S	181
	Mercury	Hg	185
	Carbon	C	193
	Iodine	I	206
2	Phosphorus	P	178
3	Carbon	C	248
	Silicon	Si	252
4	Carbon	C	248
	Mercury	Hg	254
5	Tin	Sn	303
6	Lead	Pb	406
7	Carbon	C	496
	Hydrogen	H	486
	Chlorine	Cl	479
	Bromine	Br	478
8	Hydrogen	H	656
	Deuterium	D	656
9	Fluorine	F	690
10	Oxygen	O	777

volumes of Zweig and Sherma (15), and the *AOAC Official Methods of Analysis*. Generalizing, the detectors most frequently used in pesticide analytical labs in the 1990s include NP-TSD, EC, FPD, and HECD (or OI version). Increasingly, however, the trend is in favor of less reliance on selective detectors and more utilization of GC-MS, particularly in the selective-ion monitoring mode. Not only does GC-MS afford the opportunity to analyze for virtually any organic compound at very low detection limits; it is now available in a cost range that makes it less expensive overall than the array of selective detectors needed to handle the range of pesticide structural types in commercial use, and their toxicologically important metabolites, regardless of heteroatom content.

HIGH-PERFORMANCE LIQUID CHROMATOGRAPHY (HPLC)

Liquid chromatography refers to any chromatographic process in which the moving phase is a liquid. Martin and Synge in 1941 recognized that LC could be a high-performance technique if very small particles were used as stationary phase. It was not until the late 1960s that this theoretical concept was reduced to practice, and several years later that particle packings of reproducibly high quality, and the requisite high-pressure pumps, low-volume connectors and detector cells, and reliable detector optics became available for routine use of HPLC by the analytical practitioner. Thus, applications of HPLC to pesticide residue analysis did not begin in earnest until the early to mid 1970s. In about 1980, the sales figures for HPLC equipment surpassed that of GC equipment—a trend that has continued into the 1990s. A major advantage of HPLC is that it is applicable to virtually any organic solute regardless of its volatility or thermal stability. Another advantage is that both moving- and stationary-phase compositions are variables (vs. only the stationary phase in GC). The inherent efficiency of HPLC is actually greater than that of GC, allowing for the use of very short columns. And most HPLC is practiced at ambient temperature, obviating the need for the bulky oven typical of GC systems. A major disadvantage is the lack of a readily adaptable array of selective detectors comparable to those available for GC.

For a detailed description of the theory of HPLC column separations, the reader is referred to Snyder and Kirkland (13), Simpson (34), Novotny (35), and Brown (36). Modern LC columns achieve high efficiency (high numbers of theoretical plates, low values of the height equivalent to a theoretical plate) by using uniform-size microparticles, with average particle diameters of 1–10 μm and occasionally less. The number of multiple paths available to a given solute is thus minimized, and the mass transfer into and out of the stationary phase, and between the stationary and mobile phases, is maximized. Both factors result in optimization of the efficiency equation. Solute diffusion in the mobile phase (van Deemter equation B term) is relatively unimportant in HPLC compared with GC, because of the much lower diffusion coefficients of solutes in a liquid when contrasted with a gas, reducing band broadening in the mobile phase to just a small fraction of that of GC. Thus, virtually all efficiency considerations dictate the use of small particles of stationary phase.

In adsorption HPLC [liquid–solid chromatography, (LSC)], silica (and occasionally alumina) is used as the stationary phase and an organic solvent is the mobile phase. Solutes of a wide range of polarities may be eluted by increasing the polarity of the mobile phase. As an example, Seiber et al. (37) fractionated pesticides and related compounds of varying polarities on a silica HPLC column by varying the mobile phase from hexane to methyl *t*-butyl ether (Fig. 3.8). In partition HPLC, a normal-phase mode of operation, in which the stationary phase is more polar than the mobile phase, or a reverse-phase mode of operation, in which the polarity combination is

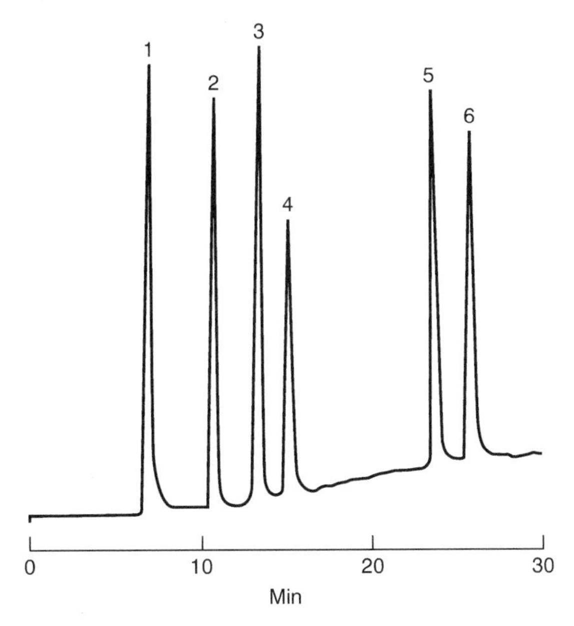

Figure 3.8. Normal-phase HPLC of six pesticides on 25 cm × 4.6 m ID, partisil silica gel using 0–100% linear gradient of hexane to methyl *t*-butyl ether over 30-min run time: (1) 50 μg DDT; (2) 20 μg methoxychlor; (3) 20 μg methyl parathion; (4) 20 μg diazinon; (5) 10 μg carbaryl; (6) 60 μg carbufuran (UV 254-nm detection) (37).

reversed, may be chosen. Many choices of stationary phases are available in both categories, exemplified in the following partial listing:

Normal Phase	Reverse Phase
Silica gel (adsorption)	C_2 hydrocarbon
Alumina (adsorption)	C_8 hydrocarbon
Cyano or alkylnitrile	C_{18} hydrocarbon
Diol or alkyl glycol	*p*-Allyl phenyl
Alkylamine	Cyclohexyl

Applications of HPLC systems to pesticide residue analysis may be found in the chapters by Moye (38) and Ivie (39). An interesting development has been the introduction of correlations of HPLC retention data with water solubilities and other properties (40), allowing for the prediction of the HPLC elution order for pesticides of far-ranging differences in polarity based on the relatively readily available physical properties (41). These correlations appear to work best for reverse-phase HPLC, but may also be used as the basis for estimating retention behavior in normal-phase systems as well. Other new developments have occurred with the introduction of highly selective stationary phases, such as the chiral phases used for enantiomer resolution and immunoaffinity phases for separation of bioactive molecules (42). Still others have occurred in extending truly high-resolution HPLC to ion-exchange chromatography, a development made possible with the introduction of microparticulate glasses with bonded ion-exchange sites (4). Finally, open-tube and microcolumn HPLC have begun to develop, although not nearly so rapidly as the comparable fields in GC. And supercritical fluid chromatography (SCFC) has been available for some time as an alternative to conventional solvent HPLC, using many of the same stationary phases as in conventional HPLC (43). Neither open-tube nor SCFC offer sufficiently significant improvements over conventional HPLC to justify the additional expense of new plumbing–pumping–detector systems, and associated hardware, at least not as of this writing.

HPLC DETECTORS

HPLC detectors are designed to monitor continuously the column eluent for the presence of solutes resolved during the chromatographic run. Virtually any difference in optical or other physicochemical characteristics between the solute and solvent mobile phase may be used as a basis for HPLC detection. In practice, the most useful to date have been refractive index, applicable to virtually any solute and thus the basis for a universal (but not very sensitive) detector approach, UV–visible absorbance and fluorescence, the most common of the HPLC detection approaches, infrared absorbance, conductivity and electrochemical, and mass spectrometric. Descriptions of detector types may be found in Willard, et al. (4), Wheuls (44), Yeong and Synorce (45), and many other references dealing with instrumental and/or HPLC techniques. LC/MS has been the subject of several books, including Brown (42) and Yergey et al. (46).

The UV–visible absorbance and fluorescence detectors and mass spectrometry are the most used detection methods for pesticides and related environmental contaminants (see Table 3.9 for a summary of commercial HPLC detectors). UV–visible absorption detectors probably constitute over 70% of detection systems in use, and include fixed-wavelength, variable-wavelength, and scanning (in real time)-wavelength versions. Advantages include highly refined optic and electrical systems, providing exceptional baseline stability and thus high signal-to-noise ratios at fairly low analyte concentrations, microflow-through cells on the order of 8 μL conventional and 2–3 μL for microcolumn HPLC, again providing low detection limits

Table 3.9. Generalized Performance Characteristics of Commercial HPLC Detectors

Type	Approximate Mass Limit of Detection (ng)	Linear Range (max)	Useful with Gradients
UV–visible fluorescence	0.1–1	10^5	Yes
Electrochemical	0.001–0.01	10^6	No
UV–visible absorbance	0.1–1	10^5	Yes
Conductivity	0.1–1	10^4	Variable
Mass spectrometry	0.1–1	—	Variable
Refractive index	100–1000	10^4	No
FTIR[a] absorbance	100–1000	10^4	Variable

[a]Fourier transform infrared.

and virtually no band spreading, and operability with virtually any solute absorbing significantly at 200 nm and above. The greater selectivity, and usually lower detection limits, occur with analytes that have moderate to strong absorbance at longer wavelengths. Higher sensitivities are generally attainable with fixed-wavelength detectors, providing the analyte of interest absorbs significantly at one of the wavelengths (254, 280, 313, 334, 365 nm) available in the medium-pressure mercury vapor lamp output. More flexibility, including the ability to optimize detection wavelength to maximum signal-to-noise levels for a given analyte/mobile-phase composition, is provided by the variable-wavelength detector. The solid-state diode array scanning-wavelength detector allows for obtaining a real-time spectrum for each solute as it elutes, in as little as 0.01 s. Some instruments may be operated to monitor simultaneously several independent wavelength signals, which is particularly useful for multicomponent detection in complex samples.

Prior knowledge of the UV–visible absorbance characteristics of the pesticide or related compound of interest can be very helpful in selecting the proper monitoring wavelength(s) (47). It is also necessary to have information on the absorbance characteristics (as well as polarities) of solvents and solvent combinations used in HPLC, for which a table of the type presented in Snyder and Kirkland (Ref. 13, pp. 248–250) is indispensable. Monitoring at very low wavelengths (200–210 nm) requires a deuterium lamp and a nonabsorbing (water, methanol, paraffin hydrocarbon) solvent, but has the advantage of approaching universality in detection and fairly high absorbance values for many analyte classes. One can approximately calculate the limit of detection for a given analyte if one knows the molar absorptivity at the wavelength selected for monitoring, the cell pathlength, cell volume, analyte molecular weight, and minimum absorbance required to produce a signal twice the noise level at that wavelength—a detector-dependent characteristic. For example, if molar absorptivity is 10,000, pathlength is 1 cm, cell volume is 5 μL, molecular weight is 500, and minimum *A* is 0.0008, the Beers law–based calculation is 0.2 ng. This assumes that all the analyte band is in the cell at once (i.e., that the chromatographic peak width is on the order of 5 μL, which will be the case only for exceptionally efficient chromatographic systems). More likely, dilution will spread the peak out over 50–100 μL or so, in which case the detection limit will be roughly 2–4 ng. When contrasted with the low-picogram-range sensitivity available with many GC detectors, the disadvantage of UV–visible HPLC for trace residue analysis becomes apparent. A typical example of application of HPLC to pesticide residues was provided by Lerch and Donald (48), who reported the analysis of hydroxylated atrazine degradation products in water using solid-phase extraction and reverse-phase HPLC with UV detection at 220 nm. Detection limits were between 0.1 and 1 ppb.

Fluorescence detectors offer greater sensitivity, to 10 pg and occasionally less, but only for those chemicals that fluoresce themselves, or may readily be converted to fluorescent derivatives. Fluorescent compounds are generally those that possess fused aromatic ring systems, or single aromatic or heteroaromatic rings with electron-donating groups such as NH_2, OH, OCH_3, or alkyl. Electron-withdrawing groups (NO_2, OCOR, etc.) are detrimental to fluorescence. Carbaryl and its breakdown product 1-naphthol, guthion and its breakdown product anthranilic acid, thiabendazole, *o*-phenylphenol, and piperonyl butoxide are examples of fluorescent pesticides. Nonfluorescent amino, phenolic, and carboxylic pesticides may be converted to fluorescent derivatives by the technique of flourogenic labeling derivatization using fluorescamine (primary amines), dansyl chloride (primary or secondary amines), or NBD (phenolics)—techniques in perhaps broader use in TLC than HPLC.

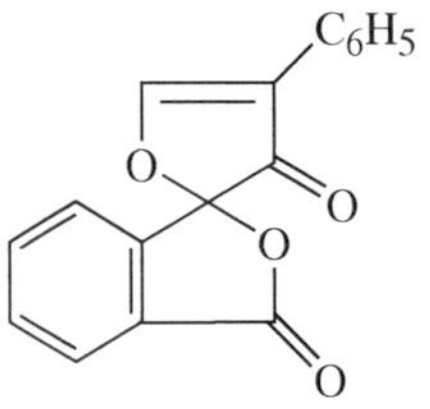

C_6H_5 O O O O SO_2Cl $N(CH_3)_2$ Cl N O N NO_2

Fluorescamine 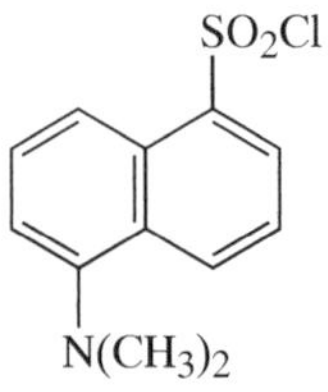 Dansyl chloride 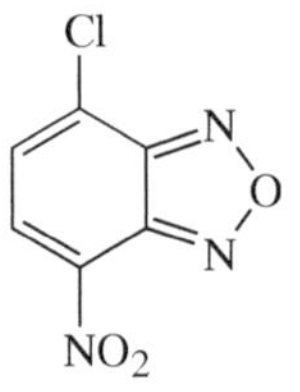 NBD

A specially useful technique of fluorometric labeling was developed by Krause (49) for carbamate insecticides. Eluted carbamates are detected at the nanogram and subnanogram levels with a fluorescent detector after in-line hydrolysis to methyl amine and subsequent reaction with *o*-phthalaldehyde

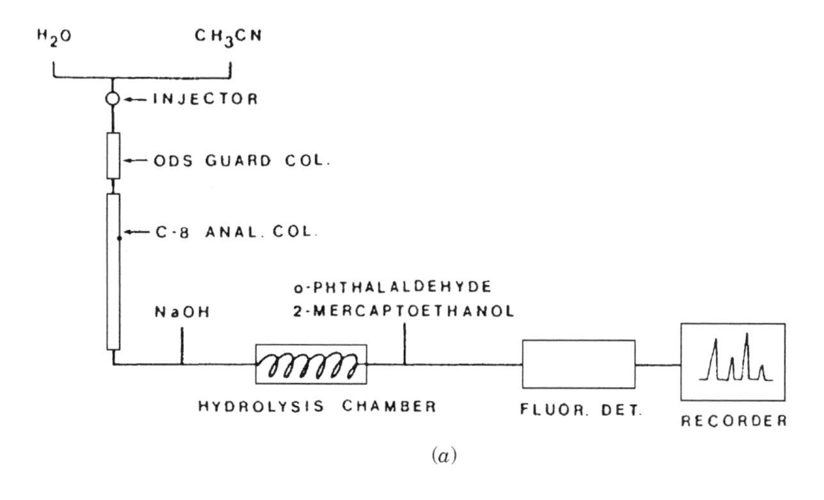

(*a*)

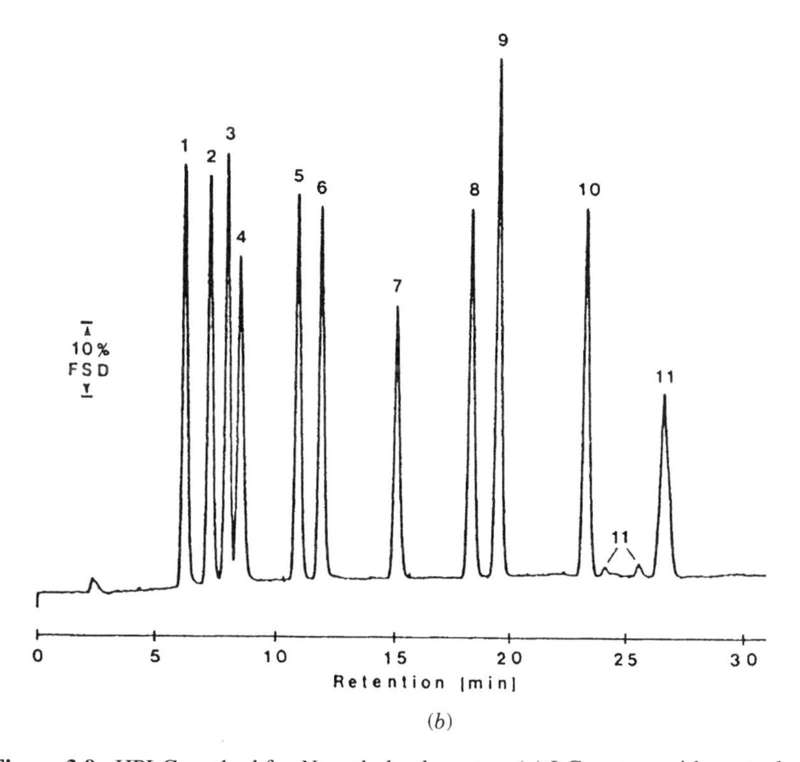

(*b*)

Figure 3.9. HPLC method for *N*-methylcarbamates: (*a*) LC system with postcolumn derivatization and fluorescence detection and (*b*) typical chromatogram for carbamates and metabolites: (1) aldicarb sulfoxide; (2) aldicarb sulfone; (3) oxamyl; (4) methomyl; (5) 3-hydroxy carbofuran; (6) methiocarb sulfoxide; (7) aldicarb; (8) carbofuran; (9) carbaryl; (10) methiocarb; (11) bufencarb (50).

Table 3.10. Retention and Responses of Pesticides that Fluoresce; Relative to Carbofuran on C_8 Column[a]

Pesticide	Relative Retention	Relative Response	Pesticide	Relative Retention	Relative Response
Picloram	0.12	0.0018	Propoxur[b]	0.98	1.02
Maleic hyrazide	0.15	0.0012	Bendiocarb[b]	1.00	0.99
Nitrapyrin metabolite	0.18	0.0003	Carbofuran[b]	1.00	1.00
Bentazon	0.19	0.0024	Karbutilate[b]	1.00	0.0089
Asulam	0.24	0.0002	Mobam[b]	1.02	1.04
Aldicarb sulfoxide[b]	0.33	1.07	Ethoxyquin	1.04	0.0013
dimethyl, 2,4-D acetamide	0.35	0.0002	Carbaryl[b]	1.06	1.44
2,4-D dimethyl amine salt	0.35	0.0002	Thiofanox[b]	1.08	0.022
β-Naphthoxy acetic acid	0.37	0.0004	α-Naphthol	1.10	0.47
Aldicarb sulfone[b]	0.40	1.13	Landrin[b]	1.14	0.90
Oxamyl[b]	0.44	1.05	Norflurazon	1.16	0.0009
Methomyl[b]	0.46	1.01	Chloramben methyl ester	1.18	0.013
Formetanate hydrochloride[b]	0.54	0.0059	Carbanolate[b]	1.18	0.95
Perfluidone	0.56	0.0002	Warfarin	1.22	0.0045
Duraset	0.57	0.0004	Dichlormate	1.23	0.0003
Thiofanox sulfoxide[b]	0.57	0.46	Methiocarb[b]	1.26	0.97
3-Hydroxy carbofuran[b]	0.60	0.99	Promecarb[b]	1.31	0.99
Methiocarb sulfoxide[b]	0.64	0.86	Azinphos-methyl	1.32	0.014
Dimethoate	0.67	0.0002	Methazole	1.36	0.11
Tranid[b]	0.67	0.84	Benomyl[c]	1.23, 1.41	
Thiofanox sulfone[b]	0.69	0.46	Bufencarb[b]	1.44	0.53
Methiocarb sulfone[b]	0.79	0.90	Azinphos-ethyl	1.49	0.0088
AIBA	0.80	0.33	Coumaphos	1.60	0.0002
Naphthyleneacetic acid	0.82	0.0002	Phosalone	1.64	0.0012
Aldicarb[b]	0.83	0.72	Mexacarbate[b]	2.12	0.60
Azinphos-methyl oxygen analog	0.84	0.0094	Aminocarb[b]	3.87	0.48
Coumafuryl	0.94	0.0005			

Source: Adapted from Kraus (49).

[a] Sensitivity of detector adjusted to provide $\frac{1}{2}$ FSD for 10 ng carbofuran. Carbofuran retention approximately 20 min.

[b] Carbamate insecticides.

[c] Multiple peaks from R_c 1.23 to 1.41, major peak at 1.41. Relative response could not be calculated. After 3 days, no peak observed. 12–70% acetonitrile in water at 30 min linear gradient.

Table 3.11. Examples of HPLC Methods for Analysis of Organophosphorus Pesticide Residues (64)

Matrix	HPLC Conditions		Analyte	LOQ[a]
	Column	Detector		
1. River water 2. Drinking water	1. Supherspher C_8[b] 2. Nucleosil C_{18} ODS[c]	1. UV-diode array 2. Variable UV	1. Chlorpyrifos, diazinon, disulfoton, fenamiphos, fenthion, isofenphos, malathion, methidathion, pyridafenthion, temephos 2. Methyl parathion, ethyl parathion, fenitrothion, diazinon, azinphos-ethyl azinphos-methyl phosmet	1. 0.002–0.088 ppb[d] 2. 0.03–0.2 ppb
Soil	LiChrospher 100 RP-18[e]	Thermospray mass spectrometry	Fonofos, fensulfothion, trichlorfon, ethyl parathion, phsomet, chlorpyrifos, fenitrothion	Parent: 20–50 ng (PI);[f] 50–70 ng (NI)[f] Oxons: 1–2 ng (PI); 50–70 ng (NI)
Blood	μBondapak C_{18}[g]	Ultraviolet; APCI-MS[h], PI, and NI modes	21 OPs	1–33 ppb (SIM);[i] 133–667 ppb (TIC)[i]
Animal tissues	Bio-Sil C_{18} HL90[j]	Thermospray mass spectrometry	17 OPs	1–5 ppm

[a] LOQ = limit of quantitation.
[b] Lacorte and Barceló (59).
[c] Driss et al. (60).
[d] Confirmed using HPLC–thermospray mass spectrometry. All other cases cited in this table did not confirm.
[e] Barceló et al. (61).
[f] PI = positive ion; NI = negative ion.
[g] Kawasaki et al. (62).
[h] APCI-MS = atmospheric-pressure chemical ionization mass spectrometry.
[i] SIM = single-ion monitoring; TIC = total-ion chromatogram.
[j] Ioerger and Smith (63).

and 2-mercaptoethanol to form the fluorophore [Fig. 3.10 (50)]. Retention and response characteristics for common insecticides, including those that react in the postcolumn derivatization as well as those that do not, are presented in Table 3.10. More information on applicability of this method for monitoring of carbamates in foodstuffs may be found in Krause (51, 52). Applications to analysis of water have been reported in Graves (53).

Another very useful technique of postcolumn fluorescence derivitization involved an adaptation of the method for carbamates so that it responds to glyphosate, a secondary amine, and its metabolite aminomethylphosphonic acid (AMPA), a primary amine (38). This method involves resolution on an aminex ion-exchange HPLC column eluted with dilute phosphoric acid, postcolumn oxidative conversion of glyphosate to glycine, a primary amine, with calcium hypochlorite solution, followed by formation of a highly fluorescent isoindole by reaction with *o*-phthalaldehyde-2-mercaptoethanol. This has become the standard method for glyphosate and AMPA analysis, replacing the very lengthy and tedious methods based on ion-exchange cleanup and derivatization for GC.

Electroreducible compounds (those containing NO_2, –S–S–, C=O, C≡N, etc. groups) and electrooxidizable compounds (with NH_2, SH, CHO, etc. groups) may be detected at picogram levels using an electrochemical detector. Extremely low detection limits may be obtained when electrochemical detection is combined with microcolumn and capillary-zone electrophoresis separations (54). Although this method would appear to be useful for pesticides that generate aromatic amines and other electrochemically active conversion products during analysis, it has not seen frequent use for residue determinations up to the present time.

Dramatic improvements in HPLC–mass spectrometry instrumentation, particularly in the development of interfaces (55) that do not overly compromise the sensitivity and information output requirements of trace analysis, have led to significant increases in the utility of LC/MS in environmental analysis (36, 45). Applications to pesticide residue and metabolite analysis have been equally forthcoming (56–58). Table 3.11 (59–63) provides examples of HPLC-based methods for OP pesticide residues using UV or mass spectrometry detection. More discussion, and examples, of the use of mass spectrometry, both GC and LC-MS, are given in a subsequent chapter of this volume.

REFERENCES

1. J. N. Seiber, Analysis of toxicants in agricultural environments, in *Genetic Toxicology*, R. A. Fleck and A. Hollaender, eds., Plenum, New York, 1982, pp. 219–234.

2. H. F. Beckman, R. B. Bruce, and D. MacDougall, Spectrophotometric methods, *in Analytical Methods for Pesticides, Plant Growth Regulators, and Food Additives*, G. Zweig, ed., Academic Press, New York, 1963, Vol. I, pp. 131–188.
3. J. M. Van Emon, J. N. Seiber, and B. D. Hammock, Immunoassay techniques for pesticide analysis, in *Analytical Method for Pesticides and Plant Growth Regulators*, Vol. XVII: *Advanced Analytical Techniques*, J. Sherma, ed., Academic Press, San Diego, 1989, pp. 217–263.
4. H. H. Willard, L. L. Merritt, Jr., J. A. Dean, and F. A. Settle, Jr., *Instrumental Methods of Analysis*, 7th ed., Wadsworth, Belmont, CA (1988).
5. D. L. Stalling, R. C. Tindle, and J. L. Johnson, Cleanup of pesticides and polychlorinated biphenyl residues in fish extracts by gel permeation chromatography, *J. Assoc. Off. Anal. Chem*, **55**, 32–38 (1972).
6. J. R. Benson, Modern ion chromatography, *Am. Lab.*, (June), 30–40 (1985).
7. J. W. Jorgenson, Electrophoresis, *Anal. Chem.* **58**, 743A–760A (1986).
8. A. P. Foucault, Countercurrent chromatography, *Anal. Chem.* **63**(10), 569A–579A (1991).
9. R. R. Walters, Affinity chromatography, *Anal. Chem.* **57**, 1099A–1114A (1985).
10. D. W. Armstrong, Optical isomer separation by liquid chromatography, *Anal. Chem.* **59**(2), 84A–91A (1987).
11. M.-L. Riekkola, P. Manninen, and K. Hartonen, in *Hyphenated Techniques in Supercritical Fluid Chromatography and Extraction*, K. Jinno, ed., *Journal of Chromatography library series*, Vol. 53, Elsevier.
12. J. C. Giddings, Field-flow fractionation, *Chem. Eng. News*, Oct. 10, 34–45 (1988).
13. L. R. Snyder, and J. J. Kirkland, *Introduction to Modern Liquid Chromatography*, Wiley, New York, 1979.
14. W. Jennings, *Analytical Gas Chromatography*, Academic Press, Orlando, FL, 1987.
15. G. Zweig, and J. Sherma, *Analytical Methods for Pesticides and Plant Growth Regulators.*, Vol. VI, *Gas Chromatographic Analysis*, Academic Press, New York, 1972.
16. FDA, *Pesticide Analytical Manual*, U.S. Department of Health and Human Services, Food and Drug Administration, Washington, DC, 1994 and prior versions.
17. R. L. Grob, ed., *Modern Practive of Gas Chromatography*, 2nd ed., Wiley, New York, 1985.
18. J. N. Seiber, S. C. Madden, M. M. McChesney, and W. L. Winterlin, Toxaphene dissipation from treated cotton field environments: component residual behavior on leaves and in air, soil, and sediments determined by capillary gas chromatography, *J. Agric. Food Chem.* **27**, 284–291 (1979).
19. K. Grob, Injection techniques in capillary GC., *Anal. Chem.* **66**, 990A–1018A (1994).

20. W. Jennings, Evolution and application of the fused silica column, *J. High Res. Chrom. Chrom. Commun.* **3**, 601–608 (1980).
21. T. A. Wehner, and J. N. Seiber, Analysis of *N*-methyl carbamate insecticides and related compounds by capillary gas chromatography, *J. High Resol. Chrom. and Chrom. Commun.* **4**, 348–350 (1981).
22. F. I. Onuska, R. J. Wilkinson, and K. Terry, Isomer-specific separation and quantitation of tetrachlorodibenzo-*p*-dioxins by HRGC and HRGC / MS, *J. High Resol. Chrom. Commun.* **11**, 9–12 (1988).
23. L. Giuffrida, A flame ionization detector highly selective and sensitive to phosphorus—a sodium thermioonic detector, *J. Assoc. Off. Agr. Chem*, **47**, 293–300 (1964).
24. B. Kolb, and J. Bischoff, A new design of a thermionic nitrogen and phosphorus detector for GC, *J. Chromatogr. Sci.* **12**, 625–629 (1974).
25. H. Jing, and A. Amariv, Pesticide analysis with the pulsed-flame photometer detector and a direct sample introduction device, *Anal. Chem.* **69**, 1426–1435 (1997).
26. M. C. Bowman, Analysis of organophosphorus pesticides, in *Analysis of Pesticide Residues,* H. A. Moye, ed., Wiley, New York, 1981, pp. 263–332.
27. H. M. McNair, and E. J. Bonelli, *Basic Gas Chromatography*, Varian Aerograph, Walnut Creek, CA, 1968.
28. J. N. Seiber, Carbamate insecticide residue analysis by gas-liquid chromatography, in *Analysis of Pesticide Residues*, H. A. Moye, ed., Wiley, New York, 1981, pp. 333–378.
29. D. M. Coulson, Electrolytic conductivity detector for gas chromatography, *J. Gas Chromatog.* **3**, 134–144 (1965).
30. R. C. Hall, A highly sensitive and selective microelectrolyte conductivity detector for gas chromatography, *J. Chromatogr. Sci.* **12**, 152–160 (1974).
31. L. A. Beaver, Detectors for gas chromatography, in *Analytical Methods for Pesticides and Plant Growth Regulators*, Vol. VI, *Gas Chromatographic Analysis*, G. Zweig and J. Sharma, eds., Academic Press, New York, 1972, pp. 39–76.
32. S. O. Farwell, D. R. Gage, and R. A. Kagel, Current status of prominent selective gas chromatographic detectors: a critical assessment, *J. Chromatogr. Sci.* **19**, 358–376 (1981).
33. P. T. Holland, and R. Greenhalgh, Selection of gas chromatographic detectors for pesticide residue analysis, in *Analysis of Pesticide Residues*, H. A. Moye, ed., Wiley, New York, 1981.
34. C. F. Simpson, *Techniques in Liquid Chromatography*, Wiley, Chichester, U.K., 1982.
35. M. Novotny, Microcolumns in liquid chromatography, *Anal. Chem.* **53**, 1294A–1308A (1981).
36. P. R. Brown, High-performance liquid chromatography. Past developments, present status, and future trends, *Anal. Chem.* **62**, 995A–1008A (1990).

37. J. N. Seiber, D. E. Glotfelty, A. D. Lucas, M. M. McChesney, J. C. Sagebiel, and T. A. Wehner, A multiresidue method by high performance liquid chromatography-based fractionation and gas chromatographic determination of trace levels of pesticides in air and water, *Arch. Environ. Contamin. Toxicol.* **19**, 583–592 (1990).
38. H. A. Moye, High performance liquid chromatographic analysis of pesticide residues, in *Analysis of Pesticide Residues*, H. A. Moye, ed., Wiley, New York, 1981, pp. 157–197.
39. K. F. Ivie, High performance liquid chromatography (HPLC) in pesticide residue analysis, in *Analytical Methods for Pesticides and Plant Growth Regulators*, Vol. IX. *Updated General Techniques and Additional Pesticides*, G. Zweig and J. Sherma, Eds., Academic Press, New York, 1980, pp. 55–78.
40. R. Kaliszan, Quantitative structure-retention relationships in chromatography, *Chromatography*, 19–29 (June 1987).
41. W. J. Lyman, W. F. Rechl, and D. H. Rosenblatt, *Handbook of Chemical Property Estimation Methods,* McGraw-Hill, New York, 1982.
42. M. A. Brown, ed., *Liquid Chromatography/Mass Spectrometry Applications in Agricultural, Pharmaceutical, and Environmental Chemistry*, American Chemical Society, Washington, DC, 1990.
43. R. D. Smith, B. W. Wright, and C. R. Yonker. Supercritical fluid chromatography. Current status and prognosis, *Anal. Chem.* **60**, 1323A–1336A (1988).
44. B. B. Wheuls, Detectors for HPLC, in *Techniques in Liquid Chromatography*, G. F. Simpson, ed., Wiley, Chichester, U.K., 1982, pp. 121–140.
45. E. S. Yeong, and R. E. Synorce, Detectors for liquid chromatography, *Anal. Chem*, **58**, 1237A–1256A (1986).
46. A. L. Yergey, C. G. Edmonds, I. A. S. Lewis, and M. L. Vestal, *Liquid Chromatography/Mass Spectrometry Techniques and Applications*, Plenum Press, New York, 1990.
47. R. C. Gore, R. W. Hannah, S. C. Pattacini, and T. J. Porro, Infared and ultraviolet spectra of seventy-six pesticides, *J. Assoc. Off. Anal. Chem.* **54**, 1040–1081 (1971).
48. R. N. Lerch, and W. W. Donald, Analysis of hydroxylated atrazine degradation products in water using solid phase extraction and high performance liquid chromatography, *J. Agric. Food Chem.* **42**, 922–927 (1994).
49. R. T. Krause, Resolution, sensitivity, and selectivity of a high-performance liquid chromatographic post-column fluorometric labeling technique for determination of carbamate insecticides, *J. Chromatogr.* **185**, 615–624 (1979).
50. L. D. Sawyer, B. M. McMahon, W. H. Newsome, and G. A. Parker, "Pesticide and Industrial Chemical Residues," Chapter 10 in *Official Methods of Analysis of Official Analytical Chemists*, 15th Edition, Association of Official Analytical Chemists, Arlington, VA, 1990.

51. R. T. Krause, Multiresidue method for determining *N*-methylcarbamate insecticides in crops, using high performance liquid chromatography, *J. Assoc. Off. Anal. Chem*, **63**, 1114–1124 (1980).
52. R. T. Krause, Liquid Chromatographic determination of N-methylcarbamate insecticides and metabolites in crops. I Collaborative study, *J. Assoc. Off. Anal. Chem.* **68**, 726–741 (1985).
53. L. Graves, U. S. EPA Method 531.1, *Measurement of N-Methylcarbanoyloximes and N-Methylcarbamates in Water by Direct Aqueous Injection HPLC with Post-column Derivatization*, U.S. EPA 600/4-85-054, rev. U. S. EPA 600/4-88-039, 1985.
54. A. G. Ewing, J. M. Mesaros, and P. F. Gavin, Electrochemical detection in microcolumn separations, *Anal. Chem*, **66**, 527A (1994).
55. P. Kebarle, and L. Tang, From ions in solution to ions in the gas phase. The mechanism of electrospray mass spectrometry, *Anal. Chem.* **65**, 972A–986A (1993).
56. R. D. Voyksner, and T. Cairns, Application of liquid chromatography–mass spectrometry to the determination of pesticides, in *Analytical Methods for Pesticides and Plant Growth Regulators*, J. Sherma, ed., Vol. XVII, Academic Press, New York, 1989, pp. 119–166.
57. R. D. Voyksner, Atmospheric pressure ionization LC/MS, *Environ. Sci. Technol.* **28**, 118A–127A (1994).
58. A. Cappiello, G. Famiglini, and F. Bruner, Determination of acidic and base/neutral pesticides in water with a new microlite flow rate LC/MS particle beam interface, *Anal. Chem.* **66**, 1416–1423 (1994).
59. S. LaCorte and D. Barceló, Determination of organophosphorus pesticides and their transformation products in river waters by automated on-line solid-phase extraction followed by thermospray liquid chromatography–mass spectrometry, *J. Chromatogr.* **712**, 103–112 (1995).
60. M. R. Driss, M. C. Hennion, and M. L. Bouguerra, Determination of carbaryl and some organophosphorus pesticides in drinking water using on-line liquid chromatographic preconcentration techniques, *J. Chromatogr.* **639**, 352–358 (1993).
61. P. Barceló, G. Durand, R.J. Vreeken, G. J. Den Jong, H. Longeman, and U. A. Brinkman, Evalution of eluents in thermospray liquid chromatography, mass spectrometry for identification and determination of pesticides in environmental samples, *J. Chromatog.* **553**, 311–328 (1991).
62. S. Kawasaki, H. Ueda, H. Itoh, and J. Tadano, Screening of organophosphorus pesticides using liquid chromatography–atmospheric pressure ionization mass spectrometry. *J. Chromatogr.* **595**, 193–202 (1992).
63. B. P. Ioerger and J. S. Smith, Multiresidue method for the extraction and detection of organophosphate pesticides and their primary and secondary metabolites from beef tissue using HPLC, *J. Agric. Food. Chem.* **41**, 303–307 (1993).
64. J. N. Seiher, J. E. Woodrow, and M. D. David, Organophosphorus esters, in *Chromatographic Analysis of Environmental and Food Toxicants*, T. Shibamoto, ed., Marcel Dekker, New York, 1998

CHAPTER

4

MASS SPECTROMETRY

JOHN P. TOTH

GENERAL ASPECTS OF MASS SPECTROMETRY

The analytical chemist may regard mass spectrometry as a group of options to be considered, along with those discussed in Chapter 3, as detection devices for GC or LC analysis. The diversity and complexity of mass spectrometric detection options merits their being treated in a separate chapter.

The principal advantage of mass spectrometers over other types of detectors is the chemical information they can provide. In general terms, mass spectrometry provides information about the masses, and in some cases the elemental compositions, of the analyte molecule and/or fragments thereof; such data are much more specifically associated with molecular structure, than, for example, FID intensities, retention times, or extinction coefficients. This specificity makes mass spectrometry particularly useful for qualitative analysis.

The most obvious disadvantage of a mass spectrometer as a detector is its cost, which, at the time of this writing, lies in the range of $50,000–$1,000,000. A second drawback is that mass spectrometric detectors are generally less sensitive than other types, although this can be overcome by using selected ion monitoring, which is discussed in greater detail below. Mass spectral sensitivities are also intrinsically unstable, making quantitative analysis difficult. This problem, and means for overcoming it, will also be discussed below in more detail.

The goal of this chapter is to provide a brief introduction to the diverse technology and sometimes confusing terminology associated with mass spectral analysis and to evaluate the current and potential utility of various options for analysis of pesticide residues in foods. More detailed treatments of each subject can be found in the references listed. Technologies that are

Pesticide Residues in Foods: Methods, Techniques, and Regulations, by W. G. Fong, H. A. Moye, J. N. Seiber, and J. P. Toth. Chemical Analysis Series, Vol. 151
ISBN 0-471-57400-7

obsolete and/or not currently applicable to such analysis are included here because the analyst will encounter references to these techniques, and it is useful to be able to recognize inappropriate as well as appropriate techniques. Moreover, techniques not currently useful may become so in the future. More detailed discussions of instrumental components, spectral interpretation and applications may be found in a number of general references (1).

Although there are many mass spectrometer designs, all share six basic components: sample inlet, ion source, mass analyzer, detector, data system, and vacuum system (Fig. 4.1).

The heart of any mass spectrometer, specifically, the component that cannot be changed short of buying a new instrument, is the mass analyzer, which allows selective or sequential detection of ions of different masses. The choice of a mass analyzer determines the mass range, resolution, sensitivity, and cost of the instrument. Although there are at least five distinct types of mass analyzer, all use electric and magnetic fields to separate ions

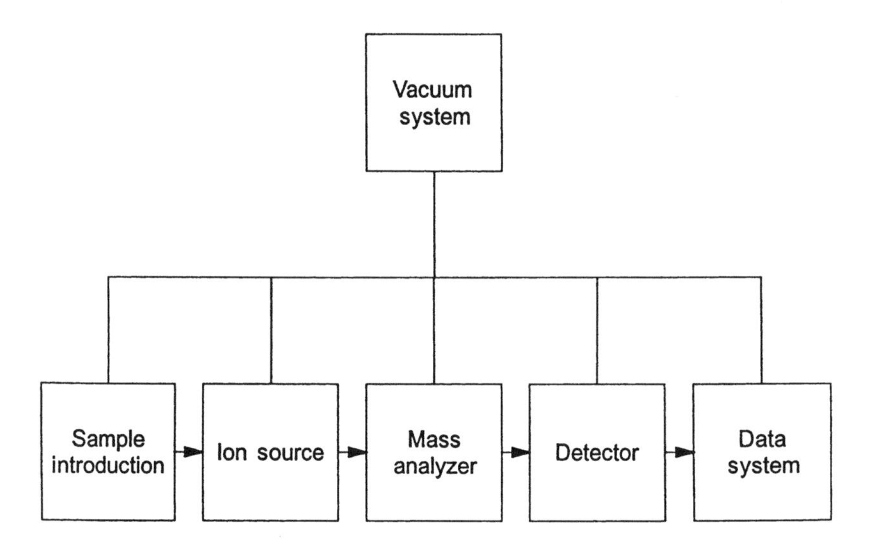

Figure 4.1. Components of a generic mass spectrometer. The *sample inlet* promotes molecules into the gas phase and introduces them to the *ion source,* where they acquire a charge. The ions produced are accelerated into the mass analyzer, where they are separated according to their mass-to-charge ratio, m/z. The detector generates a signal proportional to the abundance of ions of a given m/z value as they emerge from the mass analyzer. The *data system* is used to store the detector output–m/z data that constitutes the mass spectrum, display and analyze the acquired data and control the other components of the system. The *vacuum system* maintains the low pressure required for efficient transport molecules and ions through the system.

on the basis of their mass-to-charge ratio (*m*/*z*). Since the charge of ions produced in most mass spectrometry experiments is +1 or −1, this is tantamount to separation by mass. The presence of a charge is essential for mass analysis, however, so all instruments have an ion source in which neutral sample molecules are ionized. The function of the detector is to produce an electrical signal whose magnitude is proportional to the number of ions striking the detector. The data system serves to store and process the data and to control the various components of the instrument. For efficient transport through the system, ions and molecules must be in the gas phase at as low a pressure as possible. The system must therefore be kept under vacuum and samples introduced in the gas phase. Conventional instruments rely on sample introduction by thermal evaporation, which has limited the applicability of mass spectrometry to volatile, thermally stable organic compounds.

Since about 1980, advances in mass analyzer and detector design have focused on improving instrument performance (vis à vis mass range, sensitivity, and resolution), decreasing instrument cost, or both. During this same period innovations in sample inlets and ion sources have been aimed at extending the applicability of mass spectrometry to involatile, thermally labile analytes. The latter advances have resulted in what can be generally termed *liquid chromatography–mass spectrometry* (LCMS) techniques, although all of them can be performed in the absence of LC separation as flow injection techniques. Because there are major instrumental differences between the volatile-analyte-compatible techniques of *gas chromatography–mass spectrometry* (GCMS) and the involatile-analyte-compatible techniques of LCMS, this chapter is divided into two sections. Each section is divided into subsections describing the available techniques for sample introduction, ionization, and mass analysis beginning with the oldest and progressing to the most recently available and still-developing techniques. Following these descriptions of available hardware, basic principles of data acquisition and analysis will be presented, followed by a brief discussion of applications to pesticide analysis and reflections on current and future options for the analysis. References to pesticide analysis in matrices other than foods are included only if no food applications have yet been reported.

GAS CHROMATOGRAPHY–MASS SPECTROMETRY

Sample Introduction

The volatility and stability requisites for sample compatability with conventional mass spectrometry (MS) are the same as those for gas chromatography (GC). For this reason, the first of the so-called hyphenated techniques to be developed was gas chromatography–mass spectrometry (GCMS). In the GCMS experiment samples are thermally vaporized in a GC injection port and separated on a GC column eluting into the MS ion source, resulting in a mass spectrum for each chromatographic peak. Pure samples not requiring separation can be run using direct probe MS, in which the sample is thermally evaporated directly into the ion source.

The high flow rates of carrier gas associated with packed-column GC require the use of a separator, located between the GC column and the ion source, which removes most of the GC carrier gas while admitting the sample molecules. Both membrane and momentum separators have been used. The lower flow rates associated with capillary GC permit direct introduction of the effluent end of the column into the ion source. The ubiquity of capillary GC has rendered packed-column GCMS as obsolete as packed-column GC, but separator interfaces may still be needed when wide-bore GC columns are used for GCMS.

The standard method for GCMS introduction of sample in complex matrices is off-line solvent extraction followed by direct introduction to the GC injection port via syringe, but any of the on- or off-line extraction techniques described in previous chapters may be used for GCMS analysis.

Ionization

Electron Ionization

The most widely used ionization technique is electron ionization, formerly called *electron impact* (EI) (Fig. 4.2), in which sample molecules (M) are bombarded by high-energy (70-eV) electrons, resulting in high-energy, singly charged molecular ions $(M^+)^*$ that lose excess energy via fragmentation, producing a collection of fragment ions $(m_1^+, m_2^+, \ldots)$ characteristic of the compound.

Chemical Ionization

Since molecular ions are not always seen in EI spectra, a complementary technique, chemical ionization (CI) (Fig. 4.3) is often used to determine molecular weights (2). In a CI experiment, the ion source is charged with a reagent gas, such as methane, at high pressure ($\sim 10^{-1}$ torr), which undergoes EI ionization, producing an excess of reagent ions, such as CH_5^+. Sample molecules are subsequently ionized by the reagent ions via proton transfer. The resulting pseudomolecular ions (MH^+) undergo little or no fragmentation, providing unambiguous molecular-weight information.

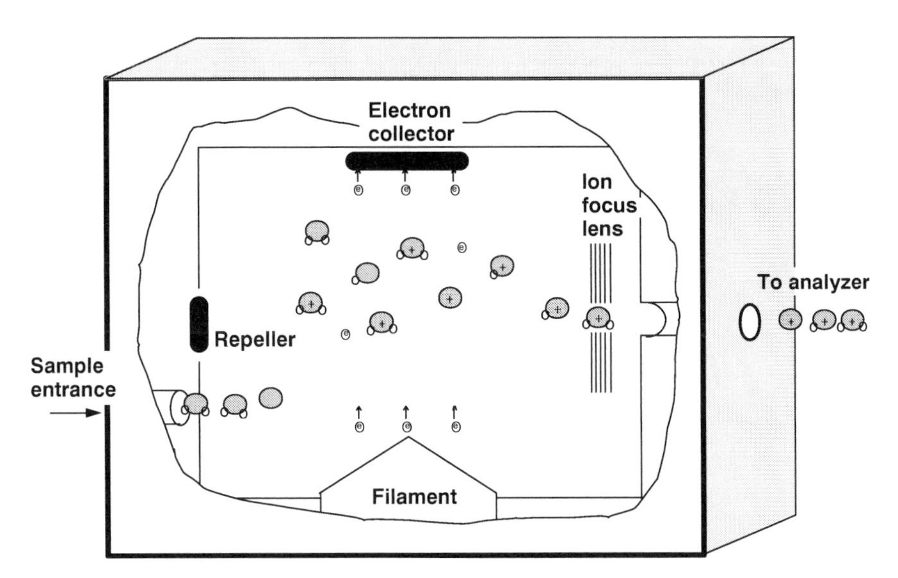

Figure 4.2. The conventional ion source operating under electron ionization (EI) conditions. Electrons emitted from a heated filament are accelerated through the volume of the ion source toward a positively charged target maintained at 70 V relative to the filament, imparting an average kinetic energy of 70 eV to the electrons. Inelastic collisions between the high-energy electrons and sample molecules entering the ion source result in transfer of energy to the molecules that relax by losing electrons, becoming molecular ions. Excess energy still residing in the molecular ions is lost by bond fission, generating fragment ions. The collection of molecular and fragment ions is accelerated and focused by the negatively charged ion focus lenses into the mass analyzer, where they are sorted by m/z value. (Adapted with permission from *What is Mass Spectrometry*? The American Society for Mass Spectrometry, 1998.)

Chemical ionization is often the ionization method of choice for trace analysis of, for instance, pesticide residues using selected ion monitoring (see paragraphs below), since the MH^+ ion is more specific for a given analyte than any lower-mass fragment ion, thus reducing chemical noise from background or coeluting contaminants.

Electron-Capture Negative Ionization

The production of a large population of low-energy electrons during CI operation provides an opportunity for a third mode of ionization, often called "negative CI" but more correctly called *electron-capture negative ionization* (ECNI) (Fig. 4.4). The mechanism of ECNI is the same as that of detection i $CH_4(g, 0.2\ \text{torr}) + e^-_{\text{High energy}} \longrightarrow CH_4^{+\bullet} + 2e^-_{\text{Low energy}}$

ii $CH_4^+ + CH_4 \longrightarrow CH_5^+ + CH_3^\bullet$

iii $CH_5^+ + M \longrightarrow MH^+ + CH_4$

Pseudo-(molecular) ion

iv $C_2H_5^+ + M \longrightarrow C_2H_5M^+$

(M + 29)

v $C_3H_5^+ + M \longrightarrow C_3H_5M^+$

(M + 41)

Figure 4.3. Chemical ionization is performed by prefilling the ion source (see Fig. 4.2) with a reagent gas at relatively high pressure, such as methane at 0.2 torr. The reagent gas ions undergo EI (i) and subsequent ion–molecule reactions (ii) to produce a high population of reagent gas ions, such as CH_5^+ and $C_2H_5^+$. Sample molecules entering the ion source cannot undergo EI due to depletion of available electrons by the reagent gas, but are instead ionized by proton transfer from the reagent gas ions producing pseudomolecular ions (iii). Reaction (iii) imparts little excess energy to the MH^+, so the latter undergoes little fragmentation and is easily identified in the mass spectrum. Further corroboration of the molecular mass is often provided by the observation of adduct ions (iv,v).

by an electron-capture GC detector; a low-energy electron is captured by an electronegative sample molecule, forming an M^- ion, the molecular anion. The molecular anion may or may not undergo fragmentation, depending on its structure. The advantage of ECNI is that ionization occurs roughly 100 times faster than for EI or CI, allowing 100-fold or greater improvement in sensitivity over the other methods. A second characteristic of ECNI is that it occurs only for electronegative molecules, such as, those containing a halogen atom, a nitro group, or an extended aromatic ring system. For other molecules ECNI does not occur at all, making ECNI highly selective for certain types of compounds. ECNI is essentially an electron-capture detector that also provides a mass spectrum (2).

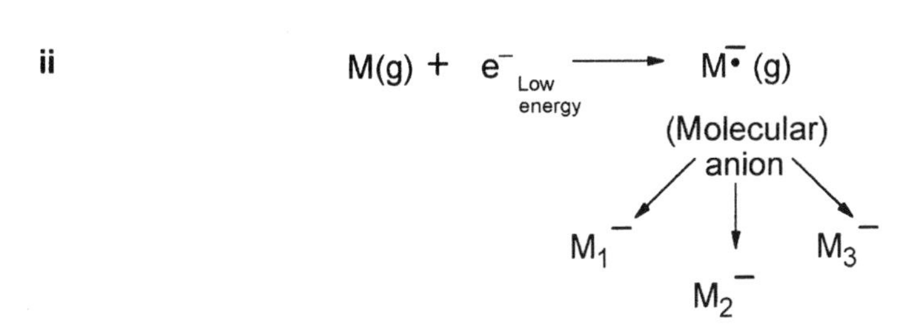

Figure 4.4. Electron-capture negative ionization, or negative CI, occurs under the same conditions as CI. Reaction (i) produces, in addition to reagent gas ions, a high population of low-energy (thermal) electrons that can interact with electrophilic sample molecules (ii) to form negative ions (molecular anions), which may or may not undergo fragmentation, depending on whether the captured electron enters a bonding, nonbonding, or antibonding molecular orbital.

The large class of halogenated and other pesticides that are amenable to ECD detection are also candidates for ECNI detection. A compilation of ECNI spectra of environmental contaminants has been published (3), and this approach has been applied to the detection of organophosphorus pesticide residues in foods (4). Polar analytes that require derivatization for gas chromatographic separation may be derivatized with any of a number of electrophilic derivatizing agents (5) that render the resulting derivatives highly sensitive to ECD or ECNI detection (6).

Mass Analysis

General Considerations

Since choosing a mass analyzer is tantamount to choosing an instrument, it is important to understand the advantages and limitations of each type of analyzer.

As stated above, mass spectrometry is useful for both qualitative analysis and target compound identification/quantitation, but the mode of mass analysis is different for these two types of experiment. When using a mass spectrometer to identify unknown constituents in a sample, it is important to be able to detect as many fragment ions as possible, so the mass analyzer must be scanned continuously from the lowest to the highest possible value; the latter is determined by the expected molecular-weight range and the former by the need to screen against contaminants. The most abundant common contaminant is atmospheric nitrogen, with a molecular weight of 28 daltons, so mass spectral scans frequently start at m/z 29, although the lower limit is often raised to m/z 45 to avoid atmospheric oxygen, argon, and CO_2. Although full-scan ECNI spectra can provide sensitivity comparable to that of an ECD, the more commonly used EI and CI ionization techniques, when used in the full-scan mode, are usually less sensitive than most GC detectors.

When using a mass spectrometer for target compound detection, considerable improvement in sensitivity and selectivity can be attained using selected ion monitoring (SIM), in which the mass analyzer is stepped between a small number of m/z values (typically one, two, or three) without scanning intervening values. Another requirement for a mass analyzer, especially in GCMS, is scan speed. The analyzer must be able to acquire an entire spectrum (or a complete set of SIM signals) in less time than it takes for a single GC peak to pass through the ion source. The rule of thumb is that a single GC peak should contain at least 10 mass spectra. For capillary GC, this requires a scan speed of approximately two scans per second ($2\,s^{-1}$).

Scan speed and the ability to do SIM were once factors that limited the choice of mass analyzers, but due to advances in electronics and data processing capabilities all the mass analyzer designs described below meet the requirements for full-scan and SIM GCMS. The important factors dictating the choice of analyzer are now sensitivity, mass range, resolution, ease of operation, and cost.

Time of Flight

The time-of-flight (TOF) instrument embodies the simplest mass analyzer design and is therefore described first as an illustration of the general principles of mass analysis. Ions are accelerated out of the ion source by a voltage pulse of magnitude V, in the kilovolt range, imparting on each ion a kinetic energy zV, where z is the charge on the ion. The ions enter a linear flight tube of length d with a detector at the opposite end. A timing device measures the time t between the initial voltage pulse and the detection of ions by the detector. The kinetic energy of a particle of mass m moving with velocity v is $mv^2/2$. From $zV = mv^2/2$ we can calculate the velocity of an ion of mass m and charge z as $v = (2zV/m)^{1/2}$. The time t required for a particle of velocity v to traverse distance d is d/v. Therefore, $t = d(m/2zV)^{1/2}$, or $t = (d^2/2V)^{1/2}(m/z)^{1/2}$. The first factor on the right side depends only on instrument conditions; the second factor is a function of the ion, specifically its mass-to-charge ratio. By measuring t, one can determine m/z.

The major advantage of TOF instruments is that ions of extremely high mass can be detected. Until recently that advantage has not been realized because of the difficulty of introducing high-molecular-weight compounds, such as biopolymers, into any mass spectrometer. Recent innovations in sample introduction have generated a resurgence of interest in TOF analyzers, particularly in conjunction with the MALDI sample inlet. The MALDI-TOF combination has not been applied to pesticide residue analysis at this time.

While analyzer designs differ widely, they have common features that are illustrated by the TOF example. Ions are accelerated by electric and/or magnetic fields; the acceleration depends on the existence of a charge z; and there is no way to measure m and z independently—only m/z can be determined.

Magnetic Sector

The double-focusing electric/magnetic sector instrument (Fig. 4.5) is one of a number of designs in which ions are first accelerated though an electric field of magnitude **E**, allowing only ions with a narrow range of kinetic

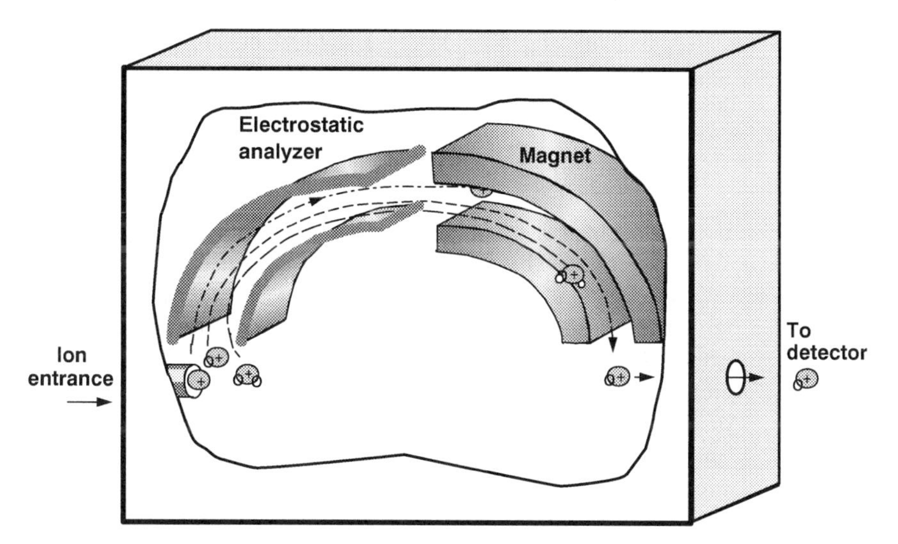

Figure 4.5. A sector mass analyzer utilizing Nier–Johnson geometry. The electric (**E**) and magnetic (**H**) fields in the two sectors force the ions into a curved path whose radius r depends on the magnitudes of **E**, **H** and the ion's mass-to-charge ratio. For a given combination of **E** and **H** only ions of a specific m/z value will have the correct trajectory to pass through to the detector. (Adapted with permission from *What Is Mass Spectrometry*? The American Society for Mass Spectrometry, 1998.)

energies to pass through. Those ions enter a curved flight tube in a magnetic field of strength **H**, which accelerates ions into a curved path of radius r. The value of r is a function of **H** and m/z. Only ions for which r matches the radius of the flight tube will get through the exit slit to the detector. By scanning values of **H**, one scans values of m/z whose paths will get them through the exit slit. The ion beam is thus focused twice, first by **E** and then by **H**.

The major advantage of sector instruments, which stems from the fact that the ion beam is focused twice, is their ability to generate high-resolution data, that is, to distinguish ions differing by less than 1 dalton, thus permitting unambiguous determination of elemental compositions of ions. Another advantage of the sector design is its ability to detect ions of a considerably higher mass than the quadrupole instruments discussed below.

The major disadvantage of sector instruments is their high cost, with prices in the \$300,000–\$500,000 range. This cost is not generally justified for routine pesticide residue analysis because the advantages of these instruments do not apply. Pesticides and their metabolites have sufficiently low molecular weights to be detectable by less expensive quadrupole instruments, and low-resolution data are usually sufficient for detection and confirmation of analytes from a list of known compounds, which is the usual goal in pesticide analysis. The high-resolution capability of sector instruments does, however, provide potentially greater selectivity for target compound detection using selected ion monitoring (see text below).

Quadrupole

In a quadrupole instrument ions are accelerated out of the ion source into a cylindrical region surrounded by four stainless-steel rods (Fig. 4.6). Adjacent rods are always oppositely charged, but the polarity oscillates on the order of 10^6 times per second (radiofrequency). This oscillation forces the ions to move in helical paths through the quadrupole, but for a given value of V, the voltage difference between adjacent rods, only ions of a given m/z value will maintain a stable path all the way though the quadrupole to the detector. By scanning V, one scans m/z.

The major advantages of the quadrupole mass analyzer are its high scanning speeds and low cost. The former was important in the development of GCMS, since the mass spectrometer must be able to keep up with rapidly eluting GC peaks. Improvements in the scan rates of other analyzers have eliminated this as an advantage for the quadrupole, but low cost is still a major advantage, with research-grade instruments in the \$100,000–\$200,000 range and smaller, EI-only benchtop models running well below \$100,000.

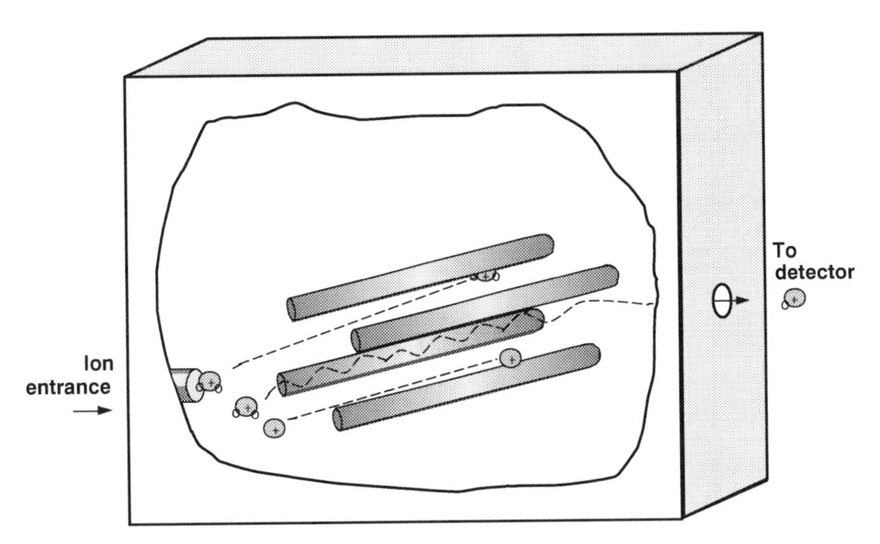

Figure 4.6. The quadrupole mass analyzer. Adjacent rods are oppositely charged, with voltage difference *V*. A positive ion emerging from the ion source will be attracted toward one of the negatively charged rods, but switching the polarity of the rods will deflect the ion's path. Rapid oscillation of the polarity ($\sim 10^6$Hz) will force the ions into a spiral path whose radius depends on the ion's mass-to-charge ratio. For a given value of *V*, only ions of a specific m/z will maintain a stable trajectory for the time it takes to traverse the rods. (Adapted with permission from *What is Mass Spectrometry*? The American Society for Mas Spectrometry, 1998.)

The principal disadvantages of quadrupole instruments are low mass range (with maximum detectable m/z in the range of 1000–2000 daltons) and low resolution (unit mass resolution, precluding unambiguous determination of elemental composition). Neither of these constitutes a serious barrier to analysis of pesticide residues, and quadrupoles have been the most commonly used mass analyzers for such applications.

Tandem Analyzers

In a tandem or MS/MS instrument, as exemplified by the triple quadrupole (Fig. 4.7), ions selected by the first analyzer, called *precursor ions* (formerly called "parent ions"), are passed into a second analyzer in which they undergo collision, usually with an inert gas, to produce fragment ions, called *product ions* (formerly called "daughter ions"). This process is referred to as *collisionally activated dissociation* (CAD) or collisionally induced dissociation (CID). The central quadrupole focuses the product ions, without separation, into the third quadrupole in which they are separated, producing a

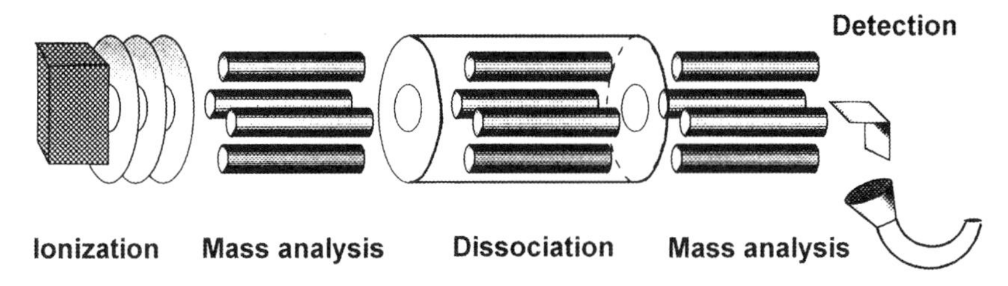

Figure 4.7. A tandem mass analyzer utilizing a triple-quadrupole configuration. Ions of a specific *m/z* value (precursor ions) are selected by the first quadrupole and enter the second, where they undergo collision with an inert gas that promotes fragmentation to product ions [collisionally induced dissociation (CID)]. The second quadrupole focuses all the product ions into the third quadrupole where they are scanned as in a standard quadrupole mass analyzer. Each chosen precursor ion thus gives rise to its own product ion mass spectrum. [Adapted with permission from R. A. Yost and C. G. Enke, *Anal. Chem.* **51**, 1251A–1264A (1979).]

product ion spectrum. Each precursor ion produced in the ion source can thus yield its own product ion spectrum, hence the term MS/MS. The three analyzers do not all have to be quadrupoles, and many analyzer permutations have been investigated, but the triple quadrupole, which is the least expensive, is the most common commercial MS/MS instrument.

The major advantage of the tandem instrument is the additional structural information provided by knowing how the fragment ions can themselves produce further fragments. It is also very useful with ionization techniques such as thermospray and fast-atom bombardment (see text below), which tend to produce MH^+ and/or adduct ions with little fragmentation.

The disadvantages of tandem instruments include their expense ($200,000–$1,000,000) and instrumental complexity. Despite the demonstrated utility of this approach for trace analysis of pesticide residues and other trace analytes, tandem instruments have not seen widespread use for routine analysis. A 1990 survey of commercial mass spectrometry laboratories revealed that of those labs that purchased equipment with MS/MS capability, only 3% ever used that capability (7).

Ion Trap

As the term implies, an ion-trap instrument separates ions of different *m/z* values in time by selectively trapping them in an enclosed space, rather than by separating them in space as in the analyzers described above. This principle is used in two quite different instrument designs, the ion cyclotron resonance (ICR) analyzer and the quadrupole ion trap. In common usage, the term *ion trap*, unless otherwise specified, is understood to refer to the quadrupole ion trap, and that terminology will be used here.

The ICR analyzer (Fig. 4.8) is a rectangular box surrounded by a very strong magnetic field. Ions entering the analyzer are forced by the magnetic field into circular paths whose radii are a function of m/z. The design of the cell permits storage of ions of a wide range of m/z values for relatively long times. The revolving ions constitute oscillating charges and are thus capable of absorbing electromagnetic radiation, with energies corresponding to radiofrequencies. The resonant frequency of a given ion is determined by the radius of its path and thus by m/z. A radiofrequency generator/detector system similar to that used in NMR is used to produce a resonance spectrum

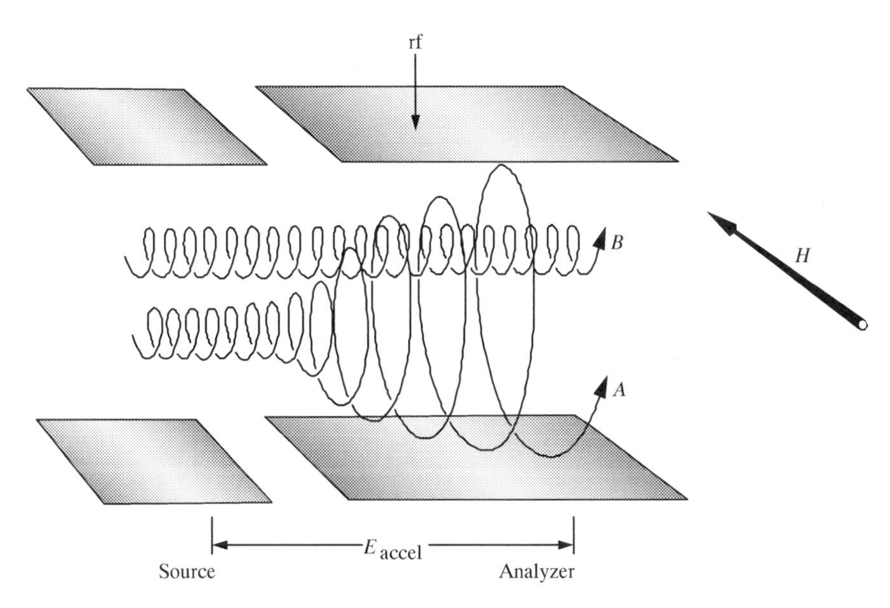

Figure 4.8. The ion cyclotron resonance mass analyzer. Ions generated in the ion source are trapped in stable orbits by an external magnetic field **H**. For a fixed-field strength **H**, the frequency of oscillation ω_{ion} for each ion is determined by its mass-to-charge ratio. Ions are accelerated by an electric field E_{accel} into the analyzer region where they are irradiated by radiofrequency energy (rf), which is scanned continuously through a range of frequencies ν. An ion, designated A, for which $\nu = \omega_A$ absorbs energy, increasing the amplitude of its oscillation. Other ions, designated B, do not absorb energy. Ions of mass-to-charge ratio m/z are detected by the absorption of rf radiation at the corresponding resonant frequency. (Adapted with permission from M. M. Bursey, Mass spectrometry, in *Chemical Instumentation: a Systematic Approach*, 2nd ed. H. A. Strobel, ed., Addison-Wesley, Reading, MA, 1973.)

in which the resonance peaks correspond to m/z values of the ions stored in the cell.

Since ions are detected by the energy they absorb, rather than by their impingement on a detector, sensitivity can be enhanced by methods developed for optical spectroscopy, such as Fourier transform analysis. Conventional spectroscopic instruments allow individual wavelengths to pass sequentially though a narrow slit, with inevitable loss of sensitivity. The higher the resolution, the narrower the slit and the greater the loss of sensitivity. Fourier transform analysis allows for simultaneous detection of all wavelengths, followed by mathematical generation of the spectrum. Applying this to ICR spectra permits very high resolution without loss of sensitivity.

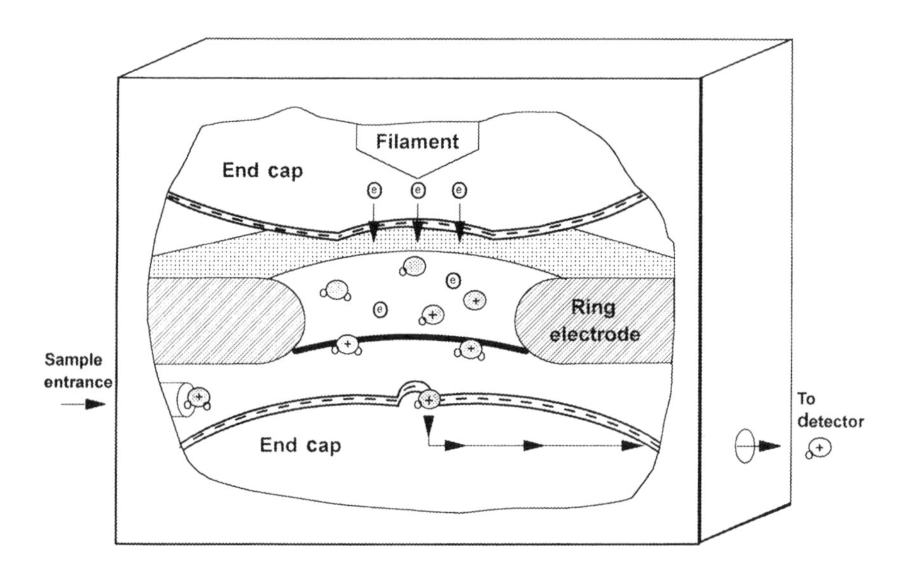

Figure 4.9. The quadrupole ion-trap mass analyzer. In this design, ions are trapped in the same space in which they are generated (by EI or CI). This space is enclosed by the two end-cap electrodes, which always bear the same charge, and the oppositely charged ring electrode. Changing the polarity $\sim 10^6$ times per second generates an oscillating field in the enclosed region analogous to that generated in the quadrupole in Figure 4.6, except that ions are trapped in circular orbits rather than spiral trajectories. The radius of the orbit depends on its mass-to-charge ratio and the voltage difference V between the ring and end-cap electrodes. Ions of a given m/z are trapped in stable orbits when V reaches the appropriate value, then ejected when V exceeds that value. Ejected ions are accelerated into the detector. (Adapted with permission from *What is Mass Spectrometry*? The American Society for Mass Spectrometry, 1998.)

The major disadvantage of FT-ICR is its high cost, which extends beyond the $\sim \$200,000$ purchase price; to produce the needed field strengths one needs a superconducting magnet that requires a steady supply of liquid helium. FT-ICR is used primarily as a research tool for studying ion–ion and ion–molecule reactions.

The quadrupole ion trap (Fig. 4.9) uses a cylindrical cell with a quadrupolar electric field to effect the trapping of ions in circular paths, as in ICR. Path radius is a function of the voltage V between the doughnut-shaped electrode and the end caps, and m/z of the ion. As in the quadrupole analyzer, only ions of a given m/z will have a stable orbit at a given V. The sample is ionized in the trap by electron impact while V is scanned. When V reaches a value corresponding to a stable orbit for a given m/z, ions with that mass-to-charge ratio are trapped. As V passes the stable-orbit value, the trapped ions go into an unstable orbit and are ejected from the trap to the detector.

The principal advantage of the ion trap for analysis is that such instruments are considerably cheaper than all other types for doing routine EI analysis of low-molecular-weight compounds. Ion traps are also capable of CI and ECNI ionization, and commercial instruments have been introduced that are capable of MS/MS.

Early commercial ion-trap designs suffered from difficulties in producing EI spectra comparable to those generated by quadrupole and sector instruments, making them incompatible with library searching using existing databases. This problem appears to have been solved in contemporary ion-trap instruments, but more recently it has been shown that ion traps operating in the CI mode can produce spectra with a combination of EI and CI characteristics, creating potential problems for multiresidue detection using CI mass spectrometry and SIM in ion-trap instruments (8).

Data Acquisition and Analysis

As discussed briefly above, mass spectrometry can be used for qualitative identification of unknown analytes, detection and/or confirmation of target compounds, and quantitation of target compounds. These three types of application require different modes of operation.

Qualitative Analysis

For qualitative analysis, the desirable characteristic is specificity, which is achieved by observing the molecular ion and as many fragment ions as possible. Electron impact is therefore the most commonly used ionization technique, supplemented, if necessary, by CI data to confirm the molecular weight. Because much of the time of a full-scan EI acquisition is spent scanning m/z values corresponding to ions of zero or low abundance, the sensitivity of this mode is not great. As a rule of thumb, a reliable identification of an unknown analyte requires an injection of $\sim$50–500 ng of that analyte.

Spectral interpretation is best done by the analyst using well-established rules of fragmentation (1c). This is not always practical due to the very large numbers of spectra that can be produced by a complex sample yielding many GC peaks and to the fact that many analysts lack formal training in mass spectral interpretation. For these reasons, most commercial GCMS data systems include an on-line library of EI spectra and a search algorithm for automated comparison of an experimental spectrum with the library spectra. The most commonly used spectral libraries are the NIST library (9), containing $\sim$75,000 EI spectra of $\sim$62,000 compounds and the Wiley library (10), with $\sim$140,000 EI spectra of 118,000 compounds. Subsets of these libraries containing, for example, spectra of pesticides and metabolites; are also available. Library searches should always be treated with caution. A computer search may find a best-fitting spectrum in the library, even when the spectrum of the actual compound does not reside in the library. The entries in any library are gleaned from the literature, and there is no guarantee of the quality of any given spectrum. Different compounds can have EI spectra that are indistinguishable (e.g., geometric isomers or homologous alkanes). Nonetheless, when used cautiously and in conjunction with other information, library searches can be an invaluable time-saving device.

Target Compound Detection

If the analyst is interested in establishing the presence or absence of a given compound without regard to any other components (i.e., target compound detection), the desirable characteristics are selectivity, that is, production of a signal unique to the target compound, and sensitivity. Both of these can be achieved using selected ion monitoring (SIM), in which the mass analyzer is stepped between a small number of m/z values instead of scanning continuously over a wide range. This improves selectivity, by collecting data only at m/z values corresponding to ions characteristic of the target compound, and sensitivity by eliminating the time spent collecting data in silent parts of the target compound's spectrum. The improvement in sensitivity over full-scan MS is typically a factor of 10^2 or better. Sensitivity and selectivity are attained at the expense of specificity, since the information contained in the ions not observed is lost; greater selectivity, however, makes SIM more compound-specific than other detectors. It has been shown that a minimum of three coeluting, structurally significant ions with retention times and intensity ratios matching those of standards is a statistical requirement for confirmation by SIM GCMS (11); in practice, expert concensus appears to prefer no fewer than four such ions for confirmation (Refs. 12, 13 and references cited therein).

The selectivity of SIM detection for a particular compound depends on the uniqueness of the ion or ions selected for monitoring. The first step in establishing a SIM procedure is to obtain a full-scan spectrum of the target compound, from a library if necessary but preferably experimentally on the same instrument to be used for SIM. The target ion or ions are chosen from those displayed in the spectrum. The desirable traits for a target ion are high mass (for selectivity) and high abundance (for sensitivity); the former is more important. For many organic compounds the most abundant ion will be a low-mass fragment ion, but these should not be used for SIM since such an ion will be common to many compounds, and selectivity will be lost. The ideal target ion would be a high-abundance molecular ion, and for this reason CI is often used for SIM. If the target compound is electronegative (e.g., a halogenated pesticide), extremely high sensitivity can be attained using electron-capture ionization and SIM detection. This is not always practical, however, since many halogenated compounds ionize under ECNI conditions to produce Cl^- or Br^- as the only ion, providing very little selectivity.

Selectivity can be enhanced by the use of a tandem mass analyzer. In SIM analysis, confidence that a candidate compound is identical with a standard is based on the fact that they possess ions with the same m/z values. Confidence is increased if these ions also have the same product ions after dissociation. Selectivity may also be improved by using a sector instrument to perform exact-mass SIM; that is, the target m/z values are specified to a sufficient number of decimal places to ensure selective detection of ions of a specific elemental composition. Methods for highly selective and sensitive detection of pesticide residues using a combination of exact-mass SIM and MS/MS have been reported; the cost of instruments capable of such analyses begin at about $500,000. Whether the advantages of such detection methods for a specific problem justify the cost of the instrumentation will be a matter of individual judgment.

Quantitation

Quantitation by mass spectrometry can be considered target compound identification taken one step further. Once a SIM method for detection of the analyte is established, quantitation is performed by comparing signal intensity of the analyte with those of a standard. This, of course, requires stability as well as selectivity and sensitivity, and this poses a problem for mass spectrometry. In a mass spectrometer, unlike an optical spectrometer, the sample passes through the optics and impinges directly on the detector, causing the response of the instrument to change from day to day or run to run. For this reason, accurate quantitation by mass spectrometry usually requires the use of internal standards. Since the instrument may display quite different sensitivities to the sample and the internal standard, a response factor R must be calculated for each analyte as $R = (A_a/A_s)(C_s/C_a)$, where A_a and A_s are the GC peak areas of the analyte and standard and C_a and C_s are the corresponding concentrations. The concentration of analyte in an unknown sample spiked with the same internal standard is then calculated as $C_a = (A_aC_s)/(A_sR)$.

The ideal internal standard is an isotopically labeled analog of the analyte; such a standard is chemically identical to the analyte, so its response should be identical, specifically, $R = 1$, and its GC retention time will also be identical to that of the analyte so the instrument conditions at the time of elution will be identical. Quantitation with such an internal standard, called *isotope dilution mass spectrometry*, is possible only if the labeled standard is available and affordable.

Because of these difficulties and the time and expense of mass spectral analysis, it is usually preferable to do quantitation by other means and use mass spectrometry only to confirm the identity of the analyte. Nonetheless, there are certain instances in which mass spectrometry is the only acceptable method of quantitation, notably when stable-isotope-labeled compounds, which can be distinguished only by their masses, need to be quantitated. Mass spectral quantitation is also a legal requirement in some regulatory contexts.

Applications and Conclusions

The most commonly used mass spectrometric technique for analysis of pesticide residues in foods and other complex matrices is GCMS with on-line or off-line extraction, EI, CI or ECNI ionization, and quadrupole mass analysis. Published pesticide residue analyses based on quadrupole GCMS are too numerous to list here. Criteria for acceptable GCMS data for pesticide residues and other trace contaminants in food and drugs were the subject of a 1989 review (13).

Although continued incremental improvements in sensitivity and detection limits may be expected, radical changes in inlets and ion sources are probably not forthcoming. The greatest focus of research in GCMS methodology at the moment appears to be in the development of the quadrupole ion-trap as the analyzer of choice. The motivation for this is primarily economical, with, at the time of this writing, at least two commercial ion-trap GCMS systems available for around $50,000. The potential for doing MS/MS with such an instrument is also a motivation, although this capability has more direct appeal for LCMS applications. Since EPA- and other officially sanctioned analytical methods emphasize performance rather than specific techniques, a number of investigators have examined the ability of the less expensive quadrupole ion traps to meet these performance criteria (14).

The use of ion traps has been reported for GCMS detection of specific contaminants in whole milk (15). Cairns et al. have studied the characteristics of chemical ionization in ion traps in the context of the development of a protocol for multiresidue pesticide analysis in foods by ion-trap GCMS (8, 16–18). Multiresidue pesticide analysis protocols have been reported using the ion-trap GCMS with the Luke extraction method (19) and supercritical fluid extraction (20, 21).

At the time of this writing, the method of choice for MS analysis of volatile pesticides and metabolites is GCMS. In the absence of a compelling need for high resolution, high mass or MS/MS capabilities, the conservative choice would be a quadrupole instrument, with a cost of ~$50,000 for an EI-only instrument to ~$150,000 for EI, CI, and ECNI capabilities. The lower cost of an ion-trap instrument makes this an attractive alternative, with the performance of EI-only instruments comparable to that of equivalent quadrupole instruments. At least one manufacturer markets an ion-trap GCMS system packaged specifically for pesticide analysis.

Ion-trap instruments with CI, ECNI, and MS/MS capabilities are available at costs competitive with those of EI/CI/ECNI single-quadrupole instruments. As with any new technology, the applicability of such instruments to a specific problem should be tested rigorously, and advanced ion-trap instruments are in that stage of development. As shown by the work of Cairns and others, the processes taking place inside the ion trap can be complex, and fundamental studies must precede any wholesale replacement of all other detectors by ion traps.

LIQUID CHROMATOGRAPHY–MASS SPECTROMETRY

General Aspects

Compounds incompatible with GC separation are de facto incompatible with conventional mass spectrometry. Chromatographic separation of such compounds has been performed routinely for decades using liquid chromatography (LC) and, more recently, supercritical fluid chromatography (SFC) and capillary electrophoresis (CE), but early attempts to introduce liquid samples to a mass spectrometer, such as moving-belt and direct liquid interfaces, had serious drawbacks. Methods developed since 1980, however, have made introduction of liquid samples routine, permitting mass spectrometers to be used as detectors for LC, SFC, CE, and flow injection. LCMS has been the subject of several reviews (22–24), the most recent of which is dedicated to applications of these techniques to pesticides and other xenobiotics in food (25).

Sample Introduction and Ionization

Moving Belt

The earliest attempt to couple liquid chromatography with mass spectrometry was the moving-belt interface in which the effluent of the LC column was deposited on a conveyor belt, which then moved through a heated region in which the solvent was evaporated and then through two stages of vacuum into a metal chamber whose tip opened into the ion source of the mass spectrometer. The solute residue on the belt is vaporized into the ion source by heating the metal tip to 300–400 °C. The incompatibility of this design with thermally labile samples has rendered it effectively obsolete.

Direct Liquid Introduction

Another early LCMS strategy involved splitting the LC effluent between the column and the ion source such that the fraction admitted to the ion source was of sufficiently low volume to maintain an adequate vacuum. Several homemade DLI interfaces have been described, but the experimental difficulties have prevented this approach from seeing widespread use.

Fast-Atom Bombardment

Fast-atom bombardment (FAB) (Fig. 4.10*a*), also referred to as liquid secondary-ion mass spectrometry (LSIMS), uses a focused beam of neutral atoms, typically argon or xenon, to sputter analyte molecules from the surface of a liquid matrix, usually glycerol, into an evacuated ion source (26). The sample introduction process is also the ionization process; proton exchange with the matrix causes analyte molecules to be ejected from the matrix as pseudomolecular ions, $(MH)^+$ or $(M-H)^-$, as well as neutral molecules. Even very high-molecular-weight biopolymers can be volatilized in this way without decomposition. In fact early FAB designs were generally limited to high-molecular-weight analytes because of the high glycerol background at low m/z values. The general utility of FAB was greatly

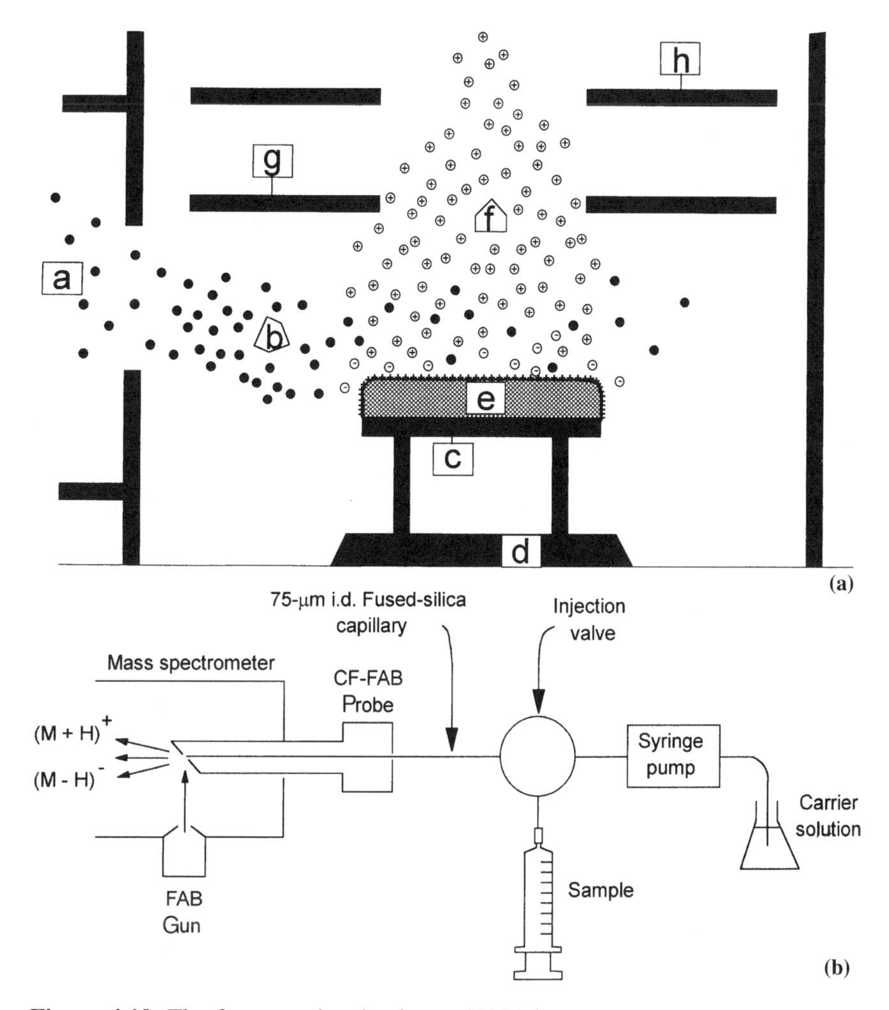

Figure 4.10. The fast-atom bombardment (FAB) ion source. (*a*) Static mode. Fast atoms are generated in the FAB gun (*a*) by first ionizing inert gas ion Xe(Slow)→Xe^+(slow), accelerating the resulting ions through an electric field, then neutralizing the accelerated ion, Xe^+(fast)→Xe(fast). The atom beam (*b*) impinges on the sample (*e*) dissolved in a low volatility liquid matrix, such as glycerin, in a sample holder (*c*) mounted on the end of a probe (*d*). The energy of the impact is transferred to the sample molecules at the surface, simultaneously ionizing (by proton exchange with the matrix) and vaporizing them. The resulting ion beam (*f*) is accelerated through lenses (*g*, *h*) into the mass analyzer. Sample molecules are continuously brought to the surface for ionization / vaporization by diffusion within the liquid matrix. (Adapted with permission from Ref. 26b.) (*b*) Continuous flow mode. Instead of manually placing a drop of matrix, with dissolved sample, in the sample holder and then inserting the probe into the ion source, the probe remains in the ion source and liquid matrix is continuously brought to the probe tip via a syringe pump; the sample is introduced to the flowing matrix through an injection valve. (Adapted with permission from Ref. 27.)

enhanced by the development of continuous-flow FAB (CF-FAB) (Fig. 4.10*b*), in which aqueous solutions of analytes could be introduced to a flowing stream of a mobile phase consisting of 5% glycerol in water (27). This permits CF-FAB to be used as an LCMS interface and for flow-injection analysis. Fast atom bombardment is used routinely with a variety of commercial systems.

Supercritical Fluid Chromatography

Supercritical fluid chromatography (SFC) uses standard GC stationary phases and supercritical fluids (see discussion of supercritical fluid extraction in Chapter 3) as a mobile phase. This technique is best suited for analytes of intermediate polarity; specifically, those more polar than GC-compatible analytes but less so than the most polar LC-compatible analytes. An attractive feature of SFC is the possibility of putting it directly on line with supercritical fluid extraction. At the other end, the SFC column is easily interfaced with MS, since the mobile phase can be allowed to return to subcritical pressure after exiting the column, thus becoming a normal gas which is easily pumped away. In spite of its advantages, SFC has not gained great popularity, perhaps because it emerged at about the same time as thermospray (see paragraphs below), which was compatible with the already familiar and widely used separation technique of LC.

Thermospray Ionization

The first truly compatible LCMS interface was the thermospray system (TSP) (Fig. 4.11), which is both a novel sample introduction system and a novel ion source (28). Its operation is based on the fact that if an ionic solution, which is electrically neutral as a bulk liquid, is rapidly converted to an aerosol of fine droplets, the charges will not be evenly distributed among the droplets, resulting in some droplets with an excess of positive ions and some with an excess of negative ions. In the TSP instrument, this rapid dispersion is effected by pumping a buffer solution, such as ammonium acetate, under high pressure (i.e., with an LC pump) through an electrically heated metal capillary. The amount of heat applied to the capillary can be adjusted so that the solution will emerge from the capillary as a fine spray, and NH_4^+ will be expelled from the positively charged droplets by electrostatic repulsion. If an analyte is injected into the flowing solution, the analyte molecules can be ionized by proton transfer from the NH_4^+ ions as in CI. This proton transfer can occur within the liquid droplets or in the gas phase, with the latter considered to be the predominant process. In addition to MH^+, other adduct ions, including MNH_4^+, are also found, with

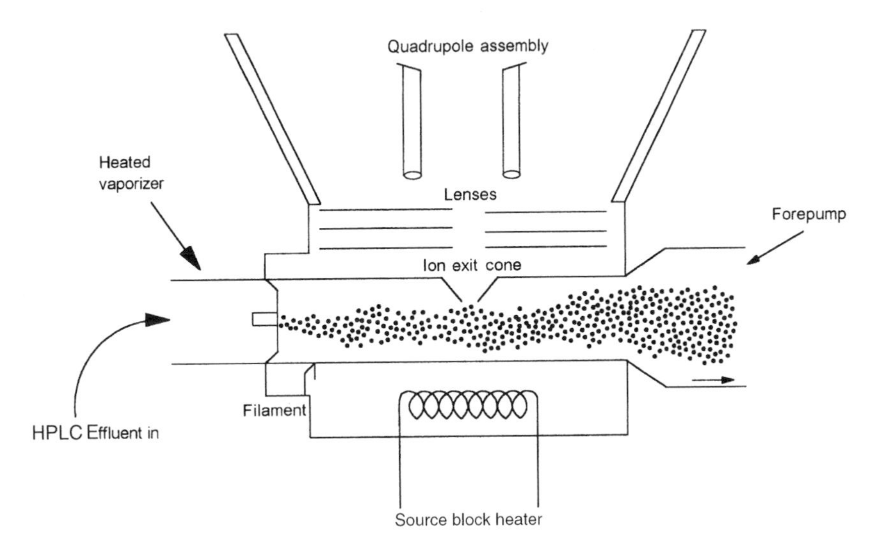

Figure 4.11. The thermospray interface/ionizer. Mobile phase, modified with a volatile buffer (e.g., ammonium acetate), enters a heated vaporizer with an inner diameter of ~10 μm where it is partially vaporized and enters the ion source as an aerosol. Rapid nebulization of the bulk liquid results in asymmetrical distribution of buffer ions (NH_4^+, $CH_3CO_2^-$) among the droplets, and the excess charge on the surface of the droplets causes ejection of buffer ions and sample molecules into the gas phase, where the sample molecules undergo ionization via proton exchange with buffer ions. Ions are deflected 90° through an exit cone and focused into the mass analyzer, while the bulk of the solvent vapor is pumped away by a mechanical forepump. (Adapted with permission from Ref. 24.)

relative abundances depending on the relative basicity of the analyte. Negative ion spectra can also be obtained as a result of proton transfer from analyte molecules to the buffer anions, such as acetate ions. If there is an LC column between the point where the sample is injected and the metal capillary, this system constitutes an LCMS system. In the absence of an LC column the system constitutes a flow-injection MS system.

Thermospray has several advantages over earlier LCMS techniques, principally its compatibility with reverse-phase liquid chromatography, tolerating up to 100% aqueous mobile phases. Another advantage is that it is compatible with flow rates normally encountered in liquid chromatography (i.e., ~1 mL/min).

A major disadvantage of thermospray is that its sensitivity is analyte-dependent and relatively poor compared to that of GCMS. Another serious drawback is that, like CI, it produces primarily MH^+ ions with little of the structurally informative fragmentation provided by EI. This is a major disadvantage for qualitative analysis, but it can be an advantage for target compound detection by SIM provided that, in addition to MH^+, some reasonably abundant and structurally significant fragment or adduct ions can be detected. An expensive remedy to the fragmentation problem is to put a TSP ionizer on a tandem mass spectrometer. A recent study aimed at the development of a TSP method for detection of 51 nitrogen- and phosphorus-containing pesticides reports the use of source and temperature variation and postcolumn derivatization to produce abundant fragmentation without MS/MS (29).

Particle-Beam Ionization

Another less expensive solution to the fragmentation problem in thermospray is the particle-beam (PB) (Fig. 4.12) interface, referred to in its earliest form as the MAGIC interface (a somewhat strained acronym not worth repeating here) a variation on the TSP interface in which solvent is stripped away from the aerosol droplets so that the analyte enters the ion source as molecular clusters that are dissociated to produce EI-like spectra (30). Several commercial variants of particle-beam interfaces are available, representing two basic designs. In the original design the aerosol droplets pass through a heated region to remove the bulk of the solvent by evaporation, and then through one or more momentum separators that permit the analyte clusters to enter the ion source while the solvent vapor is pumped away. In a second design, solvent is separated from analyte by pumping through a membrane permeable to the former.

Since the PB interface connects to a conventional ion source, ionization can be by EI, CI, or ECNI, providing much more information than TSP mass spectrometry. This advantage is offset by the incompatibility of PB with highly aqueous mobile phases caused by the difficulty in removing water from the aerosol. Since TSP is most effective for highly aqueous mobile phases, these two techniques are complementary vis à vis sample compatibility. Another disadvantage of the PB interface is its lower sensitivity relative to thermospray and electrospray (see section below).

Atmospheric Pressure Chemical Ionization

In the APCI source (Fig. 4.13), the problem of introducing a liquid at high pressure into a low-pressure ion source is side-stepped by performing the ionization at atmospheric pressure. Liquid mobile phase is nebulized and evaporated at atmospheric pressure by passing though a nozzle. Use of a standard EI or CI source is not possible since a filament would burn out at atmospheric pressure. Instead, the ambient gases are ionized by the high

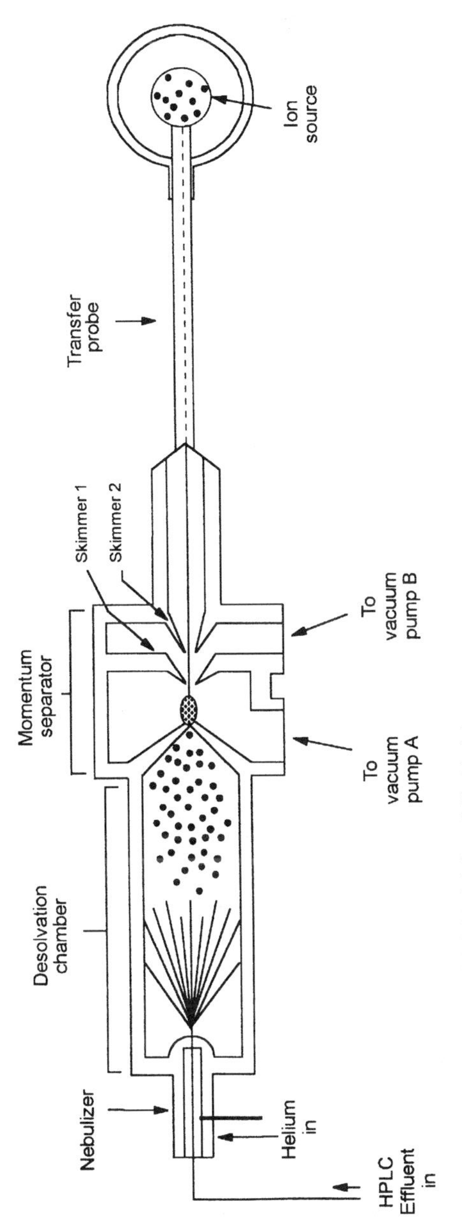

Figure 4.12. The particle-beam interface. Mobile phase is pumped through a 10-μm-ID. capillary and nebulized by a coaxial stream of helium. The aerosol droplets are desolvated in a heated chamber, and the resulting mixture of solvent vapor and gas-phase clusters of sample molecules passes through a two-stage momentum separator, where the lighter solvent molecules are pumped away and the sample clusters, by virtue of their higher momentum, pass through the transfer line into a conventional ion source, where they undergo ionization by EI, CI, or ECNI. (Adapted with permission from Ref. 24.)

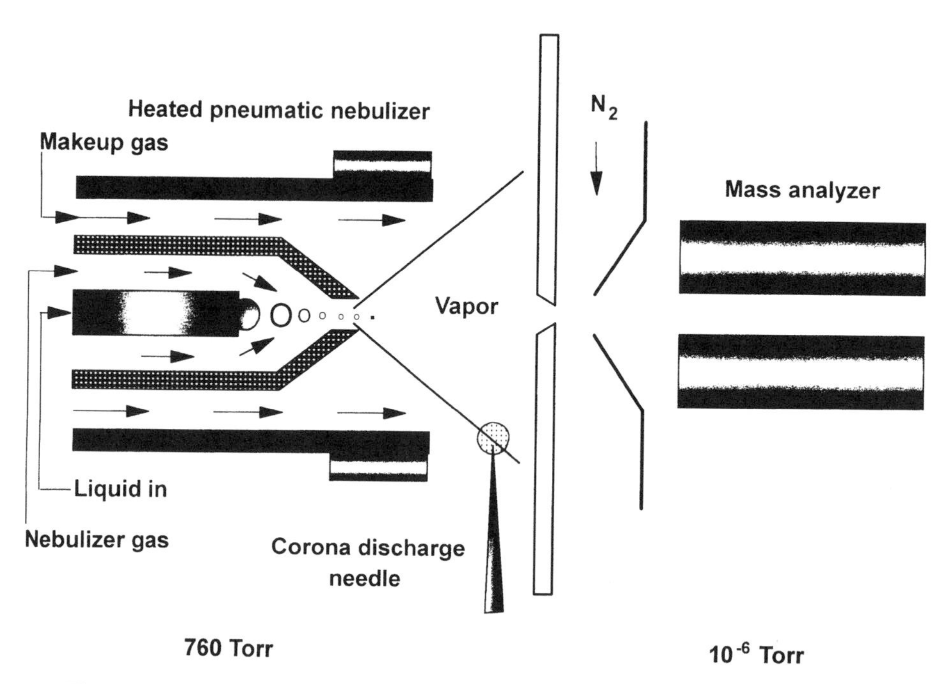

Figure 4.13. The atmospheric pressure chemical ionization interface/ionizer. As in the particle-beam interface, mobile phase is nebulized in a narrow-bore capillary by a coaxial stream of gas, in this case air or nitrogen, but is ionized immediately after exiting the capillary. Ionization at atmospheric pressure is effected by a corona discharge needle, with ambient water vapor acting as a CI reagent gas; pseudomolecular cations and anions are formed from sample molecules by proton exchange with $(H_2O)_nH^+$ or $(H_2O)_nO_2^-$, respectively. Low-mass solvent ions and neutrals are removed by a perpendicular steam of nitrogen, while the higher-momentum sample ions enter the mass analyzer, which is maintained at low pressure. (Adapted with permission from Ref. 24.)

electric field surrounding a needle-shaped electrode. The principal ions produced are $(H_2O)_nH^+$ $(H_3O^+, H_5O_2^+$, etc.) from ionization of ambient water vapor. These ions subsequently ionize the sample molecules to MH^+ by proton transfer, specifically chemical ionization. Ions are accelerated through a curtain of N_2 into the mass analyzer. The curtain gas permits maintenance of atmospheric pressure on the source side and high vacuum on the analyzer side, low-molecular weight atmospheric gases are swept away by the curtain gas, while analyte ions, by virtue of their higher momentum, penetrate through to the mass analyzer. Some commercial APCI instruments use collisionally induced dissociation (CID) in a triple quadrupole or ion-trap analyzer to effect fragmentation of the MH^+ ions. In most such systems the

APCI interface can be interchanged easily with an electrospray interface (see text below).

Electrospray Ionization

Another alternative to TSP is the electrospray ionization (ESI) interface (Fig. 4.14), which is available in several commercial variants (31, 32). In ESI, ionization takes place at atmospheric pressure, but the technique differs significantly from APCI in that nebulization and ionization of the mobile phase is effected by an electric field applied to the end of a restricted inlet nozzle. In two variations on this interface, nebulization is assisted by a stream of nitrogen introduced coaxially with the mobile phase (commonly called *ion spray*) or by ultrasonication. A major advantage to ESI is its ability to form multiply charged ions, so that high-molecular weight biopolymers such as proteins and polysaccharides can form ions that have high masses but low m/z values and can therefore be detected using inexpensive quadrupole (low-mass-range) mass analyzers. Another advantage of ESI, of more interest in this context, is its greater sensitivity than either TSP or PB.

Electrospray shares with TSP and APCI the disadvantage of producing primarily MH^+ ions. Some manufacturers offer ESI ion sources in triple-quadrupole tandem instruments; in other, less expensive designs, CID can be performed in the ionization region of the ESI source. Most recently, commercial instruments have been introduced that couple ESI and APCI (see discussion above) with ion-trap MS/MS.

Other disadvantages of ESI include the fact that it is currently compatible only with microliter-per-minute flow rates associated with microcapillary LC and that molecules of low polarity are often not ionizable and therefore not detectable. Such compounds are often detectable by APCI, however, making ESI and APCI complementary in the same sense that TSP and PB are complementary.

Capillary Electrophoresis–Mass Spectrometry

At the time of this writing, no applications of CEMS to pesticide analysis have been reported. Such applications will occur when (1) the desirability of CE as a method for separation of pesticides has been demonstrated and (2) an interface for routine use of mass spectrometry as a CE detector has been developed. The latter development has been pursued fairly extensively; electrospray is the preferred method because of its compatability with highly polar, ionic, and polyionic analytes (see Ref. 33 and references therein), but the application of CE to pesticides, as discussed in a previous chapter, is still

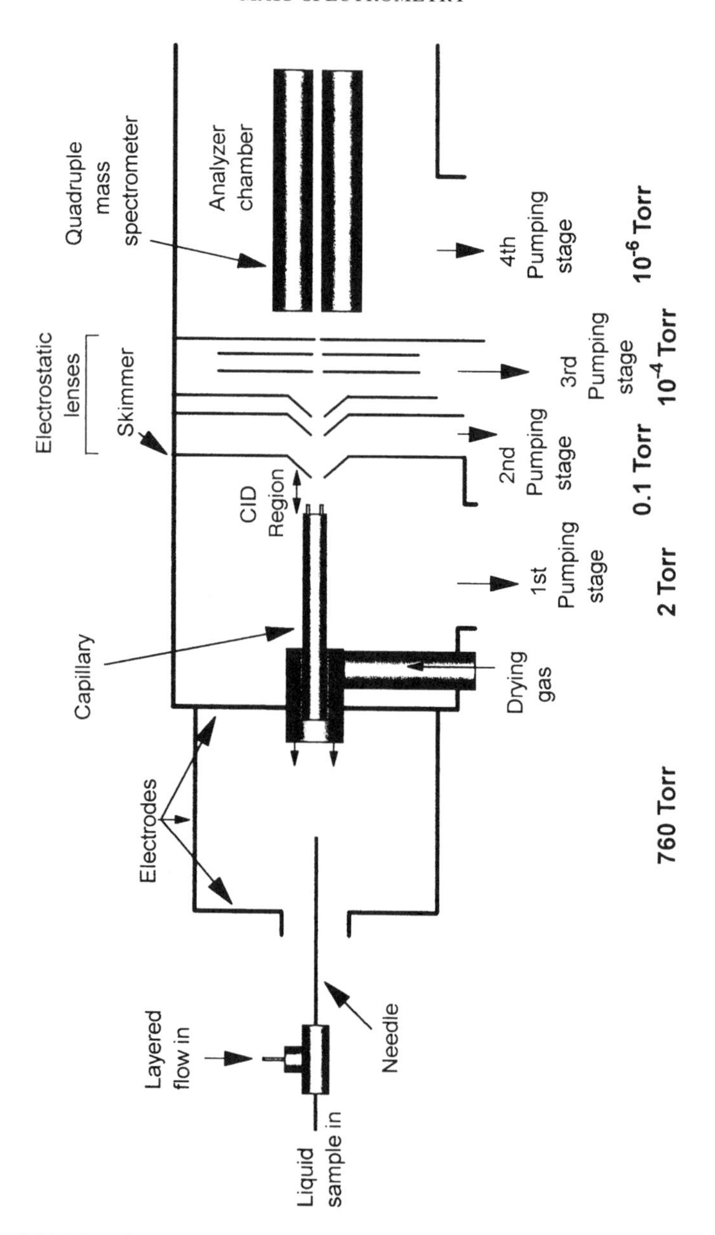

Figure 4.14. The electrospray interface / ionizer. The cylindrical ionization chamber, the walls of which constitute an electrode, is maintained at atmospheric pressure and 2–4 kV relative to the stainless-steel sample introduction needle, which is at ground.

(*contd*)

in its earliest stages of development. At least two commercial LC-ESI MS instruments offer an adaptor for coupling the ESI interface to a capillary electrophoresis device.

Membrane Introduction

Membrane-introduction mass spectrometry (MIMS), like purge-and-trap introduction, is a means of introducing organic compounds in aqueous matrices to a conventional ion source (34). In MIMS, separation of organic molecules from the aqueous matrix is effected by selective diffusion of organic molecules through a hydrophobic membrane which excludes water while admitting small organic molecules. The design of such an interface is fairly simple, with static or flowing aqueous solution at ambient pressure on one side of the membrane and a conventional high-vacuum mass spectrometer ion source on the other. The advantage of MIMS over purge-and-trap introduction is that it does not rely on a timed sequence of events (purge, trap, desorb) and is thus suitable for continuous process monitoring. MIMS interfaces have been used, for example, for continuous fermentation monitoring. Another advantage is that analysis is not limited to highly volatile compounds as in purge-and-trap. A disadvantage of MIMS is that the size-exclusion properties of membranes preclude the analysis of biopolymers and other high-molecular-weight compounds.

Matrix-Assisted Laser Desorption and Ionization

Matrix-assisted laser desorption and ionization (MALDI) is the most recently developed and commercially successful of the techniques that rely on desorption of a sample from its matrix by laser excitation (35). Since it has not been succesfully coupled to flow-injection or liquid chromatography, it cannot be called an LCMS technique but is included here because it is compatible with involatile samples. MALDI emerged in the late 1980s and became commercially available in the early 1990s. In a MALDI experiment

---

Figure 4.14. (*contd*) Molecules at the surface of the mobile phase emerging from the needle are ionized by the strong electric field within the ionization chamber, and the resulting high charge density nebulizes the solution. A coaxial flow of organic solvent introduced at the tee enhances the ionization efficiency. The charged droplets are desolvated by the drying gas (nitrogen) and accelerated through the capillary into the first pumping stage. If desired, ions can undergo fragmentation at this stage via collisionally induced dissociation (CID) by applying a voltage across the indicated region. Subsequent pumping stages reduce the pressure to the requisite 10^{-6} range for mass analysis. (Adapted with permission from Ref. 24.)

the analyte and a matrix compound that absorbs light at the frequency of the laser being used are codissolved in a suitable solvent and the resulting solution deposited on a metal target and allowed to dry. The concentration of analyte in the resulting matrix residue is $\sim 10^{-6}$ M. The dry residue is then irradiated with a laser pulse which desorbs and ionizes the sample. The function of the matrix is to absorb the energy of the laser pulse, thus preventing thermal degradation of the analyte, and ionize the sample by proton transfer. Various pulsed UV and IR lasers are used for MALDI, and the most commonly used matrix compounds are nicotinic acid and sinapinic acid, both of which are solids at room temperature. Since this technique is particularly well suited for involatile high-molecular-weight analytes, including biopolymers, it has been used primarily in conjunction with time-of-flight analyzers, giving rise to the double acronym MALDI-TOF. As previously discussed, MALDI instruments have seen little or no application to pesticide residue analysis.

Mass Analysis

In principle, any of the LCMS sample inlets and ion sources may be coupled to any of the mass analyzers described in the GCMS section. In practice, most commercial LCMS systems make use of a quadrupole mass analyzer. The major exceptions are FAB, which in its earliest inception was applicable only to compounds with molecular weights of 1000 daltons or more and thus required the use of sector instruments; and MALDI, whose most successful commercial version uses a time-of-flight analyzer. Since MALDI has not to date seen application to pesticide analysis, the TOF analyzer will not be discussed here.

Quadrupole

The most common LCMS instruments currently in use are thermospray and particle-beam instruments with quadrupole analyzers.

Magnetic Sectors

As stated above, the only reason for using a sector instrument for pesticide analysis in GCMS is a requirement for high-resolution data for elemental composition or for the exceptional selectivity of exact-mass SIM. The same applies in the case of LCMS, with the additional factor that FAB sources, intended as they originally were for high-molecular-weight analytes, are most commonly found on sector instruments.

Tandem Analyzers

The rationale for the use of tandem analyzers for LCMS is somewhat different from that for GCMS. In the latter, product ions provide additional structural information for qualitative analysis or enhanced selectivity for target compound detection. In the case of thermospray, APCI, and electrospray LCMS, collisionally induced dissociation is often the only way to generate fragment ions at all, making MS/MS essential for qualitative analysis or for SIM with more than one target ion. In commercial instruments, CID is most commonly done in triple-quadrupole or ion-trap analyzers. In the case of ESI, CID can also be performed in the ionization region of the source.

Ion Trap

As in the case of GCMS, the low cost of the ion–trap analyzer has made the development of ion–trap LCMS instruments a major focus of commercial development. As discussed above, the potential for doing MS/MS in an ion trap is more desirable for LCMS than for GCMS, and the desirability of doing it inexpensively is self-explanatory. Commercial instruments that couple LC with ion traps emerged in the mid-1990s.

Data Acquisition and Analysis

The principles governing the application of mass spectrometry to qualitative analysis, target compound detection and quantitative analysis are the same for LCMS as for GCMS. As discussed above, some form of CID is often necessary for qualitative analysis or target compound detection. Limits of detection are generally higher for LCMS than for GCMS.

Applications and Conclusions

LCMS technology is still in a state of flux, and the ideal interface has yet to be devised. The moving-belt and DLI interfaces may be safely relegated to history, although applications of the latter to pesticide residues in crop samples have been reported fairly recently (36). It is probably safe to say that CF-FAB and SFC/MS have passed their peak of popularity, at least among instrument manufacturers. The ability to detect pesticide standards by FAB has been demonstrated (37–39), and the use of FAB with high-resolution and tandem MS has been applied to the detection of specific pesticides and metabolites in water (40, 41). The utility of supercritical fluid chromatography/mass spectrometry has been demonstrated for the detection of organophosphorus pesticides in spiked fruit samples (42) and carbamate pesticides in chicken tissue after extraction by SFE (43).

At the other end of the historical curve, commercial capillary electrophoresis – mass spectrometry systems have become available recently, but no applications to pesticide analysis have been reported at the time of this writing. Membrane-introduction MS has been applied to the detection of nonpesticide pollutants in water (44), and the development of a LC-MALDI interface is at an early stage (45).

This leaves the four aerosol-based techniques of thermospray, particle-beam, APCI, and electrospray as the well-established but not yet obsolete LCMS techniques of greatest current interest to the analyst.

Thermospray is by far the most widely used of these four techniques, by virtue of its compatibility with highly polar water-soluble pesticides, reverse-phase LC, and normal LC flow rates and also because it has been available for the longest time. Studies using pesticide standards to demonstrate the utility of TSP for their detection are too numerous to list here, as are applications to environmental samples. Thermospray has been used without MS/MS for multiresidue determination of pesticides in fruits and vegetables (46), detection of supercritical fluid–extracted thiosulfinates in garlic and onion (47), and organophosphorus pesticides and metabolites in beef tissue (48). All of these studies employed a single quadrupole for mass analysis. Thermospray MS/MS has been used for the determination of sulfonamides in blood and meat by LC and flow injection analysis (49).

Recent reports of particle-beam LCMS for the detection of pesticides in complex matrices have focused on water (50, 51), but no applications to foods could be found. The use of APCI MS for the detection of triazine herbicides (52) and *N*-methyl carbamates (53) in spiked crop samples has been described. Electrospray ionization with quadrupole mass analysis has been used in conjunction with solid-phase extraction for the determination of pesticides in water (54–56). The inevitable next step has been taken with the use of an electrospray ion source coupled to an ion-trap mass analyzer for the determination of pesticides in water (57).

At the time of this writing, the choices for MS analysis of polar pesticides and their metabolites are some form of LCMS or derivatization followed by GCMS. The latter is more labor-intensive but offers greater sensitivity and the confidence associated with using a mature method. The former eliminates the necessity for derivatization but presents an array of choices, none of which is universal in its applicability to different samples. The most versatile LCMS system at this time would probably be a single mass analyzer (single-quadrupole, triple-quadrupole, or ion-trap) with two interchangable ion sources, one for use with more volatile, less polar, analytes (PB or APCI), the other for involatile, highly polar, and ionic compounds (TSP or ESI). Five years ago the clear choice would have been TSP and PB with a single quadrupole. Recent results suggest that the combination of an APCI/ESI source with an ion trap is emerging as the inevitable choice. At this time the rational choice must be based on careful consideration of the characteristic advantages and disadvantages of each option in the context of the analyst's immediate and anticipated needs.

REFERENCES

1. (a) W. McFadden, *Techniques of Combined Gas Chromatography / Mass Spectrometry: Applications in Organic Analysis*, Wiley, New York, 1973; (b) B. S. Middleditch, ed., *Practical Mass Spectrometry. A Contemporary Introduction*, Plenum Press, New York, 1979; (c) F. W. McLafferty, *Interpretation of Mass Spectra*, University Science Books, Mill Valley, CA, 1980; (d) F. W. Karasek, O. Hutzinger, and S. Safe, eds., *Mass Spectrometry in Environmental Sciences*, Plenum Press, New York, 1985; R. Davis, M. Frearson, and F. E. Pritchard, *Mass Spectrometry*, Wiley, New York, 1987; (f) J. Gilbert, ed., *Applications of Mass Spectrometry in Food Science*, Elsevier, London, 1987.
2. A. G. Harrison, *Chemical Ionization Mass Spectrometry*, CRC Press, Boca Raton, FL, 1983.
3. E. A. Stemmler and R. A. Hites, *Electron Capture Negative Ion Mass Spectra of Environmental Contaminants and Related Compounds*, VCH, New York, 1988.
4. H. J. Stan and G. Kellner, *Biomed. Environ. Mass Spectrom.* **18**, 645–651 (1989).
5. D. R. Knapp, *Handbook of Analytical Derivatization Reactions*, Wiley, New York, 1979.
6. (a) C. L. Deyrup, S.-M. Chang, R. Weintraub, and H. A. Moye, *J. Agric. Food Chem.* **33**, 944–947 (1985); (b) J. P. Toth and H. A. Moye, Negative ion electron capture mass spectrometry of fluorinated pesticide derevatives, in J. D. Rosen, ed., *Applications of New Mass Spectrometric Techniques in Pesticide Chemistry*, Wiley, New York, 1987.
7. D. Powell. *Proceedings, 38th ASMS Conf. Mass Spectrometry and Allied Topics*, 1990, pp. 1573–1574 .
8. T. Cairns, K. S. Chiu, and E. Siegmund, *Rapid Commun. Mass Spectrom.* **6**, 331–338, (1992).
9. NIST/EPA/NIH Mass Spectral Database, U.S. Deptartment of Commerce, Gaithersberg, MD, 1992.
10. *The Wiley Registry of Mass Spectral Data*, 5th ed., J Wiley, New York, 1992.
11. J. A. Sphon, *J. AOAC* **61**, 1247–1252 (1978).
12. T. Cairns and R. A. Baldwin, *Anal. Chem.* **67**, 552A–557A (1995).
13. T. Cairns, E. G. Siegmund, and J. J. Stamp, *Mass Spectrom. Rev.* **8**, 93–117 (1989).

14. (a) W. L. Budde, J. W. Eichelberger, J. W. Munch, and J. A. Shoemaker, Abstract 488; (b) R. D. Brittain, Abstract 489; (c) M. V. Buchanan, R. Merriweather, and M. R. Guerin, Abstract 490; (d) R. G. Cooks, S. Bauer, P. Wong, N. Kasthurikrishna and M. Soni, Abstract 491; (e) R. A. Hites and W. C. Schnute, Abstract 492; Pittsburgh Conference and Exposition on Analytical Chemistry and Mass Spectrometry, 1995.
15. T. M. P. Chichila and D. R. Erney, *J. AOAC Int. Intl.* **77**, 1574–1580 (1994).
16. T. Cairns, K. S. Chiu, E. Siegmund, and M. Weber, *Rapid Commun. Mass Spectrom.* **6**, 449–453 (1992).
17. T. Cairns, K. S. Chiu, D. Navarro and E. Siegmund, *Rapid Commun. Mass Spectrom.* **7**, 971–988 (1993).
18. T. Cairns, M. A. Luke, K. S. Chiu, D. Navarro, and E. Siegmund, *Rapid Commun. Mass Spectrom.* **7**, 1070–1076 (1993).
19. G. C. Mattern and J. D. Rosen, The use of ion trap mass spectrometry for multiresidue pesticide analysis, in *Emerging Strategies for Pesticide Analysis*, T. Cairns and J. Sherma, eds., CRC Press, Boca Raton, FL, 1992, Chapter 12.
20. S. J. Lehotay and M. A. Ibrahim, *J. AOAC Intl.* **78**, 445–452 (1995).
21. S. J. Lehotay and K. I. Eller, *J. AOAC Intl.* **78**, 821–830 (1995).
22. T. Cairns, E. G. Siegmund, and J. J. Stamp, *Mass Spectrom. Rev.* **8**, 127–145 (1989).
23. A. L. Yergey, C. G. Edmonds, I. A. S. Lewis, and M. L. Vestal, *Liquid Chromatography/Mass Spectrometry Techniques and Applications*, Plenum Press, New York, 1990.
24. R. D. Voyksner, *Environ. Sci. Technol.* **28**, 118A–127A (1994).
25. M. Careri, A. Magia, and M. Musci, *J. Chromatogr. A* **727**, 153–184 (1996).
26. (a) M. Barber, R. S. Bordoli, R. D. Sedgwick, and A. N. Tyler, *J. Chem. Soc., Chem. Commun.* **1981**, 325–327 (1981); (b) M. Barber, R. S. Bordoli, G. Elliot, R. D. Sedgwick, and A. N. Tyler, *Anal. Chem.* **54**, 645A–657A (1982).
27. R. M. Caprioli, *Anal. Chem.* **62**, 477A–485A (1990).
28. C. R. Blakley, J. J. Carmody, and M. L. Vestal, *J. Am. Chem. Soc.* **102**, 5931–5933 (1980).
29. D. Volmer and K. Levsen, *J. Am. Soc. Mass Spectrom.* **5**, 655–675 (1994).
30. R. C. Willoughby and R. F. Browner, *Anal. Chem.* **56**, 2625–2631 (1984).
31. C. M. Whitehouse, R. N. Dreyer, M. Yamashita, and J. B. Fenn, *Anal. Chem.* **57**, 675–679 (1985).
32. J. B. Fenn, M. Mann, C. K. Meng, S. F. Wong, and C. M. Whitehouse, *Mass Spectrom. Rev.* **9**, 37–70 (1990).
33. C. A. Monnig and R. T. Kennedy, *Anal. Chem.* **66**, 280R–314R (1994).
34. T. Kotiaho, F. R. Lauritsen, T. K. Choudhury, R. G. Cooks, and G. T. Tsao, *Anal. Chem.* **63**, 875A–883A (1991).

35. F. Hillenkamp, M. Karas, R. C. Beavis, and B. T. Chait, *Anal. Chem.* **63**, 1193A–1203A (1991).
36. B. H. Escoffier, C. E. Parker, T. C. Mester, J. S. M. Dewit, F. T. Corbin, J. W. Jorgensen, and K. B. Tomer, *J. Chromatogr.* **474**, 301–316 (1989).
37. H. Fujiwara and S. J. Wratten, *Chem. Anal.* **91**, 128–145 (1987).
38. G. M. Allmier and E. R. Schmid, *Rapid Commun. Mass Spectrom.* **1**, 42–45 (1987).
39. Y. Tondeur, G. Sovocool, R. K. Mitchum, W. J. Niederhut, and J. R. Donnelly, *Biomed. Environ. Mass Spectrom.* **14**, 733–736 (1987).
40. K. A. Caldwell, V. M. Sadagopa Ramanujam, Z. Cai, M. L. Gross, and R. F. Spalding, *Anal. Chem.* **65**, 2372–2379 (1993).
41. Z. Cai, V. M. Sadagopa Ramanujam, M. L. Gross, S. J. Monson, D. A. Cassada, and R. F. Spalding, *Anal. Chem.* **66**, 4202–4209 (1994).
42. H. T. Kalinoski and R. D. Smith, *Anal. Chem.* **60**, 529–535 (1988).
43. B. Murugaveri, A. Gharaibeh, and K. J. Voorhees, *J. Chromatogr. A* **657**, 223–226 (1993).
44. M. Gazda, L. E. Dejarme, T. K. Choudhury, R. G. Cooks, and D. W. Margerum, *Environ. Sci. Technol.* **27**, 557–561 (1993).
45. K. K. Murray and D. H. Russell, *J. Am. Soc. Mass Spectrom.* **5**, 1 (1994).
46. C. H. Liu, G. C. Mattern, X. Yu, R. T. Rosen, and J. D. Rosen, *J. Agric. Food Chem.* **39**, 718–728 (1991).
47. E. M. Calvey, J. E. Matusik, K. D. White, J. M. Betz, E. Block, M. Littlejohn, S. Naganatha, and D. Putnam, *J. Agric. Food Chem.* **42**, 1335–1341 (1994).
48. B. M. Ioerger and J. S. Smith, *J. Agric. Food Chem.* **42**, 2619–2624 (1994).
49. G. K. Kristiansen, R. Brock, and G. Bojesen, *Anal. Chem.* **66**, 3253–3258 (1994).
50. J. S. Ho and W. L. Budde, *Anal. Chem.* **66**, 3716–3722 (1994).
51. A. Cappiello, G. Famiglini, and F. Bruner, *Anal. Chem.* **66**, 1416–1423 (1994).
52. J. P. Toth and A. P. Snyder, *Biol. Mass Spectrom.* **20**, 70–74 (1991).
53. W. H. Newsome, B. P. Y. Lau, D. Ducharme, and D. Lewis, *J. AOAC Intl.* **78**, 1312–1316 (1995).
54. C. Molina, M. Honig, and D. Barcelo, *Anal. Chem.* **66**, 4444–4449 (1994).
55. S. Chiron, S. Papilloud, W. Haerdi, and D. Barcelo, *Anal. Chem.* **67**, 1637–1643 (1995).
56. C. Crescenzi, A. DiCorcia, S. Marchese, and R. Samperi, *Anal. Chem.* **67**, 1968–1975 (1995).
57. H.-Y Lin and R. D. Voyksner, *Anal. Chem.* **65**, 451–456 (1993).

CHAPTER

5

EMERGING METHODS: EXTRACTIONS AND CLEANUP

H. ANSON MOYE

SOLID-PHASE EXTRACTIONS: CARTRIDGES

Introduction

First described in 1976 as a technique for the extraction of drugs from body fluids (1), and called then "column extraction," what has since come to be known as *solid-phase extraction* (SPE) has lately experienced a resurgence for the extraction and cleanup of pesticides. No one single reason for this phenomenon can be rightly given, but rather several innate characteristics of the technique together have probably generated a new interest in the incorporation of SPE into contemporary analytical methods for the analysis of pesticides, first in water and, increasingly, in foods. Some of the advantages that SPE devices have over conventional solvent extractions and cleanup of pesticides include, but are not limited to, the following:

Better Reproducibility. Since SPE extraction devices are based on careful selection of adsorbents in terms of particle size, pore size distribution, and purity of adsorbent, they perform more consistently, from device to device, from batch to batch, and from manufacturer to manufacturer, than do conventional adsorbents, such as silica or alumina.

Elimination of Solvents. An increasing pressure is being felt by regulatory agencies in the United States to reduce or eliminate organic solvents in the analytical chemistry laboratory because they present problems by their acquisition cost, potential exposure of humans during their use, cost of proper disposal, and the potential for environmental pollution during their use and their disposal (2). Solvents typically used in the U.S. Food and Drug Administration's (FDA's) Luke method (3) and the State of California's Department of

Pesticide Residues in Foods: Methods, Techniques, and Regulations, by W. G. Fong, H. A. Moye, J. N. Seiber, and J. P. Toth. Chemical Analysis Series, Vol. 151
ISBN 0-471-57400-7

Food and Agriculture method for the analysis of pesticides in fruits and vegetables include acetonitrile, acetone, *n*-hexane, methylene chloride, and petroleum ether. These solvents, and others used in methods developed by the U.S. Environmental Protection Agency (EPA), such as benzene, carbon tetrachloride, chloroform, toluene, methyl ethyl ketone, methyl isobutyl ketone, tetrachloroethylene, 1,1,1-trichloroethane, trichloroethylene, and xylene, have been targeted to be reduced by 50% by 1995 in contract and government laboratories (2). There is also a growing concern about the release of chlorinated hydrocarbons to the atmosphere, which may deplete the ozone layer.

Speed. Recent development of the SPE disk has increased the rate at which pesticides can be extracted from water due to an improvement in device permeability. Since the polarity of the solution from which the pesticides are being absorbed can be adjusted by the proper addition of water, absorption can be made to occur quickly, reducing the contact time between the solution being processed and the SPE device.

Versatility. Many absorption modes can be brought into play with SPE devices, making the technique applicable to a wide array of pesticide molecular structures and polarities. Consequently, pesticides of a hydrophobic, hydrophilic, and ionic nature can be extracted with the devices now available.

Cost. Competition among manufacturers, mass-production techniques, and a growing demand for device quantities have together made typical SPE devices attractive from a cost standpoint. Their costs have actually diminished in the last 5 years, compared to solvent costs, which have increased.

Automation. Because of their small size, physical configuration, and solvent permeability, they are becoming increasingly popular for use in automated sample preparation devices. Recent incorporation of the devices into well-established and widely used analytical methods used for monitoring of foods should provide the impetus to develop automated sample extraction and cleanup techniques.

Freedom from Interference. Just as attention to the quality of the adsorbents and phases used in the SPE devices has contributed to excellent reproducibility in their performance, they are also extremely clean, and free from interference that might otherwise appear on the newer and highly sensitive instruments used for detection and quantitation.

Extractions in the Field. Since SPE devices are small and can be put into use with only a minimal of equipment, they are particularly suitable for analysis in the field, particularly water. They eliminate the need for transporting large volumes of water back to the laboratory under refrigerated condions, since they have been shown to stabilize pesticides under ambient conditions.

How the Technique Works

Any extraction or separation that can be accomplished by SPE is based on the concept of selective retention by the device for either the analyte, in this case the pesticide, or for potential interference that would otherwise disrupt subsequent steps in the analytical method, usually the chromatographic separations or the detections. Solid-phase extractions can be made to work in either the batch or column mode. Both rely on the ability of an adsoptive particle to adsorb either the pesticide or the interference from a solution, typically water. In the batch mode, the particles are poured into the aqueous phase and either stirred or shaken, after which they are subsequently removed by filtration. This mode has not gained favor as a technique for the extraction or cleanup of pesticides, probably because loose particles present difficulty in being fully recovered from containers, where they tend to adhere to beaker and tube surfaces. The accepted mode, technically termed *column liquid–solid extraction* (CLSE) by the International Union of Pure and Applied Chemistry (4), uses adsorbent particles that are contained in cartridges of various sizes, usually made of a plastic, such as polyethylene or polypropylene that are of extremely high purity. For simplicity, in this discussion we shall call these cartridges *SPE cartridges*, since this term is now well known among analytical chemists around the world.

Some cartridges are designed with a small reservoir at the top, in which the sample and subsequently eluting solvent can be placed, while others are designed for being attached directly to syringes or to plastic lines that can be made to carry solvent (Fig. 5.1). Both sample and eluting solvent is made to flow through the cartridge by pressure differentials, normally of only 10–15 psi (lb / in.^{2}), which is easily achievable with laboratory vacuum systems or aspirators.

Retention on the particles within the cartridge is achieved by interactions that are developed between the surfaces of the particles and the dissolved solute. Such interactions, more fully described below, can be adjusted by modification of the surface of the particle, which is done by the cartridge manufacturer, or by adjusting sample and eluting solvent type, pH, ionic strength, and volume. With few exceptions, the particles are generally porous silica that have been chemically modified, usually by the reactions of silane reagents containing various functionalites with the silanol groups on the silica surface. Particle size is typically 40 μm in diameter, and pore size is typically 60–100 Å. The result of such reactions produces particles that are

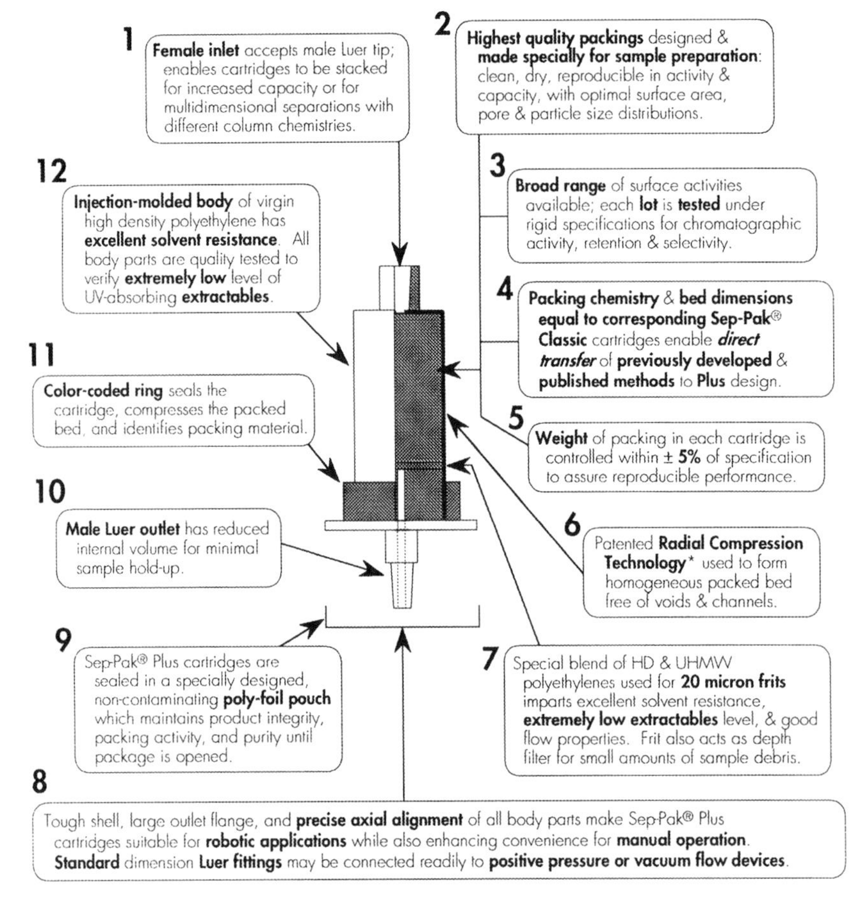

Figure 5.1. Sketch of typical solid-phase extraction cartridge, SPE cartridge (courtesy Waters Company).

chemically coated with organic molecules of various functionalities, a sort of "brush" configuration that projects into the air or into the solvent that surrounds the particle. These chemically bound molecules are thus made available for intermolecular interactions with the pesticide or interference that are present in the sample to be extracted.

Also available for interactions are cartridge particles that have not been "end-capped" after the first reaction with the bulky silane reagent, or cartridge particles that have not been exposed to silane reagent at all.

Because silanol groups are plentiful on the surface of the particle and into the pores, after bulky silane groups have been attached to the surface, there are remaining silanol groups that were shielded by adjacent "brush" molecules. Such silanol groups can be left alone, and hence become available for interactions of their own, or they can be end-capped with smaller sized silane reagents. If still allowed to persist, these silanol groups can participate in what are termed "secondary" interactions, which can affect additional retention on the particle. Figure 5.2 shows how silane reagents are attached to silica particles, and how residual silanols are still available.

A wide array of silane reagents can be used for bonding to the silica particles, providing a selection of mechanisms for solute–particle interactions. Probably the most popular particle phase for pesticide extractions is the octadecylsilane, C_{18}, which participates in what are typically called "hydrophobic" interactions, although this is a misnomer, since it has been

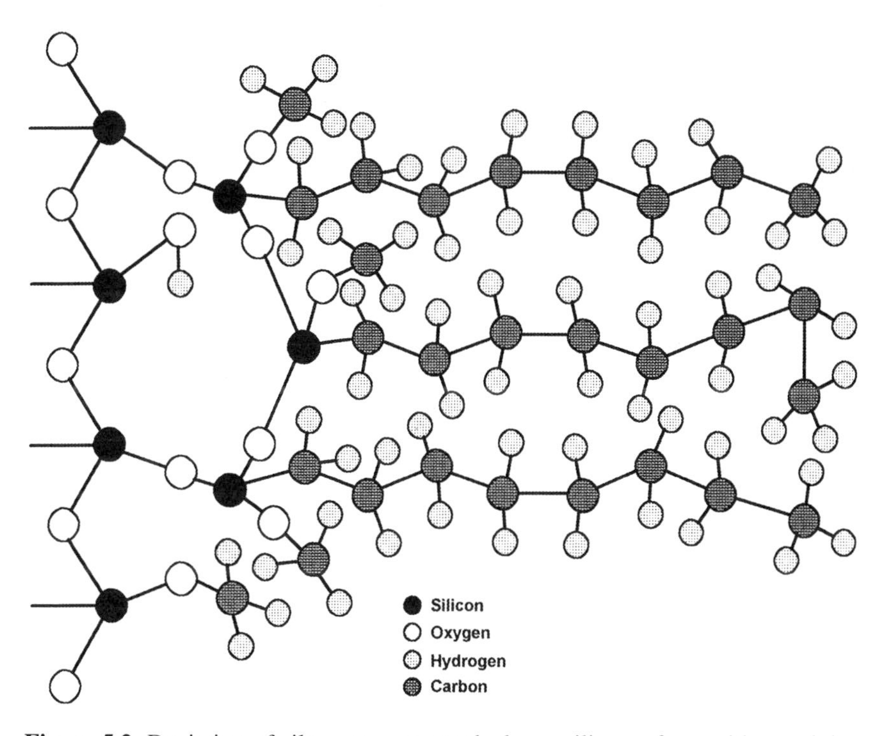

Figure 5.2. Depiction of silane reagent attached to a silica surface, with remaining uncoated silanol groups (courtesy Varian Sample Preparation Products). Three-bonded silica synthesis. Bonded silicas are formed by the reaction of organosilanes with activated silicas. A subsequent reaction called end-capping may take place, which deactivates some of the remaining silica groups on the silica.

shown that the real interactions involved are those between the water molecules themselves (5). Such interactions arise from the preferred orientation of a hydrophobic pesticide, such as resmethrin, which aligns itself with the bristle of the C_{18} brush, making it unnecessary to form a six-sided cavity in the water. Since this is energetically preferred, the pesticide would remain in this orientation until organic modifiers are called into play that, by diluting water, make the cavity formation within the water more energetically feasible. Other interactions include dispersive interactions, those that result from the close proximity of molecules containing a degree of unsaturation. Of the more commonly available stationary phases, the cyano (CN) one would enter into these types of interactions, and would respond to pesticides of an aromatic nature, such as resmethrin. One of the stronger types of interaction would be hydrogen bonding, which requires a proton donor and an acceptor. SPE particles that would participate in these interactions would include the amino (NH_2) and the diol (Diol). Because it is a primary amine, the NH_2 phase can either donate or accept a proton, whereas the Diol would be more prone to behave as a proton donor. Pesticides that could act as proton donors would be the phenols and acids; those that would act as acceptors would be the anilides, amides, triazines, and other nitrogen-containing molecules. Also available as proton donors or acceptors would be the unmodified particles, such as the ubiquitous silica or the less used alumina. Those chemically bound particles that are not end-capped would have a limited number of silanol groups that would be available for what are termed "secondary" interactions. Table 5.1 summarizes the types of particle phases that are commercially available and the types of interactions that can come into play with them.

Extraction and Isolation Strategies

Two general approaches can be taken when using SPE cartridges:

- Retention of the pesticide on the cartridge while allowing interferences to pass through and be discarded, with subsequent eluent change to elute the pesticide
- Retention of the interference while allowing the pesticide to pass through and be collected for subsequent analysis

Selection of one of these two modes will depend on the molecular structure of the pesticide and the nature of the sample that is being analyzed, along with the determinative step being used, particularly the chromatographic separation employed and the detector type. Typical interferences that are found in foods include lipids, oils and waxes, sugars, pigments, and anions of

Table 5.1. Summary of Commercially Available Solid-Phase Extraction Cartridge Phases

Packing Type	Interaction
C_{18}	Hydrophobic
C_8	Hydrophobic
Cyclohexyl	Hydrophobic
C_2	Hyrophobic, hydrogen bonding
Phenyl	Dispersive, hydrophobic
Acrylic acid/acrylamide	Ion exchange, hydrogen bonding
Acrylamide	Ion exchange, hydrogen bonding
CN (cyanopropyl)	Dispersive, hydrophobic
Diol	Hydrogen bonding
NH_2 (aminopropyl)	Hydrogen bonding (acceptor)
SCX (benzenesulfonylpropyl)	Cation exchange
PRS (sulfonylpropyl)	Cation exchange
CBA (carboxymethyl)	Cation exchange
DEA (diethylaminopropyl)	Anion exhange
SAX (trimethylaminopropyl)	Anion exchange
SI (silica)	Hydrogen bonding
Alumina N	Hydrogen bonding
Florisil	Hydrogen bonding

various types. Although there are many exceptions, most successful strategies for the isolation of pesticides from plant extracts employ the second strategy, in which the cartridge particle and the eluting solvent are chosen to elute the pesticide after application while retaining troublesome interferences.

Since fruits and vegetables consist largely of water (6), and are typically extracted with a water-miscible solvent (3,4, 7–9), the resulting solutions are typically mostly water and thus lend themselves to the second strategy listed above. Such solutions would then be amenable to what Tippins has deemed "nonpolar extractions," whereby oils and waxes, which have limited but measurable solubilities in aqueous systems, could be extracted using one of the hydrophobic cartridges, such as the C_{18}, C_8, or phenyl (10). If the water content of the extract is carefully adjusted by the addition of the extracting solvent, which is typically acetone or acetonitrile, and the pesticide has a moderate degree of polarity, with a K_{ow} of 100–1000, there ought to be good opportunity to elute it from one of these cartridges. Any opportunities the pesticide would have to form hydrogen bonds with the water, or to be ionized by appropriate pH adjustment, would further enhance the chances of elution by the water/organic extract.

If there is a concern about the presence of sugars, carbohydrates, and other anionic species, the subsequent adsorptions could be undertaken using cartridges that are retentive for them. This would necessitate a solvent exchange, usually by partitioning, so that the amount of water will be either eliminated or reduced, in order to give the "polar extraction" type of column an opportunity for retention of the interferences (10). Examples of this sort of removal of polar interferences include the use of the quaternary methylamine anion exchanger (QMA), the aminopropyl (NH_2) under acidic conditions, the SAX, or the DEA (8,11,12). Some researchers have termed this technique that of "normal phase" SPE, meaning a separation in which polar isolates are retained by a cartridge that is being developed by a nonaqueous solvent.

Use of normal-phase SPE cartridges in conjunction with a reverse-phase HPLC analytical column, such as the C_{18}, has been shown by de Kok et al. to be superior for a large variety of crop samples, when *N*-methylcarbamates are analyzed using postcolumn fluorogenic labelling with *o*-phthalaldehyde (12). They studied silica, aminopropyl, cyanopropyl, and diol cartridges, using sample extracts in dichloromethane. Of these, they found the aminopropyl cartridge to surpass all the others for its ability to remove early-eluting interferences that normally would obscure the more water-soluble carbamates. Elution of the aminopropyl cartridge was accomplished with varying percentages of methanol in methylene chloride.

The approach of coupling an SPE cleanup step that is of a mode different from that of the determinative step appears attractive when HPLC is used as the determinative step. For example, the analysis of the *N*-methylcarbamate and carbamoyl oxime insecticides is traditionally accomplished by a reverse-phase separation on a C_{18} column (13). Consequently, success has been achieved in removing troublesome interferences by such a normal-phase SPE step, using the aminopropyl (7–9,11,12) and the quaternary amine (8,11) SPE.

We must emphasize that the preceding generalizations apply to the analyses of many pesticides in one group, the so-called multiresidue approach. Frequently this may not be necessary for the task at hand, and the same SPE particle type may be employed that is used in the analytical column of the HPLC, if that is the determinative step. For example, by adjusting the percentage of acetonitrile in water used to elute pesticides from the C_{18} cartridge, it may be possible to fractionate individual pesticides from each other, to simplify the determinative step. One study using this cartridge, which included 29 insecticides of several classes, including the organophosphates, *N*-methylcarbamates, and synthetic pyrethrins showed clear separations between individual groups of pesticides, separations that seemed to follow their water solubilities (Table 5.2) (13).

Table 5.2. Elution Pattern and Recovery of 29 Pesticides from C_{18} Sorbent

Pesticides[a]	MW	Solubility[b] (mg/L)	Elution with % CH_3CN in Water (% Recovery)					
			20	40	60	80	100	Total
P1: Monocrotophos	223	1,000,000	—	—	—	—	—	—
P2: Trichlorfon (DEP)	257	154,000	17	—	—	—	—	17
P3: Dimethoate	229	25,000	27	53	—	—	—	80
O1: Metalaxyl	279	7,100	—	30	51	—	—	81
C1: Metolcarb (MTMC)	165	2,600	—	85	—	—	—	85
C2: Propoxur (PHC)	209	2,000	—	92	—	—	—	92
C3: Fenobucarb (BPMC)	207	610	—	—	89	—	—	89
C4: Xylylcarb (MPMC)	179	580	—	90	—	—	—	90
C5: XMC[c]	179	470	—	88	—	—	—	88
P4: Phenthoate (PAP)	320	200	—	—	81	12	—	93
P5: Malathion	330	145	—	—	94	1	—	95
P6: Dimethylvinphos (*E*)	331.5	130	—	—	83	—	—	83
P7: Dimethylvinphos (*Z*)	331.5	130	—	—	95	—	—	95
C6: Carbaryl (NAC)	201	120	—	97	—	—	—	97
C7: Isoprocarb (MIPC)	193	ss[c]	—	73	14	—	—	87
P8: Fenthion (MPP) SO_2	310	ss	—	24	77	—	—	101
P9: Pyridaphenthion	340	ss	—	—	93	2	—	95
P10: EPN	323	is[d]	—	—	40	53	—	93
P11: Diazinon	304	40	—	—	43	48	—	91
P12: Phosmet (PMP)	317	22	—	—	96	2	—	98
P13: Fenitrothion (MEP)	277	14	—	—	88	—	—	88
P14: Tetrachlorvinphos (*E*)	366	11	—	—	83	10	—	93
P15: Tetrachlorvinphos (*Z*)	366	11	—	—	95	—	—	95

Table 5.2. *(contd.)*

Pesticides[a]	MW	Solubility[b] (mg/L)	Elution with % CH_3CN in Water (% Recovery)					
			20	40	60	80	100	Total
C8: Benfuracarb	410	8	—	—	3	97	—	100
P17: Isoxathion	313	2	—	—	48	44	2	94
O2: Fenvalerate	420	<1	—	—	—	87	2	89
O3: Oxadiazon	345	0.7	—	—	—	97	—	97
C9: Carbosulfan	380	0.3	—	—	—	68	23	91

Source: Ref. 13.

[a] P1–17: organophosphorus pesticides; C1–9: carbamate pesticides, O1–3: others.
[b] Water solubility.
[c] Slightly soluble.
[d] Insoluble.

The most soluble organophosphates, monocrotophos and trichlorfon, could be eluted with only 20% acetonitrile in water, when 0.5 g of adsorbent was used and 5 mL of solvent was used for elution. The remaining 27 pesticides remained on the column under these conditions. When the acetonitrile percentage was increased to 40%, and another 5 mL of solvent was applied, four more pesticides—one organophosphate and two carbamates—eluted, along with one acetanilide that only partially eluted, metalaxyl. When the acetonitrile percentage was increased to 60%, 23 of the pesticides eluted; those remaining had water solubilities of 8 mg / L or less. This same study clearly showed the effects of alkyl chain length on pesticide retention, when acetonitrile was used to elute a group of nine pesticides of various classes ranging in water solubility from 0.3 to 2600 mg / L. Little differences were seen for the retention of these pesticides on the C_{18} and C_8 cartridges, although for even 100% acetonitrile the least water-soluble was somewhat retained on the C_{18} cartridge, but fully eluted from the C_8 cartridge. However, when the methylsilane, C_1, cartridge was studied, only 40% acetonitrile was required to elute all nine, with the exception of fenobucarb, which eluted with only 62% acetonitrile (Table 5.3) (13).

Other cartridges were studied as well, including the cyclohexyl (CH), phenyl (PH), and cyanopropyl (CN), all using the water / acetonitrile eluents that were described above (see Table 5.4) (13). None of these were as retentive as the C_{18}, although the CH cartridge approached it, requiring 80% acetonitrile to elute benfuracarb and carbosulfan, both carbamates. It is interesting to note here how sharp some separations can be on such SPE devices. For example, even for the least retentive CN cartridge, 100% of bufencarb elutes in the 60% fraction of acetonitrile / water, with none appearing in either the 40% fraction or the 80% fraction. Consequently, the efficiency of separation for some of the water insoluble pesticides does not necessarily depend on the retention of the cartridges.

While no definitive studies have been done to clearly define the sample load that can be tolerated by the various cartridge particles, some general guidelines can be listed. Most manufacturers claim adsorption capacities, for total solute, to be in the range of 20–100 mg / g, with some exceptions given for the polymeric coatings (Accell Plus CM; Accell Plus QMA, Waters). These speciality cartridges can adsorb as much as 200 mg / g of packing, when proteins are adsorbed (14,15). However, knowing what to expect in the way of sample coextractives may be impossible, so the reader is referred to the following sections on recent advances to understand what has been accomplished by others in the way of sample loading capacity. Most successful chemists have left a broad margin for error in loading their samples, so that interference breakthrough doesn't occur as a result of

Table 5.3. Elution Pattern and Recovery of Carbamates from Four Different Sorbents

Sorbents	MW	Solubility[a] (mg/L)	Elution with % CH_3CN in Water (% Recovery)					
			20	40	60	80	100	Total
C_{18}								
Metolcarb (MTMC)	165	2600	—	85	—	—	—	85
Propoxur (PHC)	209	2000	—	92	—	—	—	92
Fenobucarb (BPMC)	207	610	—	—	89	—	—	89
Xylylcarb (MPMC)	179	580	—	90	—	—	—	90
XMC[a]	179	470	—	88	—	—	—	88
Carbaryl (NAC)	201	120	—	97	14	—	—	97
Isoprocarb (MIPC)	193	ss	—	73	3	—	—	87
Benfuracarb	410	8	—	—	—	97	—	100
Carbosulfan	380	0.3	—	—	—	68	23	91
C_8								
Metolcarb (MTMC)	165	2600	—	82	—	—	—	82
Propoxur (PHC)	209	2000	—	94	—	—	—	94
Fenobucarb (BPMC)	207	610	—	—	88	—	—	88
Xylylcarb (MPMC)	179	580	—	89	—	—	—	89
XMC[a]	179	470	—	87	—	—	—	87
Carbaryl (NAC)	201	120	—	107	—	—	—	107
Isoprocarb (MIPC)	193	ss	—	58	31	—	—	89
Benfuracarb	410	8	—	—	25	78	—	103
Carbosulfan	380	0.3	—	—	—	93	—	93
C_2								
Metolcarb (MTMC)	165	2600	—	84	—	—	—	84
Propoxur (PHC)	209	2000	—	92	—	—	—	92

Fenobucarb (BPMC)	207	610	—	—	87	—	—	87
Xylylcarb (MPMC)	179	580	—	90	—	—	—	90
XMC[a]	179	470	—	89	—	—	—	89
Carbaryl (NAC)	201	120	—	97	—	—	—	97
Isoprocarb (MIPC)	193	ss	—	80	7	—	—	87
Benfuracarb	410	8	—	—	102	5	—	107
Carbosulfan	380	0.3	—	—	7	85	—	92
C_1								
Metolcarb (MTMC)	165	2600	20	62	—	—	—	82
Propoxur (PHC)	209	2000	7	87	—	—	—	94
Fenobucarb (BPMC)	207	610	—	62	29	—	—	91
Xylylcarb (MPMC)	179	580	—	93	—	—	—	93
XMC[a]	179	470	—	90	—	—	—	90
Carbaryl (NAC)	201	120	—	100	—	—	—	100
Isoprocarb (MIPC)	193	ss	—	86	—	—	—	86
Benfuracarb	410	8	—	—	107	—	—	107
Carbosulfan	380	0.3	—	—	13	77	—	90

Source: Ref. 13.

[a] Water solubility.

Table 5.4. Elution Pattern and Recovery of Carbamates from Three Additional Sorbents[a]

Pesticides	MW	Solubility[b] (mg/L)	Elution with % CH_3CN in Water (% Recovery)					
			20	40	60	80	100	Total
CH								
Metolcarb (MTMC)	165	2600	—	81	—	—	—	81
Propoxur (PHC)	209	2000	—	88	3	—	—	91
Fenobucarb (BPMC)	207	610	—	—	86	—	—	86
Xylylcarb (MPMC)	179	580	—	82	6	—	—	88
XMC[c]	179	470	—	78	8	—	—	86
Carbaryl (NAC)	201	120	—	80	16	—	—	96
Isoprocarb (MIPC)	193	ss	—	35	48	—	—	83
Benfuracarb	410	8	—	—	27	77	—	104
Carbosulfan	380	0.3	—	—	4	85	—	89
PH								
Metolcarb (MTMC)	165	2600	8	74	—	—	—	82
Propoxur (PHC)	209	2000	—	92	—	—	—	92
Fenobucarb (BPMC)	207	610	—	35	54	—	—	89
Xylylcarb (MPMC)	179	580	—	87	3	—	—	90
XMC[c]	179	470	—	84	4	—	—	88
Carbaryl (NAC)	201	120	—	96	7	—	—	103
Isoprocarb (MIPC)	193	ss	—	76	9	—	—	85
Benfuracarb	410	8	—	—	92	10	—	102
Carbosulfan	380	0.3	—	—	15	78	—	93
CN								
Metolcarb (MTMC)	165	2600	71	6	—	—	—	77
Propoxur (PHC)	209	2000	79	10	—	—	—	89

Fenobucarb (BPMC)	207	610	—	79	4	—	—	83
Xylylcarb (MPMC)	179	580	24	61	—	—	—	85
XMC[c]	179	470	20	64	—	—	—	84
Carbaryl (NAC)	201	120	4	93	—	—	—	97
Isoprocarb (MIPC)	193	ss	5	77	—	—	—	82
Benfuracarb	410	8	—	—	102	—	—	102
Carbosulfan	380	0.3	—	4	82	—	—	86

Source: Ref. 13.

[a] *Method*: Grain extract (equivalent to 5 g) spiked with 9 carbamates at 1.0-ppm level was applied to 7 different phase cartridges with 0.5 g each sorbent mass. The cartridges were washed with 5 ml water and eluted with 5 ml each of above solution (20–100% acetonitrile in water).

[b] Water solubility.

[c] 3,5-Xylyl methylcarbamate.

variations in food composition due to variety, maturity, and harvest interval, if fresh food is being analyzed.

Coupling Extracting Solvent to SPE Device

Most published analytical methods employing SPE devices for pesticide isolation begin with a hydrophobic interaction (reverse-phase) cartridge in the first separation, if there are more than one (see section on recent advances below). Consequently, either a water miscible solvent is used or water itself, so that interferences that are of a hydrophobic nature are adsorbed, allowing the pesticides to flow through. If liquid foods, such as beverages, are analyzed for pesticides, they can be added directly to the cartridges and eluted with appropriate solvents. Wines have been analyzed for a wide array of pesticides after extraction of aliquots with either C_8 or C_{18} (16,17); they were added directly with elution by ethanol/water, followed by elution with ethyl acetate (16) or methylene chloride.

Recent Advances

Mixed Mode Cartridges

As described below under the sections dealing with methods, multiple cartridges in sequence are frequently used at various stages throughout the method, often employing various retention mechanisms, sometimes as many as three. Together, these sequential cartridges give increasingly clean sample extracts, such as is required when the ion-trap detector is employed (11). More recently, at least one study has been dedicated to exploring the potential of a single cartridge having mixed-mode retention characteristics (18). Specifically, cartridges were filled with plastic beads composed of styrene–divinylbenze but with chains on the polymeric backbone of the familiar C_{18} group, and the less familiar sulfonic acid group, were chosen. One type of cartridge, called the MP-3, was produced by Interaction Chromatography (Mountain View, CA). The structure of the MP-3 bead is shown in Figure 5.3. The authors chose to study the retention characteristics of a group of triazine herbicides, including atrazine, and its metabolites deethylatrazine and deisopropylatrazine. Other triazines studied were simazine, propazine, ametryn, prometryn, terbutryn, and prometon.

Distribution coefficients were measured for all five triazines for the affinity of the MP-3 bead, along with a bead having only C_{18} functional groups, the MP-1 bead (18). Free-energy-of-adsorption differences between the two beads, ΔG_a, amounted to 2.59 kcal/mol for those herbicides having a primary amine moiety, and to 1.77 kcal/mol for those having a secondary

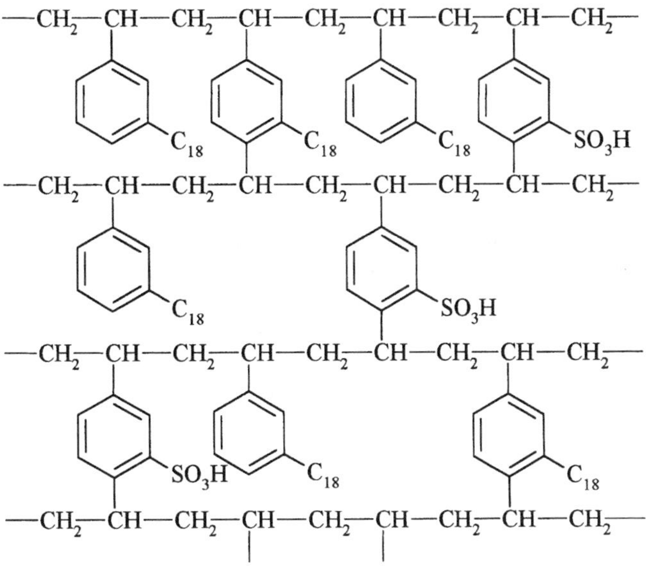

Figure 5.3. Structure of the chemically bonded-phase MP-3, produced by Interaction Chromatography.

amine moiety. This approximates energies attributable to hydrogen bonding, apparently with the sulfonic acid functionality of the MP-3 bead. Attempts to simulate the MP-3 bead, by mixing equal volumes of beads having only the C_{18} functionality and the sulfonic acid functionality, showed less recovery of the triazine herbicides from water than observed for the mixed-mode MP-3 bead; this bead consistently gave 95% recoveries of all triazines from laboratory water. It is obvious that the MP-3 bead brings interactions into play for the retention of the triazine herbicides that go beyond the typical "hydrophobic" interactions of the C_{18}-type beads, and which ultimately lead to improved recoveries for pesticides that can behave as proton acceptors. Other workers have studied the effect of sulfonation levels of the SDVB polymer bead on the retention of several phenolic compounds, including phenol (19). Retention of this compound when various sulfonated polymers were eluted with pure water was a maximum at approximately 0.6 meq/g. Porous Teflon Empore disks containing the various sulfonated resins were used for this study (see below, section on solid-phase extraction: disks).

A recently available "hydrophilic" particle packed cartridge, the Oasis HLB (Waters), has been put to use for the analysis of an array of drugs and pollutants, and may have the potential for cleanup of some of the more water-soluble pesticides (20). This particle, produced by increasing the degree of *N*-vinylpyrrolidone crosslinking in the divinylbenzene:*N*-vinyl-

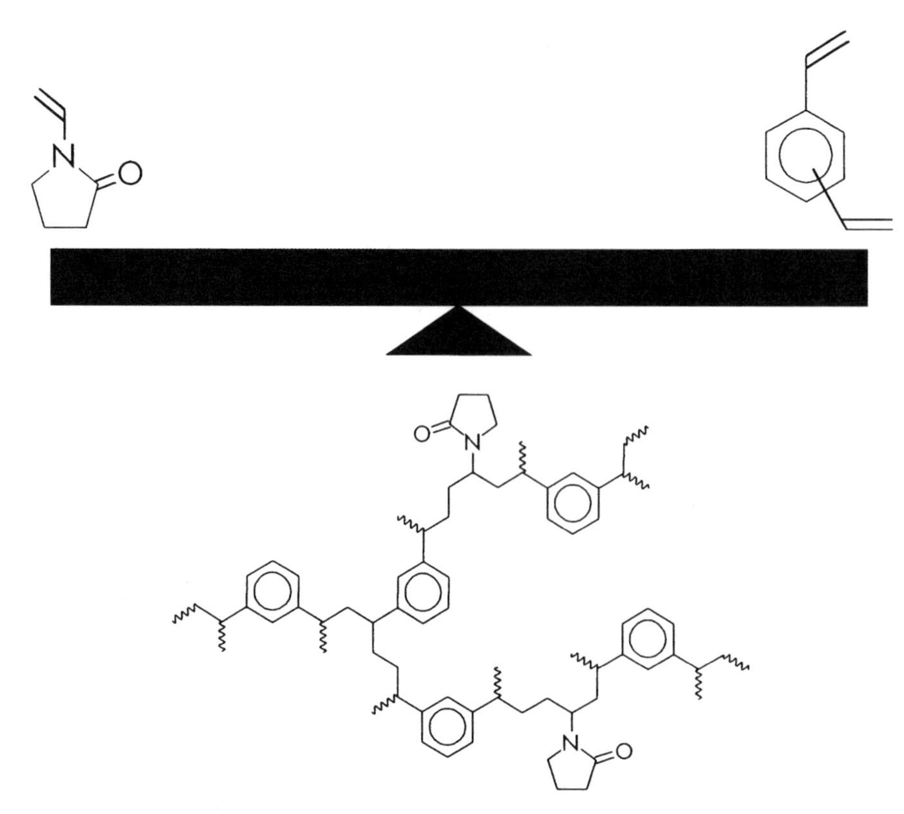

Figure 5.4. Structure of the chemically bonded phase Oasis HLB, produced by the Waters Company (courtesy Waters Comapany). Hydrophilic–lipophilic balance.

pyrrolidone copolymer above that of an older particle, Porapak RDX, has some features that should make it attractive to the pesticide analytical chemist. The chemical structure of this copolymer is shown in Figure 5.4. Porapak RDX has been shown to efficiently extract several sulfonylurea herbicides from water and aqueous soil extracts, including thifensulfuron methyl, metsulfuron methyl, chlorsulfuron, and tribenuron methyl (21). Since water is able to wet the surface of the particle very effectively, there is no need for cartridge "activation" with a water-miscible organic solvent, such as methanol, which is the case for silica-based particles having "hydrophobic" coatings, such as the octadecylsilyl groups. Not only does this simplify sample application to the cartridge; it also means that they are immune to the "deactivation" that can occur if the device is left under vacuum after "activation" before the aqueous sample is added. This insensitivity to drying out can be seen in Figure 5.5, where recovery of five

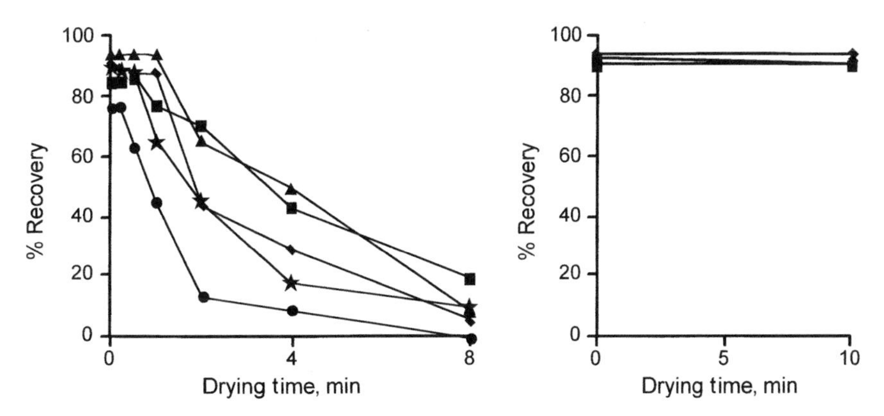

Figure 5.5. The effect of cartridge drying on the recovery of several drugs spiked in porcine serum: (*A*) 100 mg C_{18}–silica cartridge; (*B*) 30 Oasis HLB cartridge. (*a*) C_{18}, (*b*) HLB.

different drugs—procainamide, acetaminophen, rantidine, doxepin, and propranolol—is shown for various drying times after activation with methanol. While the C_{18} silica cartridge shows diminished recoveries of the drugs from water as a function of methanol drying, no losses are observed for the Oasis HLB cartridge (30).

Elution of Pesticides from Cartridges

A wide array of solvents have been used to elute retained pesticides from SPE cartridges, depending on the hydrophobicity of the pesticide and the retention mechanism on the cartridge. For nonpolar pesticides, these solvents have included methanol, acetonitrile, ethyl acetate, chloroform, and hexane. For polar pesticides, they have included methanol, isopropanol, and acetone (10). Undoubtedly, many others that have escaped notice here have been used. Choice of the appropriate solvent for complete elution may not be trivial, and may be dependent on the matrix being analyzed, and the classes of chemicals that are coextracted and retained on the cartridge along with the analyte. For example, de Kok et al. in 1987 showed the extreme sensitivity that effective cleanup has on the eluting solvent, arriving at a solvent composed of 1% methanol in methylene chloride for the elution of the *N*-methylcarbamate pesticides (12). This sensitivity was confirmed by the work of Page and French, who arrived at a 2% methanol in methylene chloride eluting solvent composition (7).

More recently, in a study examining the elution of a number of widely ranging toxicants from a C_{18} cartridge after extraction from various types of contaminated waters, including sediment pore water, it was clearly shown that a less polar solvent is required to remove the hydrophobic pesticide metabolite DDE from the cartridge bead (22). DDE, which has an estimated log K_{ow} of 6.94, was adsorbed from water at a concentration of 10 μg/L, by passing 950 mL of filtered pore water solution through 1 g of SPE adsorbent (J. T. Baker, Phillipsburg, NJ). When an elutropic series of methanol and water was passed through the cartridge, with methanol content increasing from 25 to 100%, a portion of the DDE eluted in the first, second, and third portions of 100% methanol; most remained attached to the cartridge beads. However, when the eluting solvent was switched to methylene chloride, the remaining DDE was eluted. A more polar analyte, diethyl phthalate (DEP), with a log K_{ow} of 2.57, came out in the methanol–water series of solvents. Each solvent aliquot was exactly 1.5 mL. A less polar analyte, hexachlorobenze (HCB) (log $K_{ow} = 6.42$), behaved somewhat similarly to DDE. From this study, it seems as if both DDE and HCB were retained on the cartridge by two different mechanisms. Such behavior would not be predicted from HPLC retention data alone, since both compounds elute rapidly with 100% methanol. It seems highly likely that coextractives from soil pore water may come into play here. Consequently, the analytical chemist must be continually aware of apparent losses on the SPE cartridge that might be a result of cartridge adsorption behavior modifications resulting from the presence of unknown but active coextractives.

Somewhat similar phenomena have been observed in the retention of amino group containing drugs that have been extracted by SPE from human plasma (23). When two drugs with pK_a values of approximately 8 were extracted from laboratory water with either C_{18} or cyano SPE cartridges at low concentrations, very low recoveries were observed. Further study showed that the drugs were highly dependent for their retention on both buffer strength and pH. Plasma adjusted to pH 7 with 0.05 M buffer and 0.1% triethylamine gave the highest recoveries observed, when compared to buffered water. It was also shown that small amounts (0.04%) of detergent, such as Tween 80 and Triton X-100, aided in recovery of the drugs. Such behavior seem to show that there still remained active sites on the silica particles that bound the polar drugs in an irreversible manner. The presence of plasma constituents or detergents seemed to mask them and allow elution by methanol containing 0.1% trifluoroacetic acid.

Multiresidue Methods

Incorporation of SPE cartridges into pesticide analytical methods for residues in foods has taken nearly 20 years after they were first introduced, and has lagged their incorporation into analytical methods for water and hazardous wastes considerably. While numerous methods have been developed for an array of single pesticides, or small groups related by molecular structure, by far the most significant are those that have been developed for multiple classes, typically termed *multiresidue methods.* Laboratories within the U.S. Food and Drug Administration, the California Department of Food and Agriculture, and the Florida Department of Agriculture and Consumer Services have all developed multiresidue methods utilizing either GC, HPLC, or GC/MS in the determinative steps (7–9,11). One laboratory in New Zealand, the Horticulture and Food Research Institute of New Zealand, has developed a multiresidue method using GC as the determinative step (16). This work was subsequent to somewhat similar approaches taken in the laboratory of the Department of Toxicology of the University of Cagliari, Italy, which employed reverse-phase HPLC with diode array detection (17).

The method described by the Florida workers was based on an adaptation of an extraction and cleanup method developed 15 years ago by Muth and Erro (24), who used the method described by Luke (8) to extract the *N*-methylcarbamates aldicarb sulfoxide, oxamyl, methomyl, 3-hydroxycarbofuran, propoxur, carbofuran, carbaryl, and methiocarb from an array of fruits, feeds, and vegetables (7). They incorporated the aminopropyl cartridge in the cleanup step, using 2% methanol–methylenechloride. Recoveries at the 0.05-, 0.5-, and 5.0-μg/g levels averaged 98, 90, and 91%, respectively, for all crops, with standard deviations for replicate analyses less than two standard deviations from the statistical means. Separations were achieved on both a C_8 and cyanopropyl (CN) Zorbax reverse-phase columns. Such a cartridge cleanup method allows a single chemist to prepare 15–20 samples in a single working day.

Workers at the Los Angeles, CA, U.S. FDA laboratory have recently developed an SPE cartridge based cleanup method specifically designed to be used with the gas chromatograph/ion trap detector (GC/ITD) for quantitations (11). This method was a follow-up to a previous study on the gas chromatograph/ion-trap detector for 254 target pesticides that employed no cleanup after acetone extraction, and which demonstrated low and variable recoveries resulting from the presence of interfering coextractives (25). A benchtop instrument using an ITD was used in the selected ion monitoring mode. Such an approach required that injections onto the GC/ITD representing 3.6 mg of crop be used, making pesticide masses per injection in the 50-pg range achievable. In order to present samples that are extremely clean, they selected three cartridges of different bonded phases that are applied in sequence. Those chosen were the C_{18} Sep-Pak (Waters), the Accell Plus QMA (Waters), and the aminopropyl (Waters). As reported before by Luke (8), 200 mL of acetone is used to extract 100 grams of crop by blending; for low-moisture products, 10–15 g is blended with 100 mL of water and 200 mL of acetone. Exactly 40 mL of the acetone–water extract is passed through the C_{18} cartridge, which is then eluted with an addtional 10 mL of 30% water in acetone. The collected eluate is further cleaned up by partitioning with 50 mL of acetone added to 1 mL of methylene chloride. The methylene chloride phase is then evaporated to near dryness in a Kuderna–Danish concentrator and the residue taken up in 5 mL of acetone. To this acetone solution, 10 mL of petroleum ether is added; the resulting solution is then passed through the Accell Plus QMA and aminopropyl Sep-Pak cartridges connected in tandem. Three additional 10 mL volumes of acetone–petroleum ether are used to further elute the tandem cartridges. This solvent is then evaporated to near dryness in the Kuderna–Danish, and the residue taken up in 50 mL of methylene chloride, which is further concentrated to 4 mL before analysis by the gas chromatograph / mass spectrometer.

When taken together, the preceding combination of cartridges were claimed to remove long chain fats and waxes (C_{18}), ionic species (Accell Plus QMA), and sugars (aminopropyl). Removal of such species permitted recoveries of 24 pesticides from strawberries, carrots, tomatoes, and lettuce ranging within 32.3–129% at the 0.5-μg / g level. Total-ion chromatograms for lettuce fortified at this level before and after the three cartridge cleanup are shown in Figure 5.6. It is readily apparent that pesticide peaks are simply missing in the unclean extract, along with a large elevation in baseline current. However, even with this cleanup, ions were still seen in the chromatogram at multiple retentions, especially at the masses used to quantitate chlorpyrifos ($m/z = 153$), and tetrahydrophathalimide ($m/z =$ 152,181) (Fig. 5.7).

A simpler approach was taken by a group of pesticide analytical chemists at the California Department of Food and Agriculture laboratories in Sacramento, CA (9). They studied seven chlorinated hydrocarbons, seven organophosphates, and seven *N*-methylcarbamates, and used gas chromatography with electrolytic conductivity detection (CHCs), flame photometric detection (OPs), and HPLC with postcolumn fluorogenic labeling detection (*N*-methylcarbamates). Extractions were done by blending 50 g of crop with 100 mL of acetonitrile. Since the authors claim that acetonitrile removes more coextractives from the crop than does the acetone used in the U.S. FDA–developed method described above, they were faced with utilizing sequential SPE cartridges also, although only two were found to be required in order to give clean chromatograms: the C_{18} (370 mg, Waters) and the aminopropyl (100 mg, Analytichem International). The C_{18} cartridge was employed immediately after the extraction was performed with acetonitrile, which gave large amounts of water that was removed from the fresh crop.

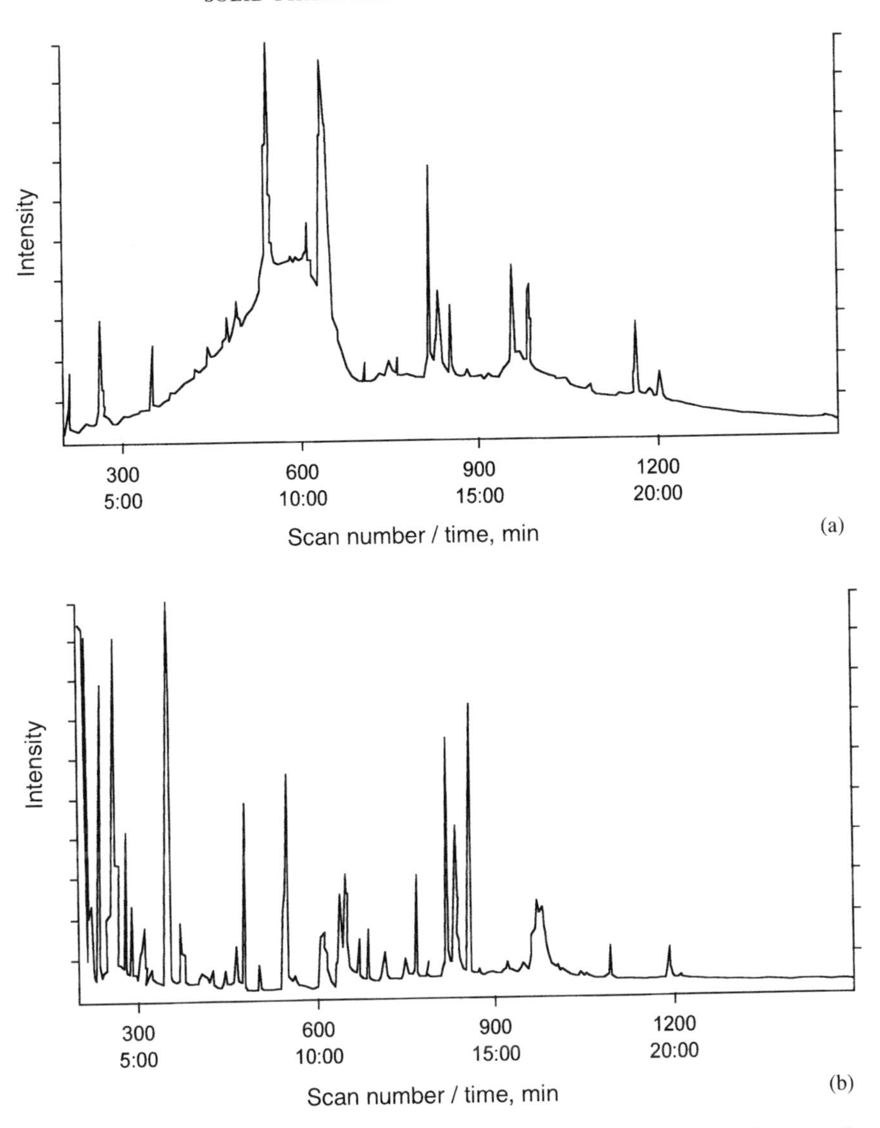

Figure 5.6. Total-ion chromatograms for a lettuce extract obtained (*a*) before sample cleanup and (*b*) after sample cleanup through three different SPE cartridges (Ref. 11).

This eluate was heated on a Kuderna–Danish to replace the acetonitrile with acetone, the acetone was further concentrated, and aliquots were taken for GC analysis for the chlorinated hydrocarbons (electron-capture detector) and the organophosphates (flame photometric detector). A separate aliquot of the

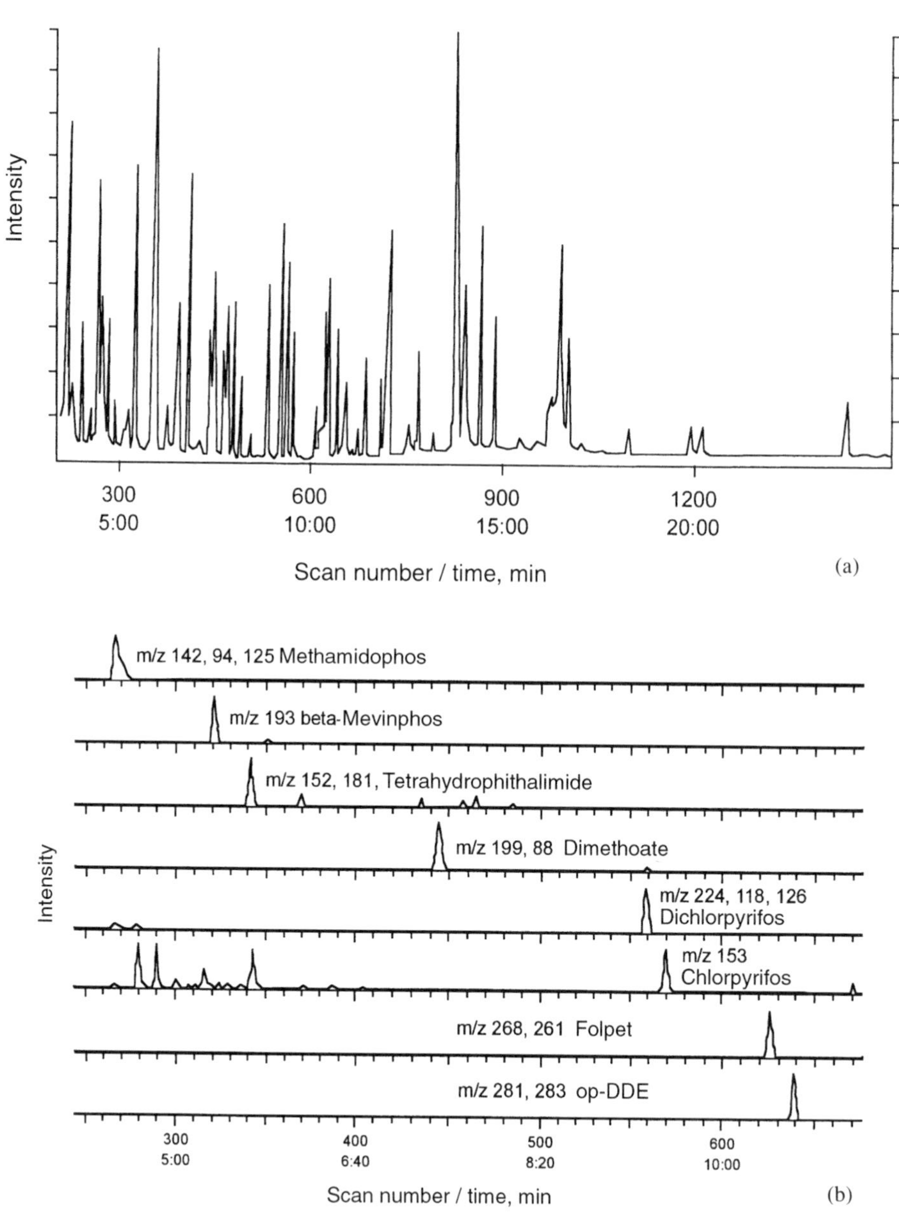

Figure 5.7. Total-ion chromatogram of a lettuce extract spiked with 24 pesticides at the 0.5-μg/g level (*a*), together with single-ion chromatograms to illustrate position of pesticides using ions selected for quantification (Ref. 11).

acetone solution was made to undergo another solvent exchange procedure, resulting in dissolution of the pesticides in 1% methanol in methylene chloride, and the sample was passed through the aminopropyl cartridge. Finally, this cartridge was eluted with an additional 2 mL of that solvent. Another solvent exchange resulted in the residue being dissolved in acetonitrile : water (1 : 1), which was analyzed for the *N*-methylcarbamates by reverse-phase chromatography and postcolumn fluorogenic labeling with *o*-phthalaldehyde.

Even though the electrolytic conductivity detector was used for the chlorinated hydrocarbons, which normally presents few problems when crop coextractives are present, it was necessary to employ a neutral phosphate buffer–saturated brine solution to partition away carboxylic acid interferences in citrus samples. Such a partitioning also greatly eliminated potential interferences in the *N*-methylcarbamate analyses. Similarly, for green onions (scallions), it was necessary to employ a sodium borohydride reduction of unidentified sulfur containing compounds to eliminate interferences (Fig. 5.8). Fortifications at 0.1 μg / g for the chlorinated hydrocarbons and *N*-methylcarbamates and at 0.2 μg / g for the *N*-methylcarbamates produced recoveries ranging from 24.5% for chlorothalonil on potato to 153.2% for dimethoate on orange fruit. Most recoveries were in the 80–95% range. Detection is also possible at the 0.01-μg / g level.

A Bond Elut C_8 cartridge was found by a group of scientists at the University of Cagliari (Italy) to be sufficient to remove potential interferences from wine so that 15 pesticides of various classes, including organophosphates, chlorinated hydrocarbons, and *N*-methylcarbamates, can be separated on a reverse-phase column (Spherisorb S_5-ODS-1; Waddinxveen, Netherlands) and quantitated on the diode array detector (17). Wine was cleaned up by passing 2 mL through the cartridge, followed by a wash with HPLC-grade water (3 × 2 mL) and 30% ethanol (3 × 2 mL), after which it was dried by vacuum. The pesticides were removed with 2 mL of methylene chloride. They found that the cartridge could be reused after it was washed with 25 mL of methanol. Fortifications at 1.00 and 0.01 μg / g gave excellent recoveries, ranging from 85% for chlozolinate to 108% for triadimefon. Limits of detection were in the 0.006–0.020 μg / g range. They found that methylene chloride was a superior solvent for removing the pesticides from the cartridge, leaving behind many interferences that were removed with more polar solvents, such as methanol or acetonitrile.

Holland et al., have developed a somewhat similar approach using a C_{18} cartridge for the analysis of 81 pesticides from wine (16). Polypropylene cartridges containing 500 mg of material were used (Alltech or Varian-Analytichem). Each cartridge was prewashed with acetone (3 × 1 mL) and ethanol : water (20 : 80, 4 mL) before the wine was extracted (10 mL).

Pesticides were eluted with ethyl acetate (1 mL). Screening was performed on an intermediate-polarity GC column (HP5, Hewlett-Packard) and confirmation on another (DB-17), with detection by electron-capture and nitrogen/phosphorous detectors. Of the 81 pesticides studied, recoveries for several of the organophosphates were less than 25% at the 0.2-μg/mL fortification level, including acephate, dicrotophos, dimethoate, methamidaphos, and omethoate. All are highly water-soluble, and probably were not retained by the cartridge. Fenpropimorph, imazalil, and pirimicarb also gave unsatisfactory recoveries. Recoveries of the remaining 73 ranged from 58% (cylothrin) to 102% (simazine). Using ethyl acetate for elution of the pesticides from the cartridges gave visible amounts of water, which could be

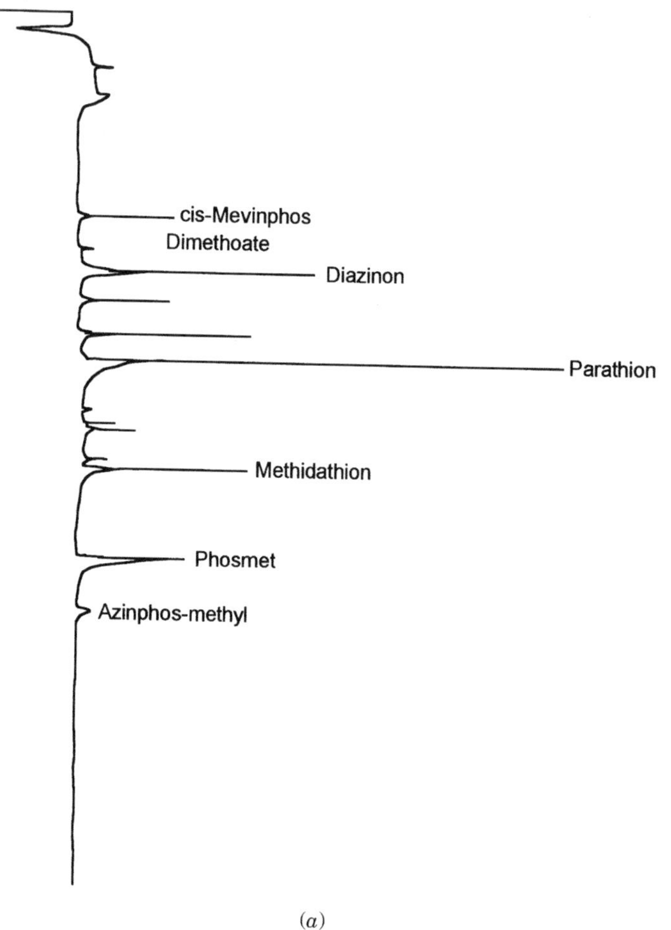

(*a*)

Figure 5.8. *(see facing page)*

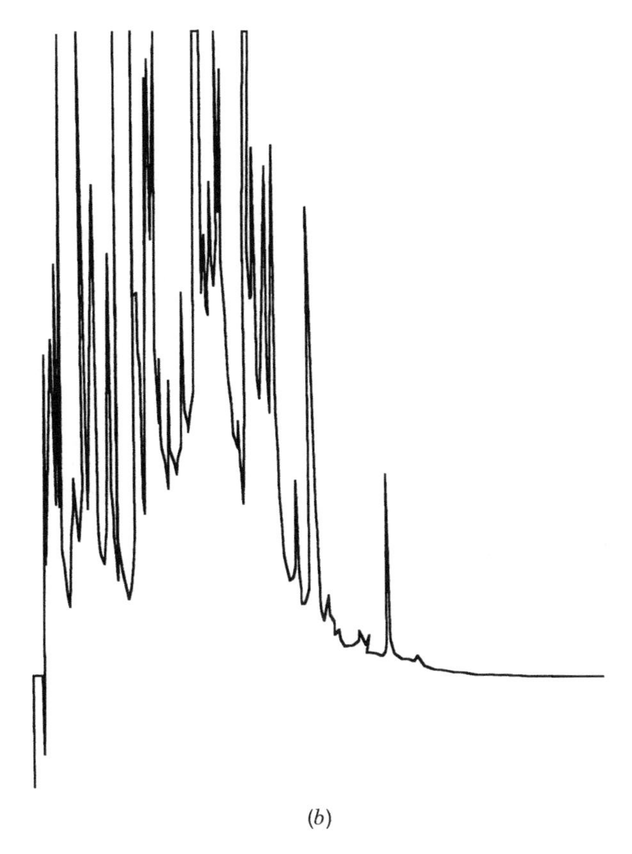

(*b*)

Figure 5.8. GC / FPD analysis of organophosphate pesticides. Comparison of the green onion sample with the addition of $NaBH_4$ (*a*), and without addtion of $NaBH_4$ (*b*) (Ref. 9).

isolated by adding isooctane to the collection tube, along with a small amount (0.3 g), of anhydrous sodium sulfate. Using larger volumes of wine, to 20 mL, did not give enhanced limits of detection, which remained at about 0.01–0.02 μg / mL.

More recently, a group of pesticide residue chemists at the Baltimore District Laboratory of the FDA developed a method that uses two SPE cartridges for the cleanup of organochlorine pesticides and polychlorinated biphenyls from nonfatty seafood products (26). Chlorinated pesticides studied included lindane, heptachlor epoxide, *trans*-chlordane, *p,p′*-DDE, dieldrin, endrin, *o,p′*-TDE, *cis*-nonachlor, and *p,p′*-DDT. PCB congeners studied were the Aroclor 1242, 1254, and 1260 groups. Nonfatty fish selected were flounder, whiting, and sea trout, while shellfish were represented by crab, shrimp, and scallops. Fortifications were made at 0.01, 0.05, and 0.40 µg/g, before extractions were done using acetonitrile and homogenization with a Polytron ultrasound device. After dilution of the acetonitrile solution with water, the mixture was passed through a 1-g Mega Bond Elut C_{18} column (Varian Sample Products), and subsequently eluted with toluene–petroleum ether (3 + 97), which was directed onto the top of a 1-g LC-Florisil column (Supelco). Elution of the pesticides and PCBs from the Florisil column was accomplished with ethyl ether–petroleum ether (10 + 90). Analysis was performed by capillary gas chromatography with electron-capture detection. Average recoveries were in the mid 90% range for all levels of fortification, and did not seem to fluctuate with fish type. Chromatograms from control fish were remarkably clean, compared to what is typically seen for such sample types.

Coupling cartridges of various adsorption modes has been found by some investigators to be more efficient in removing lipids from fat samples, particularly when one of the determinative steps is the highly sensitive electron-capture detection. Workers at the FDA's Pesticides and Industrial Chemicals Research Center in Detroit have coupled diatomaceous earth (Extrelut QE), reverse-phase (Sep-Pak C_{18}), and neutral surfaced alumina (Alumina-N) for the analysis of a mixture of organophosphate and chlorinated hydrocarbon insecticides from solid fat samples (butter) and edible oils (soybean, sunflower, corn; 27). Hexane was used for the extraction, giving a fat concentration of 0.6 g/mL, which was then deposited onto the 3-mL Extrelut column positioned directly above the C_{18} cartridge (Fig. 5.9).

Both column and cartridge were then eluted with 20 mL of a 1 : 1 acetonitrile : methanol mix saturated with hexane; additional elution of the C_{18} cartridge was accomplished with 20 mL of methanol using gravity flow. Half was retained for GC analysis using the flame photometric detector (FPD), and half was passed through the Alumina-N cartridge, which had been deactivated with 10 mL of acetonitrile containing 0.2 mL of water and prewashed with 50 mL of 20% dichloromethane in hexane. Elution of the chlorinated hydrocarbon pesticides from the Alumina-N cartridge was then accomplished with an additional 25 mL of 20% dichloromethane in hexane, using gravity flow. Most of the chlorinated solvent was removed by careful rotary evaporation, being replaced with hexane before injection onto the ECD gas chromatograph. When 1.8-g fat samples were tested by elution through the tandem Extrelut QE/C_{18} column/cartridge, the highest lipid breakthrough occurred with butterfat (47 mg) and the lowest with corn oil (19 mg). When sources of the Alumina-N cartridges were used other than from the Waters Company (Burdick & Jackson, J & W Accubond), differences in the degree of deactivation were noted, with the Waters and J & W cartridges being equivalent, while the Burdick & Jackson cartridge

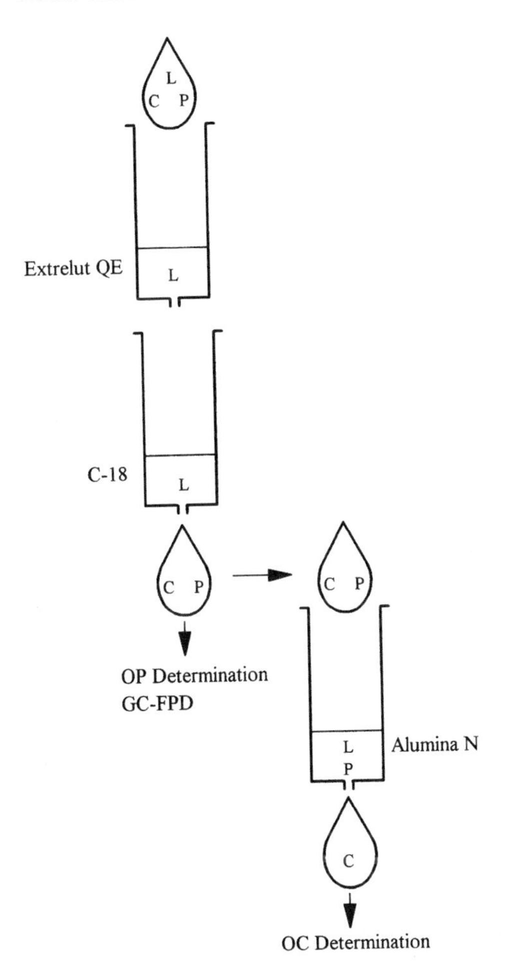

Figure 5.9. Tandem SPE cartridge arrangement using Extrelut QE and Alumina-N for the cleanup of vegetable oils and butterfat (Ref. 27). Key: L, lipids; C, OC pesticides; P, OP pesticides.

required much less deactivation. Chromatograms of the cleaned up fats and oils, when subjected to ECD gas chromatography, were remarkably free from interference (Fig. 5.10), and recoveries were greater than 89% for all pesticides including the chlorinated hydrocarbons (0.15–1.14 μg / g) and the organophosphates (1.07–4.20 μg / g).

Tandem SPE columns have also been used for the cleanup and analysis of the fungicide Dinocap (Karathane) in apples, grapes, and pears (28). A well-characterized solvent, acetone, was used for the extraction of this pesticide

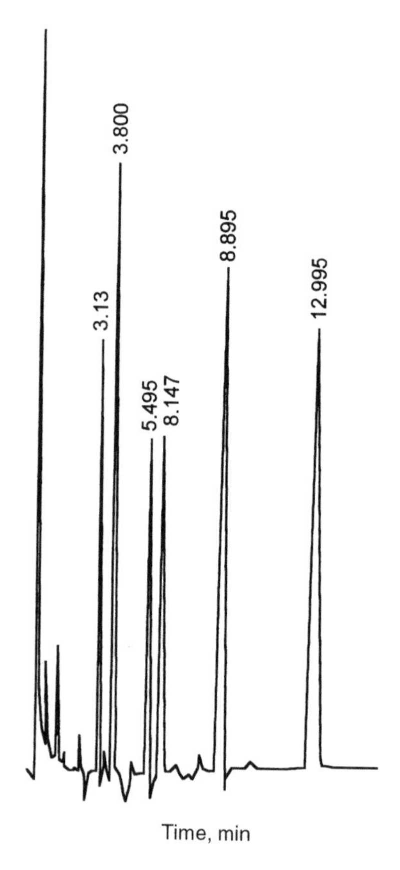

Figure 5.10. Chromatogram with GC-ECD of cleaned up olive-oil extracts for aldrin, heptachlor epoxide, dieldrin, endrin, *p,p'*-DDT, and methoxychlor (Ref. 27).

from these foods, one known to present the analyst with a challenging cleanup problem because of its propensity for also extracting a large array of potential interferences. In this study, a full 100-g sample was blended with 200 mL of acetone, the particulates filtered away, and the supernatant passed through the C_{18} Sep-Pak (1 g), and the supernatant discarded. The SPE cartridge was then eluted with 1 mL of petroleum ether followed by 5 mL of chloroform, both fractions combined, evaporated to dryness with a stream of dry nitrogen, and the residue dissolved in 1 mL of petroleum ether. This eluate was then applied to the silica SPE, which was subsequently developed with a 3×2 mL portion of ethyl ether/petroleum ether (10/90), the eluates combined, evaporated to dryness and redissolved in 1 mL of mobile phase (56% acetonitrile–44% water). Analysis was conducted on a C_{18} column with a UV absorbance detector set at 245 nm. Recoveries were almost all above 85% at fortifications of 0.05, 0.10, and 0.50 µg/g, with good coefficients of variation, ranging from 3.7 to 9.4% for triplicate analyses. Even after the tandem SPE cartridge cleanup, many potential interferences appeared on the chromatograms, requiring that the mobile phase selected elute the three isomers of Dinocap at 30 min and later, making for relatively low sample throughputs during an analytical day.

The effectiveness of a single cartridge, Sep-Pak silica (Waters), for cleanup of vegetable residues that contain lower amounts of lipid was well demonstrated in light of the use of the more universal UV absorbance detector coupled with a reverse-phase separation on a cyanopropyl column (29). Although a diode array detector was used, only one wavelength (202 nm), was employed, for the simultaneous analysis of aldicarb, aldicarb sulfoxide, and aldicarb sulfone. Homogenized white potatoes (20 g) were extracted by shaking for 30 min with dichloromethane (20 mL) while being dried with sodium sulfate (10 g), after which the solids were removed by centrifugation. After the Sep-Pak silica cartridge (50 mg) was deactivated with 5 mL of dichloromethane, the potato extract was passed through under negative pressure in about 10 min, and this fraction was discarded and the aldicarb and metabolites were eluted with 2 mL of acetonitrile. For HPLC the acetonitrile was diluted with water to give 1 : 2 acetonitrile–water, and injected onto the cyanopropyl column, which was developed with 30% acetonitrile in water at 1.2 mL / min. Aldicarb eluted in 3 min, aldicarb sulfoxide in 2.3 min, and aldicarb sulfone in 2.1 min, with full baseline resolution, with no interfering peaks from potato coextractives. The somewhat low limits of detection for aldicarb at the 5-μg / kg level (60%) could probably be improved by performing multiple extractions, although better recoveries were observed for the sulfoxide and sulfone.

Automation of the SPE cleanup step has been shown to greatly increase sample throughput, as demonstrated for the analysis of carbendazim, the major benomyl fungicide breakdown product, and another leading fungicide, thiabendazole (30). Such automation has greatly enhanced the pesticide monitoring capability of the Food Inspection Service of the Netherlands, freeing up analysts for other parts of the analytical method, particularly extractions and chromatograph operation. The device employed, a Gilson ASPEC model 401, is capable of moving the disposable extraction cartridge (DEC) to a drain position where it is preconditioned with solvent, applying the sample extract to the cartridge, washing the cartridge with eluant, moving the DEC back to a collection position, eluting it with the appropriate solvent, adding pH adjusting solution to the collection vial (a procedure that must be written into machine language by operator), mixing the eluate in the collection vial, and injecting an aliquot into the HPLC system. Separations were achieved on a variety of polymeric reverse phase columns, with the Shodex DE-613 determined to be superior for numbers of theoretical plates and peak shape. Detection was either with a UV absorbance detector set at

280 nm or with a fluorometric detector, with excitation at 235 nm and emission above 280 nm. A photograph of a successor to the ASPEC 401, the ASPEC XL, is shown in Figure 5.11. Fifteen-gram samples of chopped fruit or vegetable were extracted with 30 mL of acetone by homogenizing with ultrasound (Polytron, Brinkmann) for 30 s. The polarity of the extract was adjusted by adding 30 mL petroleum ether and 30 mL dichloromethane, extraction was performed again by ultrasound, and the solids were separated out by centrifugation. Exactly 3 mL of the extract was placed in the ASPEC sample vial and gently evaporated to dryness at 60°C with a stream of dry nitrogen, and redissolved in 2 mL of methanol. This solution was placed in the ASPEC sample rack, and cleaned up as described above on a 500-mg diol-bonded silica cartridge. Large differences were seen in the performances between cartridges from various manufacturers; the phase supplied by Varian (Bond Elut) was the only one that retained the benzimidizole compounds in the methanol solutions used, retaining those compounds in a reproducible manner even from batch to batch. In a 2 year study, their method showed a remarkable number of positives for various types of fruits and vegetables, with a total of over 2000 samples analyzed in a 2-year period (Table 5.5).

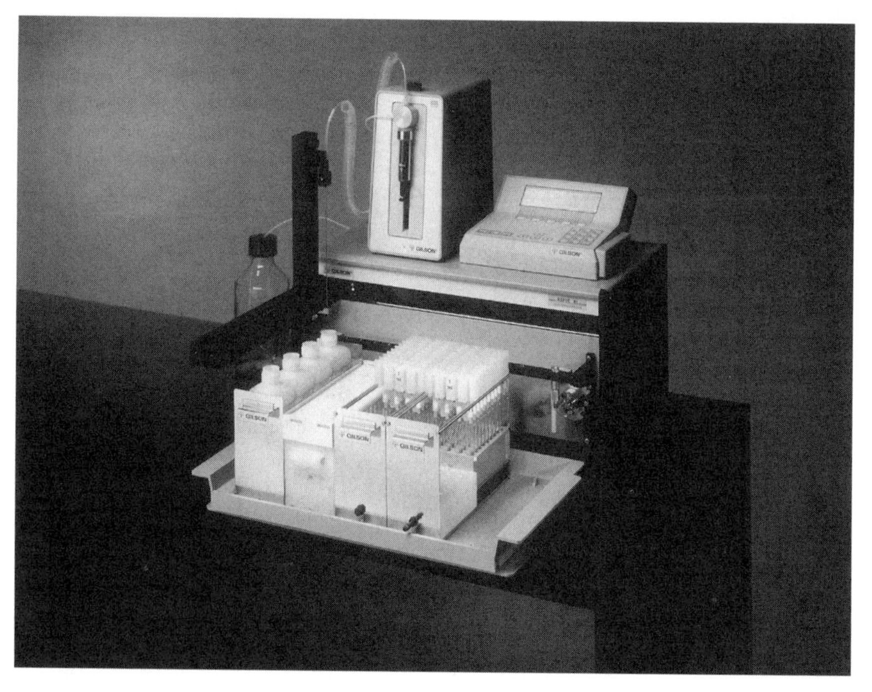

Figure 5.11. The Gilson ASPEC XL solid-phase extraction system, Gilson, Inc. Middleton, Wisconsin, USA (courtesy Gilson).

Table 5.5. Benzimidazole Fungicide Residues in Fruits and Vegetables from January 1992 to December 1994

Product	Number of Samples Analyzed[a]	Number of Residue Findings[a]		EC Maximum Residue Limit for Carbendazim (mg/kg)	Dutch Maximum Residue Limit for Thiabendazole (mg/kg)
		Carbendazim	Thiabendazole		
Vegetables	64	2	—	3	0.1
Bean	151	3	—	3	0.1
Cabbage	40	1	—	3	0.1
Celery	44 (2)	1	2 (2)	3	0.1
Cucumber	39	1	—	3	0.1
Endive	87	1	—	3	0.1
Kale	46 (7)	9 (7)	—	0.1	0.1
Lettuce	146	2	—	3	0.1
Potato	43	—	3	3	5
Spinach	73	1	—	3	0.1
Sweet pepper	61	4	1	3	0.1
Tomato	74	8	3	3	2
Fruits	—	—	—	—	—
Apple	101	28	27	2	10
Banana	24	—	13	1	3
Grapes	44	11	—	3	0.1
Kiwi	11	1	—	0.1	0.1
Mango	5 (1)	2	3 (1)	0.1	0.1
Melon	23 (2)	3	5 (2)	0.5	0.1
Nectarine	14	5	—	3	0.1
Peach or plum	44	2	—	3	0.1
Pear	19	1	—	2	10
Pineapple	5 (1)	2 (1)	1	0.1	0.1
Strawberry	273	24	—	3	3
Citrus fruits	—	—	—	—	-
Grapefruit	33	1	26	5	6
Lemon	12	—	7	5	6
Mandarin	27 (1)	2	23 (1)	5	6
Orange	66	8	46	5	6
Subtotal	1569 (14)	123 (8)	160 (6)	—	—
Other products	555	1	1	—	—
Total analyzed	2124 (14)	124 (8)	161 (6)	—	—

[a] Number of samples with benzimidazole fungicide residues exceeding residue tolerances is given in parentheses.

Although not making use of laboratory automation for performing SPE on crop extracts, another laboratory has been very successful in developing an effective method for the carbendazim breakdown product of benomyl, the active ingredient in Benlate fungicide (31). Making use of a highly sensitive fluorometric detector, with excitation set at 285 nm and emission at 315 nm, as little as 0.05 ng of carbendazim could be detected with good precision. Extraction was performed on 10 g of grains, 5 g of straw, or 50 g of fruits and vegetables, and was accomplished with methanol of various volumes. After filtering through two Whatman GF / A glass fiber filters, the methanol extract was subsequently rotary-evaporated, leaving behind an aqueous phase, which was extracted with acetonitrile saturated hexane. This gave another aqueous acetonitrile phase, which was extracted with dichloromethane, dried over sodium sulfate, and placed on a Sep-Pak Plus silica cartridge (Waters, 690 mg). Crop interferences were eluted with dichloromethane : ethyl acetate (90 + 10 by volume; 10 mL), which was discarded. Carbendazim was eluted with 15 mL of dichloromethane : ethyl acetate (40 + 60 by volume), which was evaporated to dryness, and the residues were picked up in 10 mL of HPLC mobile phase (hexane : ethanol, 90 + 10). Wine (25 mL) was extracted with 3 × 50-mL portions of dichloromethane, which were combined, evaporated to dryness, picked up in 20-mL dichloromethane, and chromatographed on the Sep-Pak as above. HPLC separations were conducted "normal phase," on a Spherisorb NH_2 column, 25 cm × 3.2 mm, at a flow of 1 mL / min. Recoveries for grape, wine, wheat grain, wheat straw, peach, sunflower grain, and rape grain generally ranged above 85%, at fortifications ranging from 0.01 to 2 μg / g. Standard deviations for replicate analyses, where n ranged from 9 to 11, were exceptionally good, usually below 5%. Comparisons were made between cleanup on the SPE cartridge and small columns packed with Kieselgel 60 Extrapure, 70 / 230 mesh (Merck). Crop interferences for wheat grain were far more problematic for the Kieselgel 60 cleanup than for the SPE cartridge, with large peaks appearing all the way out to 15 min on the chromatogram. Such effective cleanup was found by the authors to be necessary, due to the fragility of the Spherisorb NH_2 column, which tended to plug and give tailing peaks when exposed to such coextractives. It greatly reduced the frequency with which the column needed reconditioning with a solvent regime.

As pointed out above, there is a great deal of variability in the performance of the same type of SPE cartridge when products from various manufacturers are compared. This was the case in a recent method that was developed for the analysis of 23 triazine herbicides in various raw agricultural products (32). A single-step cleanup was devised using a strong cation-exchange cartridge of sulfonic acid functionalized plastic bead, Supelclean LC-SCX (Supelco), 500 mg, in a 3-mL tube configuration.

Methanol (200 mL) was used to extract fresh fruit and vegetables (100 g), or methanol : water (250 mL, 10% water) for silage and wheat (25 g). After the methanol extract was separated from plant tissue by filtration, it was diluted with water and partitioned with methylene chloride, which was dried with sodium sulfate, and evaporated to dryness by rotary evaporation. The residue was dissolved in 10 mL of hexane, 1 mL was applied to the top of the SPE cartridge, and the cartridge was then washed with acetone (2 mL) and methylene chloride (2 mL), giving an eluate that contained potential interferences that were discarded. Elution of the cartridge was accomplished with 2.5 mL of 1N NH_4OH in methanol (1 + 3) into a separatory funnel containing a two-phase system, pH-6.5 phosphate buffer (10 mL) and methylene chloride (10 mL). After elution the triazines were partitioned into the methylene chloride, which was exchanged for 2 mL of acetone after evaporation to dryness. Analysis was by gas chromatography and was conducted with a DB-17 or DB-1 capillary column and a nitrogen–phosphorus (N-P) detector. Such an extremely simple one-step SPE cleanup provided nearly interference-free chromatograms for sweet corn at the 0.1-μg / g fortification level (Fig. 5.12), which led to very good recoveries, usually above 80% for corn, apples, celery, wheat, silage, and liquid milk.

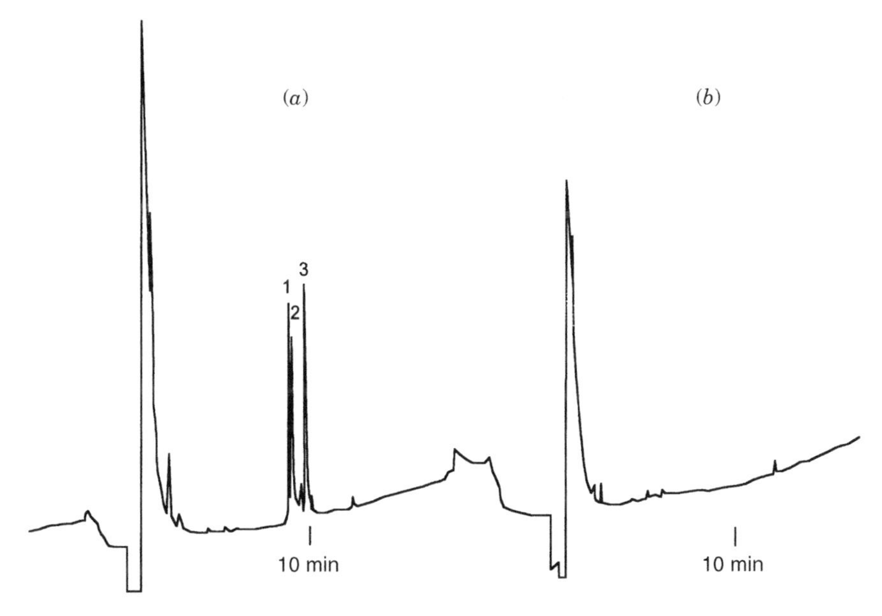

Figure 5.12. (*a*) Chromatogram of 10-mg injection of corn extract spiked with 0.10 μg / g each of desethylatrazine (1), desisopropylatrazine (2), and atrazine (3); (*b*) chromatogram of 10 mg of unspiked corn sample (Ref. 32).

Several other SCX cartridges from other manufacturers were tried without success using the prewash and eluant conditions reported here. However, the Supelclean cartridges were found to be very close to overloading, using the crop and solvent conditions reported, as evidenced by the fact that using acetone for the extracting solvent rather than the methanol caused the triazines to elute in the acetone and methylene chloride washes of the cartridge.

Solid-phase extractions can be coupled with some simple but ingenious chemical derivatizations of the pesticide to produce a pesticide residue analytical method that has extremely low method limits of detection, approaching 0.002 μg/g. This was done for one of the avermectin

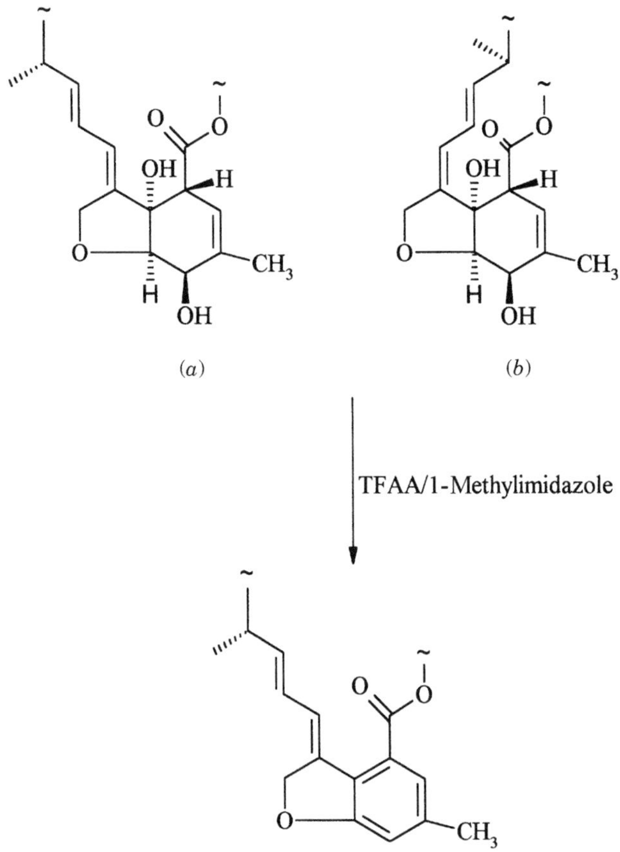

TFAA/1-Methylimidazole

Figure 5.13. Reaction of avermectin B_{1a} (*a*) and 8,9-*Z*-avermectin B_{1a} (*b*) with trifluoroacetic anhydride (TFAA) to yield a single derivative (Ref. 33).

insecticides, avermectin B_1 and 8,9-*Z*-avermectin B_1, the two toxicologically significant degradates of a mix of compounds called "Abamectin" (33). Treatment of both of these compounds with a dehydrating agent, such as trifluoroacetic anhydride (TFAA), results in a single highly fluorescent product (Fig. 5.13), which can be separated on a reverse-phase HPLC column (Chromegabond MC18; ES Industries) and measured at an emission of 470 nm after excitation at 365 nm. Fresh apples were analyzed for both Abamectin compounds by extracting a 5-g aliquot with 3 mL acetonitrile, 9 mL water, and 15 mL hexane in an ultrasound extractor (Polytron), giving a two-phase system, with hexane on top. The hexane fraction was dried with sodium sulfate, and transferred to the top of a tandem Mega Bond Elut SPE cartridge arrangement (two cartridges, 1.0 g each, 6-mL reservoir; Mega Bond Aminopropyl, Varian). Interferences were washed from the SPE with sequential rinses of hexane (4 mL), toluene (3 mL), and methylene chloride (15 mL), which were discarded. Avermectin B_1 and 8,9-*Z*-avermectin B_{1a} were eluted from the SPE with 10 mL of 1 : 1 acetone-methylene chloride. Removal of the solvent under a stream of dry nitrogen allowed for reconstitution with acetonitrile (0.5 mL), and derivatization with TFAA and methylimidazole for only 3 min at room temperature produced derivatives that gave chromatograms free of background interferences (Fig. 5.14).

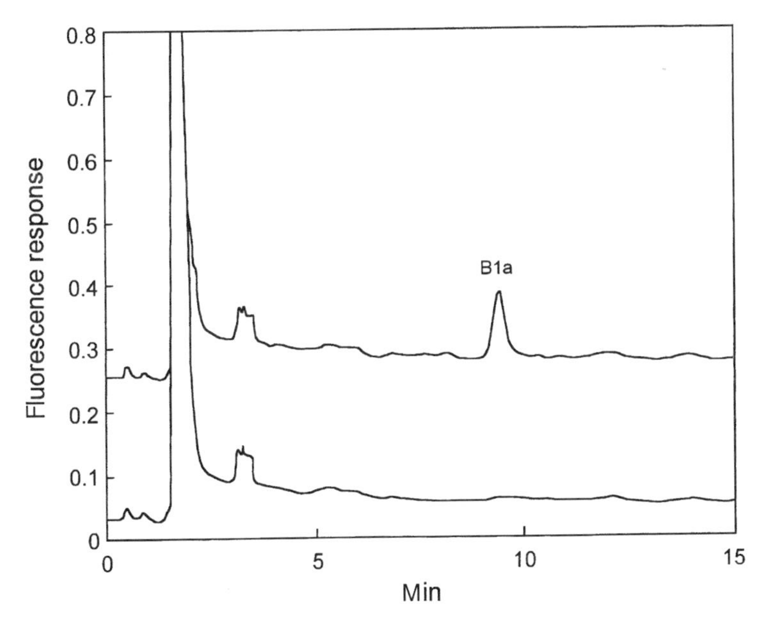

Figure 5.14. Typical HPLC chromatogram for apples fortified at 2 μg/Kg avermectin B_1 (top) and apple control (bottom) (Ref. 33).

Recoveries for both Golden Delicious and Red Delicious averaged above 80%, even for fortifications as low as 0.002 μg / g. Coefficients of variation (CV) generally were below 7% for all fortifications.

SOLID-PHASE EXTRACTIONS: DISKS

Introduction

An innovative new concept in solid-phase extractions that would further extend the capabilities of cartridge- and column-type SPE was introduced in 1990 by scientists of the 3M Company in the United States (34). They were able to incorporate an irregularly shaped HPLC-type reverse-phase bonded HPLC silica particles into a porous Teflon matrix in such a manner that the particles could be held tightly enough, yet spaced apart enough, to allow a high water permeability; they called such devices "Empore disks." Such particles, of a nominal 8-μm-diameter silica having average pore sizes of 60 Å, were spaced closely enough to allow for high linear velocities of water through the disks without exhibiting analyte "breakthrough" and consequent losses. Most of the disk's weight, >90%, is due to the weight of the particles, making for high sample capacities. Figure 5.15 shows how these particles are held by the porous Teflon fibers inside the disk matrix. At their inception, the disks were available with only octyl (C_8)- or octadecyl (C_{18})-bonded phases on the silica particles, but have since been offered with several types of silica and plastic bead particles. Such choices of particle types have greatly expanded the types of applications the disks can address for the extraction of organics and ionic species from aqueous samples. While the 3M Company selected Teflon as the containing matrix for their disks, spun glass was selected by the ANSYS company for their single Spec disk also containing irregularly shaped silica, but has an average diameter of 30 μm, and available only in the C_{18}-bonded phase. Spun glass was also selected by the Supelco Co. for their ENVI-8 and ENVI-18 disks. These disks are available only in the C_8 and C_{18} chemistries. There are other commercially available disks as of this writing, but they are of a highly specialized nature for such applications as protein purifications after various types of activation mechanisms, and will not be discused here.

Both the Teflon and the glass fiber disks offer some distinct advantages over the cartridge- or column-type SPE device, including the following: (1) much faster flow rates without breakthrough of analyte, giving shortened analysis times; (2) no opportunity for analyte loss through the disk due to "channeling" phenomena; (3) rapid elution of analyte from the disk, also shortening analysis time; (4) freedom from some types of interferences that

Figure 5.15. Photograph of a cross section of the Empore membrane, magnification 227× (courtesy 3M Co).

have shown up in the plastic encased cartridges and columns; and (5) ability to extract disks with eluting solvent in a variety of ways. Disks of the Empore type can be purchased with at least three diameter sizes—2.8, 4.7, and 9.0 cm—while the Spec disk is available only in the 4.7-cm size. The ENVI disks are available in only 4.7 and 9.0 cm diameters. Larger disks are more expensive than smaller disks, obviously, but offer advantages in terms of water throughput, which is particularly attractive when dirty water must be processed, which sometimes leads to disk plugging; larger disks offer more horizontal surface area and therefore are less prone to plugging than the smaller ones.

How the Technique Works

Whereas cartridges or columns can be activated, loaded, washed, and eluted in extremely simple vacuum manifolds that are generally inexpensive and require very little lab bench space, the use of extraction disks is somewhat more complicated and therefore ultimately expensive. For extraction of water samples, the disks must be placed in a filtration apparatus that compresses the disk between two optically flat surfaces to effect a seal and render it both gas- and liquid-leak-free. This device is typically held tight by a circular clamp, while the bottom extension of the device is placed in an airtight stopper that is inserted into a suction flask (Fig. 5.16). When a disk is installed and pressure is applied with the spring-loaded clamp, a seal is made to the outer portion of the disk surface, both top and bottom. Those disks which are very "hydrophobic" in nature, such as the C_8- and C_{18}-bound silica particles, must be "activated" by passing through a small volume of a water-miscible solvent; methanol is typically used for this. To ensure activation, the disk is allowed to rest in a layer of methanol for several minutes before the water sample is passed through under vacuum. Filtration of the water sample to remove particulates is typically accomplished before extraction with the SPE disk; this can be done using one or more filters of various pore sizes, the last of which is a 0.45-μm filter. Disks that have "hydrophilic" particles may not need to be activated, as we have seen above for certain types of cartridges, such as the Oasis LHB. After the water sample is passed through the disk, a strong vacuum is pulled for a length of time until the disk is completely dry. This is typically done for 20 min or more, depending on the quality of disk filtration device and the vacuum available. Alternately, the disk may be removed when only partially dry and placed in a desiccator overnight, to ensure complete dryness. Once removed from the filtration apparatus, the disk should be extracted by shaking with solvent in a capped tube, due to difficulties in realignment of the disk in the apparatus, particularly if the disk was deformed during handling, which is often the case.

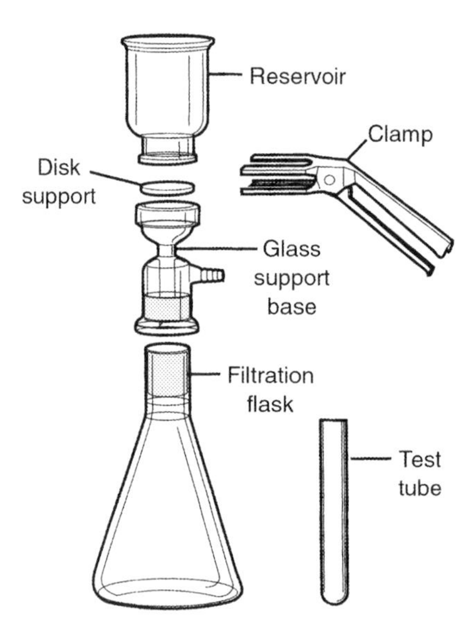

Figure 5.16. Diagram of a typical extraction (filtration) apparatus used for disk-type SPE devices (courtesy 3M Co).

If sample throughput is a concern, extraction manifolds can be purchased from a number of suppliers (Varian, Millipore, Baker, Aura Industries) that can process up to six disks at a time, with capabilities of shutting the vacuum off for each individual extraction "bell." Of particular note is the one offered by Aura Industries (Fig. 5.17), which has several interesting features. Setting it apart from many of the other extraction manifolds is the capability of operation in the "closed" or "suction" bell mode where water is pulled through Teflon tubing from the sample bottles directly into the closed disk holder and on out the vacuum line. It can also be operated in the "open" bell mode, the conventional manner, in which the water is poured into the bell and pulled through the disk by the aid of the vacuum manifold. And it can be readily adapted for use with SPE cartridges, by simple insertion of a syringe adapter kit into the glass receiving tube. We have found the manifold to be considerably faster than other manifolds in the extraction of water and even "dirty" aqueous solutions, due to its extreme tightness at the disk–bell interface, making full use of whatever vacuum is available. As it is made of Teflon and glass throughout, it is essentially contaminant-free.

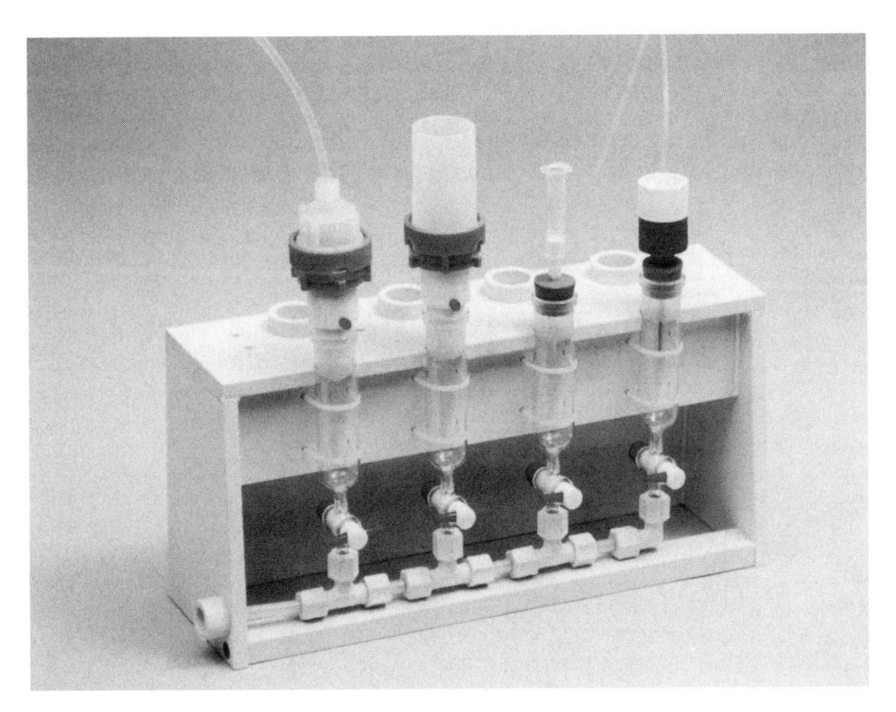

Figure 5.17. Photograph of Speedman four-station extraction manifold, AURA Industries, Staten Island, New York (Courtesy AURA Industries).

One of the more recent advances in disk technology is the introduction of the *Speedisk* and its associated extraction hardware. A distinguishing feature of this disk is that it is composed of a thin bed of "microparticles" held in place by glass fiber filters afixed to high-molecular-weight polyethylene collars that can be placed into appropriate extraction manifolds for water sample extractions (Fig. 5.18). These disks are available in about the same selection of adsorptive particle types as the Empore disks.

Advantages of the Technique

When compared to the cartridge or column technique, the distinguishing characteristic of the disks is their speed, particularly for use with a filtration device that provides a good seal at the disk. It is essentially impossible to pass water through the disks too fast, such that the analytes "break through" and are not captured, at least for the reverse-phase disks; this may be somewhat different with the ion-exchange disks. Their ease of storage and transport also distinguish them somewhat from the cartridges and columns, as they can be placed in ziplock bags, which can be sent by mail in flat envelopes.

Disadvantages of the Technique

Although not as noticeable with the 9-cm-diameter disks, the disks are more prone to plugging than the cartridges when "dirty" samples are processed.

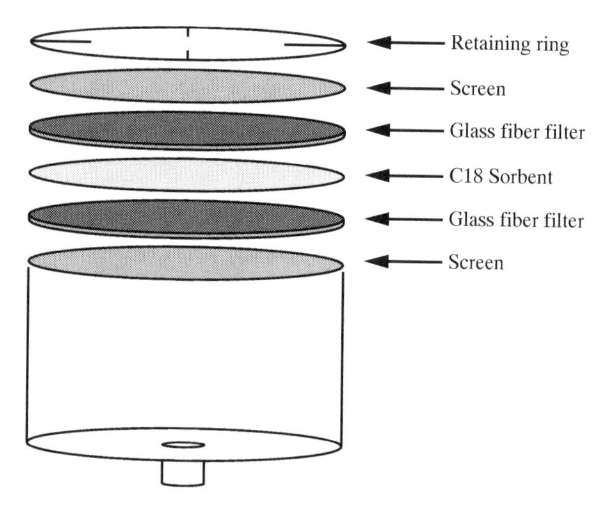

Figure 5.18. Diagram of the Speedisk extraction disk, Mallinckrodt-Baker (courtesy Mallinckrodt-Baker).

This can be mitigated to some extent by the use of proper prefilters, beginning with a multigrade type, and moving on to submicron pore sizes. If the disks are dried in the extraction manifold by pulling large quantities of air through them, as recommended by at least one of the manufacturers, then laboratory air quality becomes an issue, since the disk will trap whatever particulates are present in that air. Once again, use of a prefilter during the drying process can mitigate air-quality problems. Drying in a desiccator eliminates this problem altogether, but then the chemist is faced with the task of placing the disk back in a disk holder for elution with organic solvent if that type of elution is selected. Extraction in a capped tube, as mentioned above, eliminates this problem altogether.

Water Analysis

From their inception, the disk SPE devices were conceived with the intention of applying them to water analyses, beginning with groundwater and moving on to surface waters (35–37), and employing an array of chromatographic determinative steps, but usually coupled with GC or HPLC. It became apparent during this early work that, as with the cartridges and columns, the disks offered an attractiveness associated with their ability to stabilize pesticides after the water extraction, and in fact this was a major factor in their use in a remote water sampler that was designed for the sampling of surface and groundwaters over extended periods of time in hot and humid environments (36). Laboratory-grade water fortified at 0.1, 1.0, and 10 μg/L with herbicides, including alachlor, bromacil, ametryn, prometryn, and terbutryne, extracted with C_{18} Empore disks, and left in various temperature and humidity environments for up to 30 days gave mixed results for pesticide stability on the disks. Alachlor and bromacil herbicides survived up to the 30-day storage interval, even in the presence of a water saturated atmostphere held at 40°C. In contrast, the other herbicides were severely degraded or lost under these conditions. A wider selection of pesticides was studied for their stability on the same disk type, but held on nearly dry disks in a less humid atmosphere and at an array of lower temperatures (38). Of the 12 pesticides studied, alachlor again appeared to be the most stable, surviving with little losses out to 180 days at a temperature of 4°C. The fungicide captan was lost very quickly, however, even when stored at −20°C. Surprisingly, the organophosphate insecticide methyl parathion, which has been shown to be subjected to hydrolysis, survived fairly well up to 180 days at 4°C, although it degraded significantly in bottled water held at that temperature. It is obvious, then, that the Empore disks have the ability to stabilize at least some pesticides for extended periods of time under a variety of temperatures and humidities. This makes them attractive for use at remote

Table 5.6. EPA Methods for Pesticide Aanalysis

Method #	Analytes	Empore Disk
506	Phthalates and adipates	C_{18}, 47 mm
507	Nitrogen/phosphorous pesticides	C_{18}, 47 mm
508.1	Organochlorine/nitrogen pesticides	C_{18}, 47 mm
515.2	Chlorinated herbicides	SDB, 47 mm
525.2	Semivolatile organics	C_{18}, 47 mm
548.1	Endothall	Anion X, 47 mm
549.1	Diquat and paraquat	C_8, 47 mm
550.1	Polycyclic aromatic hydrocarbons	C_{18}, 47 mm
552.1	Haloacetic acids and dalapon	Anion X, 47 mm
553	Benzidines and nitrogen pesticides	C_{18} / SDB, 47 mm
554	Carbonyl compounds	C_{18}, 47 mm
555	Chlorinated acids	C_{18}, 47 mm
608	Pesticides	C_{18}, 90 mm
1613B	TCDD	C_{18}, 47 mm
8061	Phthalate esters	C_{18}, 47 mm

locations where sample storage problems exist before water analyses can be undertaken.

In the United States, the disks have been approved for an array of EPA methods for the analysis of pesticides and other classes of compounds. Table 5.6 summarizes these methods for which disks have been officially approved.

Recent Advances

Food Analysis

Limited work has been done to date on the analysis of pesticides in foods employing SPE disk technologies. Late in 1991, a piece of work was described in which acetone was used for the extraction of seven different pesticides from fruits and vegetables: chlorothalonil, vinclozolin, dichlofluanid, dicofol, captan, procymidone, and endosulfan (39). Extractions were accomplished on 50 g of fresh plant tissue by blending with 100 mL of acetone followed by a reextraction with 50 mL; the volumes were combined and diluted to 250 mL with distilled water. Exactly 50 mL of this extract was diluted to 500 mL with water, and that volume was passed through a 4.7-cm C_8 disk. Although fortification levels weren't given, recoveries were claimed to be within 89–101%. Quantitations were performed on a gas chromatograph equipped with an electron-capture detector.

The Italian studies described above were expanded by the southeastern FDA laboratory (40). They chose to study an assortment of organophosphate and chlorinated hydrocarbon insecticides, 18 in all, with fortifications to 5 high-moisture-content foods, tomato, potato, celery, cauliflower, and strawberry, at the 0.1–0.4 μg / g level. They used only one 100-mL portion of acetone to extract 50 g of crop sample by blending at high speed for 2 min. Acetone content of the extract was reduced by blowing down 25-mL aliquots to 10 mL and diluting back to 20 mL with water. A 4.7-cm C_{18} Empore disk was preconditioned with acetone, rather than the more typical methanol, and then pure water (3×25 mL) before the crop sample was applied. Further elimination of interferences were accomplished by simply washing the disk with 5 mL of 10% acetone in water. Pesticides were then eluted from the disk by employing 5-mL rinses of ethyl acetate–acetone ($50 + 50$), which were combined, dried with sodium sulfate, and analyzed by gas chromatography. One aspect of this study that is extremely significant, when the data are examined closely, is that capture of pesticides by the Empore disks in the presence of crop coextractives occurred best when a fairly large percentage of acetone was used in the extract, specifically 15–25% by volume, which seems contradictory to the theory of reverse-phase separations on HPLC columns. That theory states that 25% acetone in water should elute these pesticides during sample application to the disk, using the volumes that were used here, giving reduced recoveries due to breakthrough. The authors of this study claimed that the presence of large amounts of acetone probably enhanced the permeability of the disk to the pesticide, and subsequent absorption onto the disk. It can be seen from Table 5.7 that quite respectable recoveries from the crops using the acetone extraction and dilution with water were realized, with recoveries rarely falling below 80% and generally above 90%. Also of note is the very good recoveries for those pesticides that might be expected to be lost by hydrolysis with the added water, particularly diazinon, methyl parathion, malathion, chlorpyrifos, and ethion. Lowest recoveries were observed for the chlorinated hydrocarbon pesticides, particularly chlorothalonil.

There has been a recent significant expansion of the particle types available in the disks now, expanding the original C_8 and C_{18} offerings to include strong anion and cation exchangers, styrene–divinylbenzene reverse phase, styrene–divinlybenzene reverse phase sulfonated, chelating, and most recently carbon; the 3M Company has been the leader in this respect.

Future Directions

Of the disks now currently available, the reverse-phase mode types have seen the widest use for the extraction of pesticides from water and water / organic

Table 5.7 Recoveries of Pesticides Spiked at Low Levels[a] into Five High-Moisture Crops by C_{18} SPE Disk Extraction

Pesticide	Spike Level (ppm)	Recovery[b] (%)				
		Tomato	Potato	Celery	Cauliflower	Strawberry
Diazinon	0.0888	100.6 (1.5)[c]	106.1 (0.9)	104.2 (1.4)	96.7 (3.4)	97.4 (0.7)
Methyl parathion	0.1140	100.7 (2.2)	105.0 (2.6)	109.3 (4.1)	96.5 (3.4)	105.0 (0.7)
Malathion	0.1456	104.1 (1.9)	109.5 (2.0)	107.2 (3.9)	98.1 (0.1)	106.1 (1.1)
Chlorpyrifos	0.1176	99.3 (1.8)	105.8 (2.7)	100.1 (3.3)	94.4 (1.0)	100.2 (1.5)
Ethion	0.1700	99.4 (1.5)	105.4 (2.3)	100.1 (0.2)	93.7 (0.6)	104.3 (1.2)
DEF	0.2348	99.9 (0.2)	103.8 (3.0)	104.2 (2.5)	92.1 (0.9)	105.7 (0.2)
Quintozene	0.1172	80.8 (1.5)	87.1 (1.1)	92.1 (12.2)	84.6 (2.0)	78.3 (1.7)
DCPA	0.1316	94.1 (0.6)	96.9 (2.0)	96.6 (1.8)	99.5 (7.2)	94.0 (1.2)
Endosulfan I	0.1180	94.2 (0.2)	99.0 (0.1)	83.4 (2.7)	89.5 (0.2)	93.4 (1.8)
Dieldrin	0.1152	93.7 (0.1)	100.9 (1.6)	91.8 (0.8)	90.1 (0.7)	94.9 (3.9)
Endosulfan II	0.1308	92.3 (0.1)	102.4 (0.2)	86.7 (4.1)	92.5 (2.3)	97.6 (2.1)
Endosulfan sulfate	0.1824	83.4 (0.2)	104.2 (2.2)	84.3 (0.4)	92.7 (2.3)	98.3 (2.8)
Dicloran	0.1984	86.7 (0.2)	87.8 (4.4)	—[d]	83.3 (1.2)	76.7 (3.8)
Chlorothalonil	0.1424	87.8 (0.1)	84.4 (1.3)	119.9 (4.3)	17.4 (0.6)	72.2 (0.7)
Vinclozalin	0.1624	90.5 (0.2)	91.7 (0.6)	89.8 (2.6)	92.8 (0.3)	89.2 (1.8)
p,p-Dicofol	0.3856	87.7 (0.4)	88.6 (0.4)	94.4 (10.1)	92.0 (0.04)	94.6 (1.2)
Captan	0.3940	79.9 (0.4)	91.1 (1.1)	74.1 (0.5)	94.9 (3.4)	102.1 (1.6)
Procymidone	0.2476	80.3 (0.3)	94.3 (3.0)	89.1 (5.0)	88.4 (5.4)	87.5 (2.5)

[a] Spike levels correspond to 50–150% of the GC detector response of 0.3 μg/g chlorpyrifos.
[b] Average recovery of duplicate runs.
[c] Values in parentheses are standard deviations of duplicates.
[d] —, large amount of dicloran found in blank (recovery not quantitatable).

modifier solvent systems. While the work reported to date has used acetone for the extraction of pesticides from fresh foods, there is the opportunity for the use of water at elevated temperatures, since the dielectric constant, viscosity, and surface tension of that liquid all diminish with increasing temperatures, giving increased pesticide solubilities, faster kinetics for extraction, and greater plant tissue membrane permeability. The extraction scheme would then employ a hot-water (85°C) extraction, a cooling to reduce pesticide solubility, and an extraction with a reverse-phase disk. Elution from the disk would then be in the conventional manner.

As confidence grows among analytical chemists who use the disks, and as sales grow, there will be the impetus to develop disks with adsorbents of a wider selection of absorption modes. Although carbon particle disks are now available, little has been done to develop methods for their use, even with water samples. This particle type ought to allow the chemist more flexibility in his choice of extraction solvents, maybe even requiring less water addition to the extracting solvent to achieve adsorption to the disk. Particles with "affinity chromatographic" bonded phases could also be developed, although this would remove some of the "universality" characteristics of the disks.

There exist opportunities for on-line extractions, where hot water could be passed through the food to be extracted that is contained within a porous "teabag" device, either as recycle or as a single pass. After extraction is complete, the water would be cooled and passed through an extraction disk, the disk would be removed, dried, and eluted with the small volume of organic solvent.

A bright future undoubtedly exists for the disk extraction devices, if the manufacturers can expand particle-type availability, eliminate the activation step requirement, add a degree of structural rigidity, reduce their costs, and reduce their tendency to plug with dirty samples.

SUPERCRITICAL FLUID EXTRACTIONS

Introduction

How extremely valuable "supercritical fluids" (SFs) are to a rapidly advancing technological world became apparent in the 1970s when a series of patents were issued for a number of processes for the extraction of a variety of organics from foods (41). Starting the process was a patent issued to Dr. K. Zosel, a German, for the decaffeination of green coffee using CO_2 (42), a process that he discovered required soaking the beans in water before soaking them in the SF. This patent preceeded a number for the extraction of such chemicals as flavors from hops (43), cholesterol from butter, and unsaturated oils from fish (44). It has since grown into being employed by other disciplines, such as the remediation of contaminated soil, removal of nicotine from tobacco, and the cleaning of electronic parts, and the use of many fluids other than CO_2 has come about, including ammonia (NH_3), nitrous oxide (N_2O), pentane, and even water (41).

A rapidly growing realization of the hazards associated with the use of conventional organic solvents for extraction has spurred scientists into pursuing SF extractions (SFE) for the extraction of many analytes from foods, environmental samples, animal tissues, and pharmaceutical preparations (45). Costs and hazards associated with finding places to dispose of organic solvents, such as hexane, acetonitrile, and acetone, have driven many into looking into the more innocuous SF liquids, such as CO_2, water, and nitrous oxide.

Adding to the cost and safety features of SFE are several other features that have driven its exploration (46). It is fast, as a result of rapid mass transfer of analyte from the matrix being extracted into the SF, which means that frequently quantitative extractions can be achieved in a dynamic mode, requiring only minutes at times. Solvent strength, or solvent polarity, can be easily controlled by temperature and pressure of the SF, giving a degree of selectivity to the technique. Since most SFs are gases at ambient temperatures and pressures, they are easily removed by evaporation into the atmosphere after the extraction process, leaving the analyte behind in a small volume of organic solvent or on a solid-phase adsorbent, giving improvements in method limits of detection. Many of the SFs have relatively low critical temperatures (see text below for a discussion of this observation), meaning that ordinary ovens or thermal heaters can be employed to heat the liquid under pressure for SFE, and pumps that are already on hand for HPLC use can frequently be employed for delivering the SF to the sample being extracted. Other pieces of hardware making up the SFE extraction device are simply made, and are few, making the overall extraction operation easy to deal with in the laboratory. Conceptually, it is a very simple device, composed of a fluid delivery system, consisting of a pump or a pressurized cylinder, an extraction vessel that is generally enclosed in an oven, a restrictor that keeps the extraction vessel at a controlled high pressure, and a collection device, which is usually an organic solvent or an SPE device (Fig 5.19).

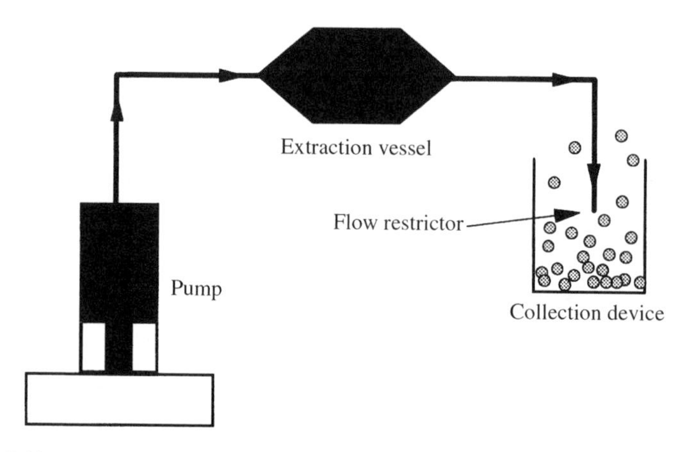

Figure 5.19. Diagramatic representation of a supercritical fluid extraction instrument (Ref. 46).

How the Technique Works

What are Supercritical Fluids?

How a substance physically responds to various temperature environments can be succintly described by what is known as a *phase diagram*, (Fig. 5.20). As both temperature (hozizontal axis; abscissa) and pressure (vertical axis; ordinate) are increased, there are *regions* on the diagram that represent the various phases that are present under those conditions, starting with a solid, a gas, transitioning to a liquid, and eventually to the *supercritical region*, which can be seen in the upper right part of the diagram. Also apparent is the *critical point*, (CP), which represents the intersection of the critical temperature line, and the critical pressure line. The *critical temperature* is the temperature beyond which a gas cannot be converted to a liquid by increasing the pressure, and the *critical pressure*, similarly, is the pressure beyond which a liquid cannot be converted to a gas by increasing the temperature. Consequently, the *supercritical fluid* region of the diagram

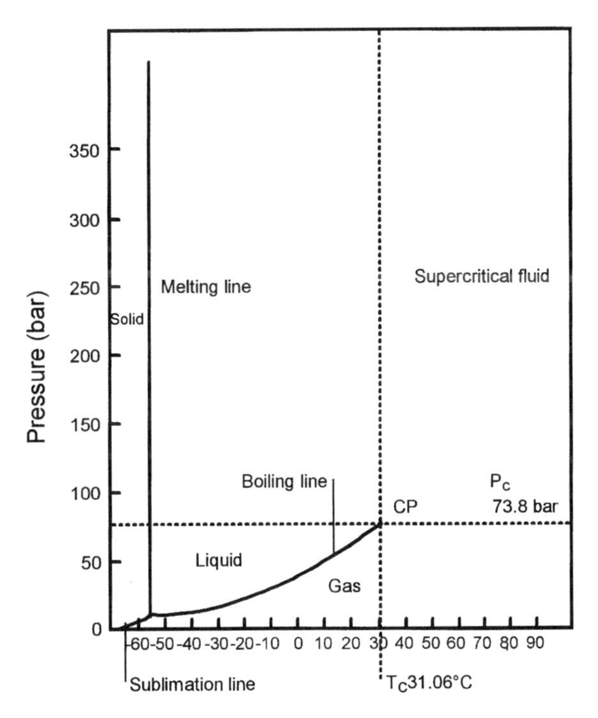

Figure 5.20. Typical phase diagram for a material exhibiting supercritical fluid characteristics.

represents a state of matter whose density, among other properties, is greatly dependent on its temperature and pressure, and lies intermediate between those of liquids and gases. Similarly, its surface tension is much lower than that of liquids. Thus, supercritical fluids can easily penetrate very small pores, which normally would be diffucult for liquids. And, since diffusion coefficients for solutes dissolved in supercritical fluids are quite high, mass transfer is more favorable, allowing for solutes to diffuse away from a solid–SF interface, and speeding up extraction kinetics.

One of the most important characteristics of any extracting solvent is the dielectric constant, ε, a measure of the *polarity* of the solvent. Water, a very polar solvent, for example, has a dielectric constant of 80 at 20°C, and hexane, a very nonpolar solvent, has a dielectric constant of 1.9 at 20°C. For SFs, both the dielectric constant and density vary strongly with pressure at constant temperature, which can be seen for CO_2 in Figure 5.21. While the polarity of SFs can be adjusted with temperature and pressure, it can also be adjusted with the addition of small percentages of low-molecular-weight modifiers, such as methanol and other miscible alcohols.

One way of measuring the polarity of a solvent molecule is to use the *solvent strength parameter*, π^*, of Kamlet and Taft (47), who observed changes in absorption spectra of a number of "probe" solutes that had been

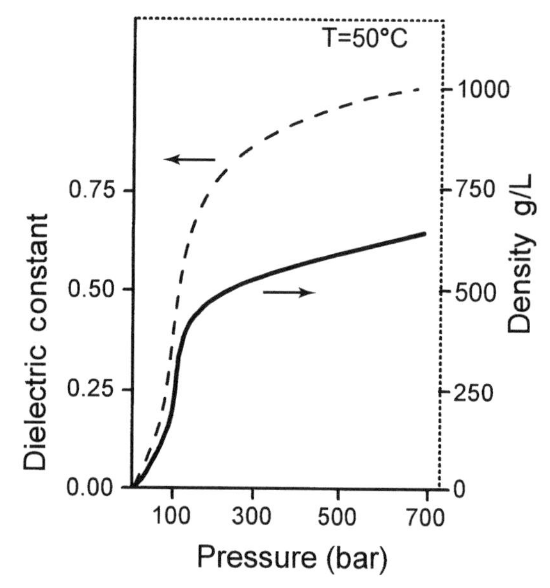

Figure 5.21. Effect of pressure on the dielectric constant and density of supercritical CO_2 at a constant temperature [courtesy Springer-Verlag; P. Hubert and O. G. Vitzthum, *Angew. Chem. Int. Ed. Engl.* **17**, 711 (1978)].

shown to be sensitive to the polarity and polarizability of the solvent environment around them. In this system of polarity classification, cyclohexane was arbitrarily assigned a π^* value of 0.0, and a very polar solvent, dimethyl sulfoxide, a π^* value of 1.0. Consequently, the larger the number, the more polar or polarizable a given solvent will be. Since solvent–solute interactions are essentially nonexistent in the vapor phase, gaseous solvents will have π^* values less than 0.0. Positive values, then, mean that the solvent is more polar than cyclohexane. How the addition of small percentages of 2-propanol affects the polarity of CO_2 is shown in Figure 5.22. With no propanol added the solvent strength reaches approximately 0.07 at approximately 500 bar; adding 10% 2-propanol boosts the polarity to approximately 0.25, a usable solvent polarity for the extraction of many types of compounds of intermediate polarity.

The following three generalizations can be made about increasing the solvent power of an SF:

- Solvent power increases with increasing temperature at constant pressure.
- Solvent power increases with increasing pressure at constant temperature.
- Solvent power increases with increasing percentage of organic modifier at constant temperature and pressure.

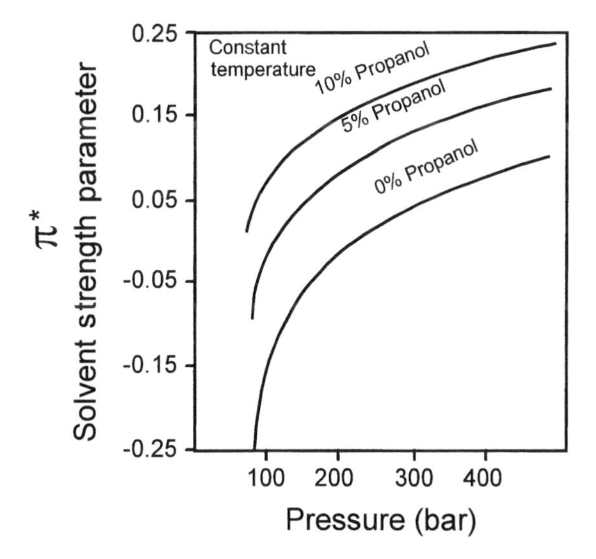

Figure 5.22. Effect of organic modifier, 2-propanol, on the solvent strength parameter of supercritical CO_2 (Ref. 47).

Table 5.8. Threshold Pressure of Substituted Tetracyclic Steroids

Steroid Functionality	Threshold Pressure (bar)
Alcohol OH (1)	80
Carbonyl (1)	80
Alcohol OH (1) + carbonyl (1)	90
Alcohol OH (1) + phenolic OH (1)	100
Alcohol OH (2)	120
Alcohol OH (3)	150
Alcohol OH (2) + carboxylic acid (1)	200
Alcoholic OH (4)	Not soluble
Alcoholic OH (3) + carboxylic acid (1)	Not soluble

Source: Ref. 48.

How can pure CO_2 be expected to extract polar compounds? Since bulky alkanes can be easily extracted with pure CO_2, it is convenient to think about the effects of various numbers and types of functional groups that might in fact inhibit the movement of a solute into that SF. Stahl used this concept to estimate the "threshold pressure" of many substituted tetracyclic steroids, determining what pressure was required to begin to see movement of these compounds into CO_2 at 40°C (48). Those data are summarized in Table 5.8. When the coumarin molecule contained only one hydroxy group (alcohol), it had a "threshold pressure" of only 80 bar; adding another hydroxy group increased it to 120 bar. When a different type of functional group, in this case a carboxylic acid group, was again added, the threshold pressure was increased to 200 bar. Those coumarin molecules that had four functional groups of any kind were not measurably soluble in SF CO_2. There are limits, then, to the types of molecules that can be solubilized into that SF, without considering kinetics, or speed of extraction.

The Extraction Process

There are several other considerations to keep in mind other than optimizing solute solubility in the SF during the extraction process. Solute solubility won't account for much if it is somehow bound to the matrix and cannot be released for transport by diffusion and convective processes into the bulk of the SF. Once it is in the SF, it is eventually swept from the extraction vessel into the collection device. It is well known that in conventional adsorptive modes of liquid chromatography, generally called *normal-phase chromatography*, small percentages of polar solvent, frequently less than 1%, compete with the analyte at adsorptive sites and elute it down the column, without actually changing the analyte's solubility in the mobile phase. Such is the case with chromatography on solid surfaces, as on alumina, silica, and other normal-phase columns. An excellent example of how this phenomenon can be one that determines ultimate extraction efficiencies was demonstrated for the extraction of polychlorinated bibenzo-*p*-dioxins (49). In that study, both supercritical CO_2 and N_2O, SFs of very similar solvent parameters, were applied to the extraction of fly ash. The N_2O gave good recoveries from untreated ash, whereas before CO_2 could perform at all the ash had to be pretreated with acid. Water, which is a very polar molecule and capable of both donating and accepting protons, has been shown to have pronounced effects on extraction efficiencies with numerous SFs (see text below), and for relatively dry sample types behaves as a competitor with the analyte on adsorptive surfaces, particularly soils and other matrices that have very polar surface groups. It has also been shown to swell pores of both organic and inorganic matrices, rendering them more open for enhanced diffusion and convective processes.

Needless to say, any diffusion process can be speeded up by shortening diffusion pathways, which for an SFE experiment is accomplished by cutting, chopping, blending, comminuting, or grinding the sample up to small particle sizes, so as to speed up mass transfer between the matrix and the SF. There are tradeoffs, however, in that backpressures can compete with the action of the restrictor, particulates can break loose and clog the restrictor, and clogging of extraction vessel frits can occur.

Sample Preparation

Since contemporary SFE instrument extraction vessels are of a limited size, typically 5–10 mL, sample preparation is a critical aspect of the technique, for the following reasons: (1) homogenous and representative samples must be taken to avoid biasing of results; (2) particle size is extremely important, and determines sample homogeneity, extraction efficiencies, extraction times, and instrument plugging; (3) control of water is a requirement, since its presence affects extraction efficiencies and instrument plugging due to icing at the restrictor; and (4) addition of a modifier of proper type and amount may improve extraction efficiencies for certain types of samples.

In their regulatory program for assessing the pesticide content of foods, the FDA specifies that 20 pounds of sample be comminuted with a vertical cutter, followed by subsampling 100 g before analysis (50). Combining comminution with the use of frozen samples that are kept frozen thoughout the sample preparation procedure measurably improves the accuracy that can be realized for replicate sample analysis (51). A recent study on pesticide homogeneity in white potato has clearly shown the advantages of mixing frozen samples for the analysis of several pesticides after extraction by SFE (52). This study and others (53–55) describe how keeping the sample cold during mixing and subsampling can improve analytical accuracy, and give techniques for employing dry ice or liquid nitrogen in doing so. These studies have shown that for produce it is desirable to analyze samples larger than 10 g, but that for more homogenous samples 10 g is statistically representative. An approach that has been used by several authors to ensure sample homogeneity is to make a conventional extract of a large sample, 25–100 g, and then take an aliquot that is deposited on a surface, such as sodium sulfate (56), Hydromatrix (see text below, also Ref. 57), or conventional SPE absorbants (58,59). These approaches appear to be unreasonable, when the purposes of SFE are considered, since solvents are not eliminated and shortening analysis time has not been accomplished.

Control of Sample Water Content

As stated above, water has a pronounced effect on the extraction efficiency and SFE extraction instrument performance. A certain amount of water in any solid sample is advantageous, and actually speeds and improves extractions (60); however, when it is present in dominant quantities, it can slow and worsen extractions (61). A recent study by Lehotay and Lee demonstrates how relative amounts of water and water adsorbent (Hydromatrix), without crop, can affect the extraction of pesticides of various water solubility; supercritical carbon dioxide was used for this study (Fig. 5.23) (62). Pesticide solubility ranged from 0.002 to 8000 mg/L; dichlorvos (DDVP) was the most soluble and fenvalerate was the least soluble. Excess (4 : 1) water seems to improve the extraction efficiency for dichlorvos, while it reduces the extraction efficiency for fenvalerate, which is extracted best without any water being present at all. For the intermediate water solubility pesticides (carbaryl, phosalone, endosulfan) some water is necessary for best extractions, although more than a 1 : 1 ratio reduces efficiency. Water, then, is necesary for the extraction of all but the most water insoluble pesticides.

Controlling the amount of water available to the extracting SF then becomes an important consideration for extractions. There are basically two approaches for gaining such control: (1) adding an anhydrous porous adsorbent and (2) partially drying the sample by exposure to heated air in an oven. For practical considerations, the first approach is now more widely accepted, mainly because the second approach has the potential for pesticide loss via volatilization or thermal degradation. Consequently, most investigators have focused on selection of an adsorbent, which they hope will be highly adsorptive, inexpensive, inert, chemically clean, and of consistent composition. Several of these absorbents are now regularly used by SFE

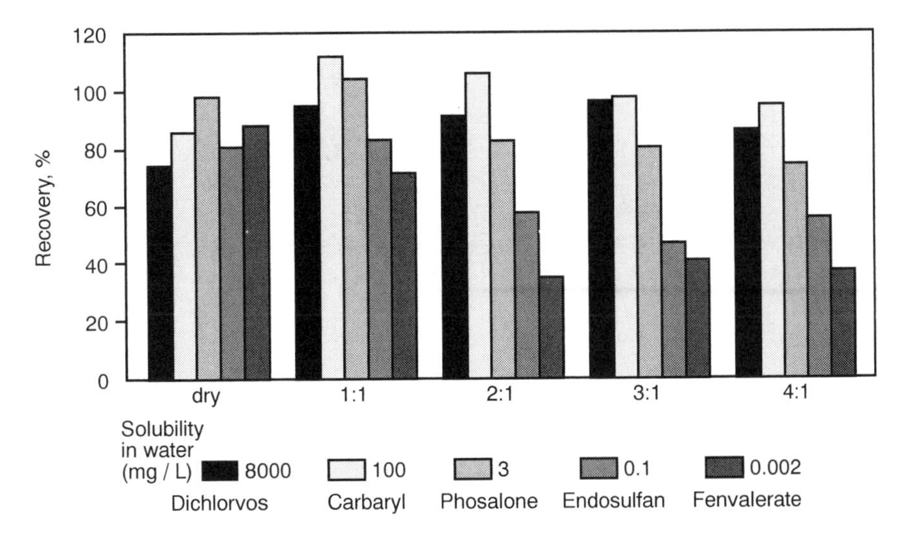

Figure 5.23. Effect of water on the extraction of pesticides with various water solubilities from cellulose powder, CF-1, using SF CO_2 (Ref. 62). Water : drying agent ratio.

users (see Table 5.9). Of those listed, the Hydromatrix product has been the most extensively studied for the extraction of pesticides, and, like many of the others, it retains the phosphoramides, such as methamidophos, acephate, and omethoate. Some of these are impractical, because they contain fines that can plug the restrictor on the SFE instrument, including magnesium sulfate, cellulose, and finely ground molecular sieves. Others are impractical because they do not absorb water extensively and have measurable heats of hydration, which can cause decomposition of analyte. Since magnesium sulfate does not adsorb the phosphoramide pesticides, and Hydromatrix doesn't adsorb the more polar pesticides, they were selected as drying agents in a USDA pesticide monitoring program that required two analyses be performed for each sample, one with extractions using Hydromatrix and the other with extractions using magnesium sulfate (61).

Use of Modifiers

At this writing there is considerable controversy about the effectiveness of modifiers in improving on the extraction of pesticides from foods, at least with CO_2 as the SF. Numerous examples exist in the literature for environmental samples, particularly soils and sediments, where it is apparent that SF CO_2 that has been modified with up to 5% organic solvent, usually methanol, is superior to unmodified SF CO_2 (63). While some authors have

Table 5.9. Comparison of Selected Drying Agents for Use in SFE

	Saturation Ratio (Water : Drying Agent)	Cost	Heat of Hydration	Density (g/mL)[c]	Consistency	Notes
Celite 545	3 : 1[a]	1.2	None	0.36	Compact	Retains phosphoramides
Hydromatrix	3 : 1[a]	3.8	None	0.29	Pelletized	Retains phosphoramides
Cellulose, CF-1	4 : 1[a]	5.0	None	0.21	Fluffy	Retains phosphoramides
Alumina	1 : 1[a]	2.4	None	0.85	Compact	Retains polar pesticides
Florisil	1.5 : 1[a]	9.5	None	0.50	Compact	Retains polar pesticides
Molecular sieves[d]	0.5 : 1[a]	4.0	Moderate	0.83	Variable	Unknown effects on pesticides
Magnesium sulfate	1.05 : 1[b]	5.0	High	0.62	Powdery	Hinders nonpolar pesticides
Sodium sulfate	1.27 : 1[b]	2.2	Low	1.4	Grainy	Hinders nonpolar pesticides

Source: Ref. 6.

[a] Determined by addition of water at room temperature.
[b] Determined by calculation using molecular weights.
[c] Amount of dry material packed in an extraction vessel divided by volume.
[d] Molecular sieves 5A ground with mortar and pestle.

claimed little benefits for using modifier additions directly to samples for the extraction of foods (64,65), at least one recent study from Canada showed pronounced effects for the addition of methanol directly to the SF before entering the extraction vessel (66). For wheat that was spiked with pirimiphos-methyl and extracted with modified SF CO_2 at 350 atm and 125°C, extraction efficiencies increased steadily from 0 to 30% methanol, reaching 100% for a 90-min extraction in the dynamic mode (SF flowing throughout the sample for nearly the entire period).

Static versus Dynamic Extractions

There are essentially three modes for using SFs to extract organics from foods: (1) dynamic, (2) static, and (3) dynamic after static. In the dynamic mode SF is continuously flowing through the extraction vessel after which it goes through the restrictor to the trapping device. Typically two to four vessel volumes of SF are required to realize recoveries above 90%. When large volumes of SF are passed through the vessel over extended periods recoveries approaching 100% are theoretically achievable, but practical considerations make this unattractive, including increased extraction of potentially interfering coextractives, the potential of collecting interferences present in the SF, and increasing difficulty in removing the analyte from the trap with time if an SPE device is used. Yet, this mode is used most frequently among SFE chemists, particularly those seeking maximum recoveries. The static mode requires the employment of fluid shutoff valves at the entrance and exit of the extraction vessel, so that the exit valve can be shut after the vessel has been purged of air, the vessel is pumped to the appropriate pressure by activating the SF pump, and the valve at the entrance is closed. After the appropriate interval, the exit valve is opened and one or two vessel volumes of SF are pumped through to the collection device. Increasing the flushing volume to two to four vessel volumes brings into play the combined static and dynamic modes, which is probably the most economical for high recoveries.

Trapping

After the SF exits the restrictor, a large expansion of the material occurs, absorbing energy and dropping the temperature of what is now a gas with various amounts of particulates in the form of droplets or crystals (aerosols) being propelled by the rapid gas expansion. For good recoveries to be realized it is necessary to interrupt the movement of these droplets and crystals while allowing the SF gas to escape. This can be accomplished in one of three ways: (1) focusing the aerosols onto an inert surface, such as the interior walls of a small vial; (2) passing the gas–aerosol mixture through a bed of highly porous SPE material; and (3) bubbling both through a solvent that dissolves the aerosols while allowing the gas to pass on through. There are merits and disadvantages to each. Trapping in a small vial, even with external cooling with ice, does not give acceptable recoveries for most analytes, including pesticides, and is not recommended. Use of liquids, with the restrictor immersed directly into the solvent, was one of the first collection techniques used. It suffers from having to replinish the solvent periodically, if dynamic extraction is conducted over extended periods (>15 min), and if there is a large Joule–Thompson effect, the restrictor may still plug, since the solution cannot be effectively externally heated. This technique has recently lost favor with SFE chemists. With the increasing availability of highly purified SPE particles having well-characterized pores of great surface area, the passing of SF gases through tubes packed with these has shown among the highest recoveries yet seen for SFE.

One recent study (67) compared the performance of four SPE cartridge-type collection devices packed with ODS, Diol, Porapak-Q, and Tenax for the collection of 50 pesticides that had been extracted from three different foods, and determined that ambient trapping was satisfactory, with the ODS giving overall better trapping efficiences. Acetone was determined to be appropriate for eluting the pesticides from the cartridges.

Pumps, Extraction Vessels, and Restrictors

While SFE instrument pumps are usually of one or two designs (the dual-reciprocating piston or the syringe), most are available with the former arrangement (41). Because most of the SFE fluids are highly compressible, SFE pumps must be of a more sophisticated design than those used in HPLC, so that reproducible flows can be maintained at any given pressure. Because SFE fluids are more difficult to feed consistently into pump reciprocating pump heads, frequently the heads will need to be cooled to temperatures around − 15°C, which makes for more expensive pumps. However, since most SFE is performed at constant pressure (the most critical parameter for controlling analyte solubilities), many pumps are designed with less attention to the elimination of pulsations due to the reciprocating-piston design. This design, in addition to having no downtime for reservoir refilling, does not have to be as rigorously cleaned as does the syringe pump when fluid-type changeover is needed. To assist filling of reciprocating pump heads with CO_2, there are cylinders available that supply that fluid under a head pressure of helium. This is not recommended, however, since helium solubility in CO_2 is such that the density of the SF is reduced at a given pressure, making that resulting fluid less effective in analyte extractions from matrices (68). Some instruments have the capability of adding the modifier to the SF before it enters the extraction vessel, although this is usually an option at extra cost, while others require the use of cylinders of the SF that have the modifier already included.

Commercially available SFE instruments can typically be operated up to pressures of 10,000 psi, with flows up to 500 mL / min, although usually flows are limited to less than 100 mL / min. This pressure is more than adequate for analytical extractions that are usually conducted at 5 mL / min or less. Vessels are made of 316 stainless steel, with ratings up to the maximum pump pressures and beyond, frequently as high as 120,000 psi, giving plenty of margin for overpressure, although all instrument manufacturers either enclose their vessels with safety shields or provide warnings about the necessity of the use of such. At least one manufacturer (Isco) makes available an instrument with a patented extraction vessel design, one that allows the vessel to be placed inside a larger and more robustly designed chamber, which is then pressurized to the same pressure as the vessel. This allows for extraction vessel construction using less expensive polymeric material that is designed so as not to absorb CO_2, so that there is no SF carryover. Extraction vessel shape—length versus diameter—has been shown to be inconsequential regarding extraction efficiencies, as long as volumes are held constant, along with pressures and flow rates. Vessels are contained in ovens that are heated in a selectable manner up to about 200°C.

Most SFE instrument manufacturers offer a choice between fixed and variable restrictor types. Since the fixed restrictor is usually fashioned from a single piece of small internal diameter capillary tubing (25–100 μm) it is the least expensive, and can be altered in length to provide needed pressure drops. However, it is prone to plugging if attention is not given to techniques for prevention. All experienced SFE users appreciate the need for heating of the fixed restrictor throughout its length, in order to eliminate partial plugging and irreproducible recoveries, and most manufacturers offer this capability. Various approaches have been used by manufacturers to provide manual and automatic variable restrictors, and it remains a critical part of instrument design that the user needs to carefully examine before decisions are made about purchase; the ideal situation is the demonstration by the manufacturer of their instrument's performance with real-world samples of the user's selection.

Table 5.10 summarizes the performance characteristics and available features of a number of the currently available SFE instruments for analytical use available at this writing in the United States. As can be seen, they range widely in both features and price. Essentially all are capable of performing extractions in both static and dynamic modes; the only instrument incapable of static-mode operation compensates for that by having the capability of

Table 5.10. Specifications for Seclected SFE Apparatus

Product	Speed 680 BAR	SFE 723	7680T SFE	SFX 2–10	SFX220	SFX 3560	DGMS	SFE-400	AutoPrep 44	PrepMaster
Manufacturer	Applied Separations 930 Hamilton St. Allentown, PA 18101 610-770-0900	Dionex 1228 Titan Way Sunnyvale, CA 94088 408-737-0700	Hewlett Packard 2850 Centerville Rd. Wilmington, DE 19808 302-633-8000	Isco 4700 Superior Lincholn, NE 68505 402-464-0231	Isco 4700 Superior Lincoln, NE 68505 402-464-0231	Isco 4700 Superior Lincoln, NE 68504 402-464-0231	Marc Sims SFE 1012 Grayson St. Berkeley, CA 94710 510-843-1306	Supelco Supelco Park Bellefonte, PA 16823 814-359-5459	Suprex 125 William Pitt Way Pittsburgh PA 15238 412-826-5200	Suprex 125 William Pitt Way Pittsburg, PA 15238 412-826-5200
Dimensions[b] ($w \times d \times h$, in.)	16 × 20 × 36	14 × 22 × 17	18 × 21 × 33	13 × 10 × 10	13 × 10 × 10	22 × 18 × 30	36 × 20 × 28	19 × 19 × 23	38 × 26 × 29	17 × 22 × 22
Automated	□	■	■	□	■	■	□	□	■	■
Autosampler capacity	NA	NA	8	NA	NA	25	NA	NA	44	NA
Can be coupled to on-line analyzer/detector	□	□	■	■	■	■	■	□	□	■
Maximum system pressure (psi)	10,000	10,000	5560	7500 or 10,000[c]	7500 or 10,000[c]	7500 or 10,000[c]	6000	6000	7400 or 10,000[c]	7450 or 10,000[c]
Maximum pump flow rate (mL / min)	500	18	4.0	90 or 40[c]	90 or 40[c]	90 or 40[c]	15	1700 d	7	7 or 14[b]
Extractor vessel capacity (mL)	2.5–300	0.5–24	7.0	0.5–10	0.5–10	0.5–10	5–300	1–10	0.5–10	0.150–50
Maximum number of simultaneous extractions	2	8	1	2	2	1	2	4	1	2
Maximum oven temperature (°C)	200	150	150	150	150	150	100	200	150	150
Restrictor type	Variable	Fixed	Variable	Fixe, variable	Fixed, variable	Fixed, variable	Variable	Fixed	Variable	Variable
Restrictor flow rate (mL / min)	100–5000[d]	250–1200[d]	0.5–4.0	0.5–10	0.5–10	0.5–10	0–15	0–2	1.0–7.0	1.0–7.0
Static and dynamic flow	■	□	■	■	■	■	■	■	■	■
Flow orientation	Vertical	Horizontal	Vertical	Vertical	Vertical	Vertical	Vertical	Vertical	Vertical	Vertical
Collection mode(s)	Sorbent, solvent	Solvent	Cryogenic, solvent	Sorbent, solvent	Sorbent, solvent	Solvent	Cryogenic, solvent	Solvent	Sorbent, solvent	Sorbent, solvent

Source: Ref. 69.

[a] Key: ■ = yes; □ = no; NA = not applicable.

[b] Dimensions may reflect stacking or side-by-side positioning of components.

[c] Pressures and flow rates depend on the model of pump purchased with the extractor.

[d] Expressed as volume of expanded gas.

extracting up to eight samples simultaneously (Dionex SFE 723), and it uses the cooled solvent collection mode rather than solid sorbents. Three different approaches are taken by those manufacturers whose instruments are limited to the number of simultaneous samples that can be extracted so that increased sample throughput can be realized. They have either chosen smaller-sized vessels that can be manipulated by a robot and are connected to the pump and collector by compression fittings (Hewlett-Packard 7680T; Suprex Autoprep 44), used inexpensive thin-walled vessels that are placed inside a pressure chamber by robot (Isco SFX 3560), or employed fingertight end fittings to minimize vessel changeover time (Suprex Prepmaster). Choices would be determined by how much automation can be afforded and how much analyst time might be available, which would vary between laboratories (69).

Recent Advances

Interest by federal agencies within the United States to provide improved multiresidue methods while striving to achieve the EPA requirements for the reduction in the use of organic solvents in the analytical laboratory has driven their scientists to the forefront of SFE method development for the extraction of pesticides from foods. Notable among these have been the groups at the USDA Agricultural Research Service Northern Regional Research Center at Peoria, Illinois (animal tissues, oils, fats, grains), their Beltsville Agricultural Research Center at Beltsville, Maryland (fruits and vegetables), and the USFDA Total Diet Research Center at Lenexa, Kansas (processed foods, meats, cheese). While the groups in Peoria and Lenexa have emphasized the need to employ relatively large sample sizes that necessitate larger shop-made apparatus of large extraction vessel capability (98 mL) (70,71), the group at Beltsville has worked with a variety of commercially available instruments (52,61,64,72–74) utilizing smaller sample sizes. Together they have shown that SFE can provide time-saving, cost-reducing, environmentally friendly approaches for the analysis of pesticides in foods, and also have prompted the instrument manufacturers to extend and refine their product line. The most comprehensive review to date for the use of SFE on pesticides in foods has been written by Lehotay (61), and the reader is referred to that article for a complete analysis of the SFE literature. For the purposes of this book, we will highlight the most significant advances to date, without putting them into a historical perspective that can be done by following the literature paper trail.

While there are many exceptions as to what classes of pesticides can be extracted by SFE, there are far more success stories. Those limitations seem to be confined to those pesticides that, because of their chemical structure, have either an inordinantly high water solubility ($> 10^5$ mg / L) or multiple sites, usually nitrogen atoms, that can participate in proton donor–proton acceptor interactions with plant cell constituents, possibly lignins and cellulose, which are well-established proton donors. And there seems to be some sort of relationship between those pesticides that don't respond well to analysis by gas chromatography and those that can't be extracted by SFE (46).

A recent encouraging study from Europe shows how some surprises can occur when SFE is undertaken (75). Three pesticides with widely different water solubilities were extracted from green pepper, tomato, and cucumber that had been preanalyzed by conventional and official multiresidue methods. These included methamidophos (water solubility, 2 kg / L), procimidone (4.5 mg / L), and endosulfan (0.32 mg / L), which were all extracted at 4500 psi and 50°C, with 15 mL of SF CO_2 used for each sample. Analyses were performed by capillary gas chromatography with a flame photometric detector (FPD), which allowed direct analysis without column or cartridge cleanup immediately after extractions were performed. Fortifications of the three pesticides were made to peppers immediately before extraction; in addition, samples having "field-incurred" residues were made available from the Spanish monitoring program, which had been preanalyzed using conventional multiresidue methodology by an association of growers and exporters of fruits and vegetables in Almería, Spain (COEXPHAL). They also analyzed some pepper samples that had been preanalyzed by the Swedish National Food Administration, and which had been found to be positive for methamidophos, procimidone, and endosulfan. Their studies showed the following: (1) SF CO_2 gave higher recoveries for methamidophos than had been reported using conventional extraction methods; (2) very low recoveries for the two pesticides (procimidone and endosulfan) having lower water solubilities, were observed, (3) the addition of methanol, 200 μL/sample, increased recoveries of methamidophos from pepper, tomato, and cucumber; and (4) the presence of moist magnesium sulfate on the crop tissue greatly enhanced recoveries of methamidophos from spiked glass wool.

While such encouraging results were obtained for the water-soluble methamidophos insecticide, a much less water-soluble insecticide, imidacloprid—a nitroimidazolidine (water solubility, 0.5 g/L)—was found to be untouched by SF CO_2 (76). Comparisons were again made between conventional methods on samples that had been previously analyzed by the Spanish cooperative (COEXPHAL) and those that were reanalyzed following extraction with SF CO_2. In addition to imidacloprid, pesticides studied were methiocarb, chlorothalonil, chlorpyrifos, endosulfan (I,II), and endosulfan sulfate. As reported previously (75), methamidophos recoveries were higher using SF CO_2 than those obtained using the conventional extractions. This did not hold true for endosulfan (summed residues), methiocarb, or chlorpyrifos, which consistently gave lower recoveries with the SF CO_2 than with the conventional methods. Water solubility, then, does not seem to be an issue in these studies. How various water-absorbing materials affected recoveries of these pesticides was also studied, with definite enhancing effects seen for magnesium sulfate and the Isco Wet Sample Support (Isco Wss), when compared to Florisil. Imidacloprid was not recovered from any of the drying agents in neat form nor from the crops, upholding the hydrogen-bonding possibilities mentioned previously.

How water or moisture control affects pesticide recoveries is clearly seen when the results are examined of studies using SF CO_2 for the extraction of the degradation product of benomyl fungicide, MBC, and thiabendazole fungicide (TBZ) from potato, apple, and banana (65). This study clearly showed that the addition of modifier (methanol), if done on a routine basis at concentrations greater than 5%, can actually reduce recoveries over those observed without modifier. This is somewhat surprising because of the moderate water solubility of TBZ (50 mg / L), which might be expected to benefit from the use of a modifier, more so than the less water-soluble MBC (8 mg / L) which may have benefited from 3% methanol, although the results were ambiguous because of relatively large standard deviations of replicates.

When any prospective extraction technique becomes a topic of investigation, one of the most important questions to be asked is whether it will be suitable when coupled with already developed and accepted determinative steps. One of the more recently employed methods is mass spectrometry, whether it be the conventional MID mode of electron impact, or the more recently emerging ion trap detection (ITD), or multiple ionizations (MS / MS), as discussed in Chapter 2 or 3 of this book. Four recent pieces of work emerging from the USDA's Beltsville laboratories clearly point to the utility of coupling SFE off line with mass spectormetry as the determinative step (52,57,72,73).

How sample preparation for SFE—including sample chopping, freezing, controling moisture, and hold time—affects recoveries of 40 pesticides from potato representing many classes, including CHs, OPs, triazines, and carbamates, was shown in a series of experiments that additionally showed the technique's compatibility with mass spectrometry (52). Two different SFE instruments were used in this study: one that was capable of holding up to eight extraction vessels (thimbles) that were rotated into place within the instrument and extracted sequentially (Hewlett-Packard 7689T) and one that held two extraction vessels that were manually replaced and reloaded, also in a sequential manner (Suprex Prepmaster). Quantitations were performed by GC / ITMS (ion-trap mass spectrometer; Finnigan MAT model ITS40) with multiple ion detection or by GC with electron-capture detection or nitrogen–

phosphorous detection. Table 5.11 summarizes the various experimental procedures that were incorporated into this study. Experiments 1 and 2 showed that pesticide losses occur when fortifications are made to fresh vegetable tissue at room temperature, which are then allowed to be exposed to open air. Losses were much less in experiments 3 and 4, even though fresh potato at room temperature was used without dry ice but with Hydromatrix (experiment 3), and with dry ice and with Hydromatrix (experiment 4). After 5-h exposure to air, losses of dichlorvos amounted to 92%, compared to 72% for samples held in closed extraction vessels for that time. Other organophosphate insecticides that showed large losses included diazinon, phorate, malathion, and disulfoton. Best recoveries for the 40 pesticides studied were almost always seen with the sample preparation scheme used in experiment 4, with the exception of the fungicide diphenylamine, which gave best recoveries with those conditions used in experiment 3 (17%). Consequently, minimum pesticide losses (maximum recoveries) were observed using a 1 : 1 Hydromatrix/sample ratio and blending the sample with dry ice. These recoveries averaged above 90% for most of the pesticides studied.

This extraction study was utilized in a subsequent piece of work intended to demonstrate the utility of SFE and GC/ITMS for the monitoring of 46 pesticides chosen for analysis in the USDA's Pesticide Data Program (74). Making up this group of 46 pesticides were 11 chlorinated hydrocarbons, 21 organophosphates, 3 pyrethroids, 3 carbamates, and 8 others in various other classes, including triazines, phthalimides, and substituted anilines. Samples of potato and broccoli originating from the Pesticide Data Program that had been found to contain measurable residues were analyzed by the proposed method, and commercially obained potatoes and broccoli were analyzed after appropriate fortifications were made. Extractions were done in the two commercial SFE instruments used previously (73), without organic modifier, and at SF CO_2 densities of 0.3 g/mL (100 atm), 0.5 g/mL (130 atm), and 0.85 g/mL (320 atm). A higher density (0.95 g/mL) was also studied, but difficult-to-manage crop background interferences were seen with it, and it was discarded. For all pesticides studied in the laboratory fortification experiments, which were made at 0.5 μg/g, the 0.5-g/mL density gave maximum recoveries except for dimethoate, carbaryl, mevinphos, atrazine, dicloran, captan, and iprodione. For these several, a density of 0.85 g/mL was required. Several pesticides, methamidophos, omethoate, and captan, recoveries were very low, ranging within 0–66%. For the 43 remaining, recoveries generally were above 85%, with a few exceptions. In addition to these laboratory fortification studies, potato check samples were analyzed that had been previously analyzed by six state laboratories using conventional and official methods; each laboratory had analyzed the same check sample for iprodione and ethion a minimum of 7 times. To simulate

Table 5.11. Procedures for Sample Preparation Used in Different Experiments

Parameter	Experiment 1	Experiment 2	Experiment 3	Experiment 4
Initial sample size	2.3 kg	2.3 kg	50 g	50 g
Spiking level(s)	0.413 μg / g, HCB; 0.434 μg/g, lindane	0.413 μg / g, HCB; 0.434 μg / g, lindane	0.4 μg / g, 37 pesticides	0.2 or 0.4 μg / g, 40 Pesticides
Sample processing	Chopped, frozen (100 g subsample); rechopped frozen (20 × 50 g subsamples)	50 g subsample from Experiment 1 blended with 100 g HMX	Blended with 100 g HMX	Blend with 50 g HMX + dry ice
Packing of extraction vessel(vessel volume)				
768T	2.1 g potato + 1.4 g HMX (7 mL)	1.3 g potato + 2.7 g HMX + 1 mL H_2O (7 mL)	1.3 g potato + 2.7 g HMX + 1 mL H_2O (7 mL)	2 g potato + 2 g HMX (7 mL)
Prepmaster	3 g potato + 2 g HMX (10 mL)	—	1 g potato + 2 g HMX + 1 mL H_2O (5 mL)	1.5 g potato + 1.5 g HMX (5 mL)
Number of extractions				
7680T	48	9	14	15
Prepmaster	32	—	6	6
Method of analysis	GC / ITMS, GC-ECD	GC / ITMS, GC-ECD	GC / ITMS	GC / ITMS

Table 5.12. Results of Interlaboratory Comparison of Analysis of Potato Check Sample

	Iprodione, (μg / g)		Ethion, (μg / g)	
Lab No.	Solvent[a]	SFE[b]	Solvent[a]	SFE[b]
1	1.1	1.3 ± 0.1[c]	0.13	0.077 ± 0.006[c]
2	0.87	1.2 ± 0.1[c,d]	0.11	0.06 ± 0.02[c,d]
3	1.4	1.6 ± 0.2[e]	0.11	0.06 ± 0.02[e]
4	1.4	0.98 ± 0.04[c]	0.11	0.057 ± 0.002[c]
5	1.4	1.16 ± 0.05[e]	0.099	0.049 ± 0.004[e]
6	1.6	1.0 ± 0.1[c]	0.13	0.040 ± 0.008[c]
7	1.1	—	0.13	—
Average	1.27	1.21	0.117	0.057
Standard deviation	0.25	0.23	0.013	0.012
RSD[f] (%)	20	19	11	22
Actual concentration	1.6 μg / g			0.12 μg / g

[a] Regulatory solvent-based extraction method (1,2).
[b] SFE using two different instruments; analyses were performed in triplicate on different days, and results are means ± standard deviations.
[c] SFE with 7680T.
[d] Result of method of standard addition.
[e] SFE with Prepmaster.
[f] Relative standard deviation.

these replicate analyses, they were analyzed using the SFE GC/ITMS method 6 times on sequential days, with the results appearing in Table 5.12. Sample fortifications were made by the lead laboratory in this program, and for these samples, the fortifications were done at 1.6 μg/g. With only one exception, these samples were found to be lower than the fortification level using the conventional solvent extraction method. Residues found by the SFE method were similarly lower, with one exception at 1.6 μg/g. Lowest RSDs were observed for ethion using the conventional solvent extraction method, 11%; the highest RSDs were observed for the SFE method for that pesticide. Both sets of data, conventional and SFE, showed about the same range in residue values and standard deviations. Consequently, the SFE method seems to be equivalent, except for consequent reductions in analysis time and solvent usage.

Conclusions

Solid-phase extraction devices continue to expand to intermediate and polar bonded phases, offering the pesticide chemist growing choices about separation modes. Their ability to stabilize pesticides during storage, thereby making it possible to defer analyses until instrumentation is available, is a real asset, in addition to reducing the volume of refrigerated space needed for such storage. Although not a topic of this chapter, SPE cartridges and disks lend themselves well to automation, due to their small size, manageable shapes, and relatively low cost. Although SPE automation has been adopted by some laboratories, almost all have applied it to water analyses. Unfortunately, several manufacturers have not found the SPE automation instrument market profitable, and have discontinued their products.

The Speedisk SPE product from J. T. Baker has only recently become available at this writing, and consequently, little is known about its performance. Since it is essentially a very short column, with particles that are held in place by the cushioning action of glass fiber filter elements, it has the potential of being less prone to plugging by dirty samples. And since it carries its own disk holder with it, it is not prone to distortions when being moved from one vacuum manifold to another, as are the Teflon matrix disks, which could give rise to errors in analyte elution due to masking of previously exposed disk areas.

Having cartridges and disks now available that require no prewetting with organic solvents is a real advantage. Prewetting not only requires additional laboratory time but also enhances the potential for disk contamination, and presents real problems if the disks are to be used in remote areas for the analysis of water. Not having that requirement should also open up some possibilities for the design of remote, unattended, water samplers.

If preliminary evidence is an indication, there is the potential for the use of subcritical water for the extraction of some pesticides from fruits and vegetables (77). Such an approach, where the polarity of water and ultimately its solvent strength is controlled by temperature and pressure, would then increase the utility of SPE cartridges and disks, since organic solvents could be eliminated, reducing costs, analysis time, and the potential for human exposure. It is conceivable that subcritical water could even be used to selectively elute interferences and pesticides from the SPE devices, further increasing their versatility for extractions and cleanups.

Recently marketed multiple sample and automated SFE instruments have greatly appreciated the value of that technique for extracting pesticide residues from foods. As their use proliferates, some of their components will undoubtedly become more refined, such as the automated restrictors, the robot-controlled extraction vessel changing mechanisms, and the precision with which SF density is controlled. The question of sample size capability still remains, with some analysts unwilling to accept results from samples less than 25 g, calling into question data from sample weights typical of the instruments now available to the analyst (1–3 g). And since the ability to handle 25 g of sample now requires special user-constructed instrumentation, which is inherently expensive and even unobtainable by some, that type is unacceptable by many laboratories. There now seems enough SFE data available to allay the fears of those concerned with the accuracy question related to small subsamples; handled properly, 1–3 g of sample is more than adequate to guarantee residue accuracy.

Of more concern is the issue of special preparation techniques that still seem to be required, which vary for sample type and pesticide type. Water control during the extraction process still requires that a solid adsorbent material be blended with the sample before placement in the extraction vessel, but the optimum sample to adsorbent ratio varies with sample type and even pesticide type. Consequently, what may be optimum for one food–pesticide combination will not be for another, with compromises being required. Some pesticide types seem to adhere to the adsorbent in varying degrees, depending on pesticide and adsorbent composition, giving rise to variable recoveries. One of the major research thrust areas ought to be that of approaching the water-control issue from a new angle, by exploring adsorbents that have a high affinity for water but not the pesticide, or by exploring new technologies that can retain water in the extraction vessel but allow for pesticide extraction.

The fact that SPE and SFE both reduce or eliminate organic solvents will continue to offer incentives for their development. Exploring water–solvent combinations for the extraction of foods followed by further dilution with water so that SPE devices can be employed should bring about further reductions in solvent usage. This approach, in combination with use of the more polar adsorbents, should become more attractive as cartridge and disk sizes are increased, allowing for larger sample preparation, and as prices drop with increases in sales. Since CO_2 has convincingly been shown to be effective for the extraction of even moderately water-soluble pesticides, even without organic modifiers, there is an impetus to provide extraction instruments with capabilities to operate at pressures higher than the 10,000-psi maximum now available. Newer approaches for dealing with these increased pressures while still keeping extraction vessel costs down will keep these instruments within reach of pesticide residue laboratories.

REFERENCES

1. J. Breiter, R. Helger, and H. Lang, *Forens. Sci.* **7**, 131–140 (1976).
2. R. Ellis, Symposium on New and Emerging Approaches to the Development of Methods in Trace-Level Analytical Chemistry, Berkeley Springs, WV, Aug. 17–21, 1992, Food Safety and Inspection Service, Un. St. Dept. Agriculture, Washington, DC.

3. M. A. Luke, J. E. Froberg, G. M. Doose, and H. T. Masumoto, *J. AOAC* **64**, 1187–1195 (1981).
4. Anonymous, Recommendations on nomenclature for chromatography, IUPAC, *Pure Appl. Chemi.* **37**, 447–462 (1974).
5. C. Horvath, and W. Melander, *J. Chromatogr. Sci.* **15**, 393–404 (1977).
6. H. A. Moye, Opportunities for pesticide residue analytical method development: the potential for aqueous extractions of pesticide residues from fruits and vegetables, in *Conference Proceedings Series, Eighth International Congress of Pesticide Chemistry*, American Chemical Society, Washington, DC, 193–203 (1995).
7. M. J. Page, and M. French, *J. AOAC Intl.* **75** (6) (1992).
8. M. A. Luke, The evolution of a multiresidue pesticide method: *J. Am. Chem. Soc.* 174–182 (1995).
9. S. M. Lee, M. L. Papathakis, H. M. C. Feng, G. F. Hunter, and J. E. Carr, *Fresenius J. Anal. Chem.* **339**, 376–383 (1991).
10. B. L. Tippins, Vol. **334**, 273–274 (1988).
11. T. Cairns, M. A. Luke, K. S. Chiu, D. Navarro, and E. Siegmund, *Rapid Commun. Mass Spectrom.* **7**, 1070–1076 (1993).
12. A. de Kok, M. Hiemstra, and C. P. Vreeker, *Chromatographia* **24**, 469–476 (1987).
13. Y. Odanaka, O. Matano, and S. Goto, *Fresenius J. Anal. Chem.* **339**, 268–273 (1991).
14. P. D. McDonald, *Waters Sep-Pak*® Cartridge Applications Bibliography, 5th ed., Waters, Division of Millipore, Milford, MS.
15. N. Simpson, and K. C. Van Horne, *Sorption Extraction Technology Handbook*, Varian Sample Preparation Products, Harbor City, CA, 1993.
16. P. T. Holland, D. E. McNaughton, and C. P. Malcolm, *J. AOAC Intl.* **77**(1) (1994).
17. P. Cabras, C. Tuberoso, M. Melis, and M. G. Martini, *J. Agric. Food Chem.* **40**, 817–819 (1992).
18. M. S. Mills, E. M. Thurman, and M. J. Pedersen, *J. Chromatogr.* **629**, 11–21 (1993).
19. P. J. Dumont, and J. S. Fritz, *J. Chromatogr. A* **691**, 123–131 (1995).
20. E. S. P. Bouvier, D. M. Martin, P. C. Iraneta, M. Capparella, Y.-F. Cheng, and D. J. Phillips, A novel polymeric reversed phase sorbent for solid phase extraction, *LC/GC Maga.* (in press).
21. M. S. Young, A polymeric solid phase procedure for determination of sulfonylurea herbicides in environmental samples, *Environ. Sci. Technol.* (in press).
22. E. Durhan, M. Lukasewycz, and S. Baker, *J. Chromatogr.* **629**, 67–74 (1993).
23. C. Y. L. Hsu, and R. R. Walters, *J. Chromatogr.* **629**, 61–65 (1993).
24. G. L. Muth, and F. Erro, *Bull. Environ. Contam. Toxicol.* **24**, 759–765 (1980).
25. T. Cairns, K. S. Chiu, D. Navarro, and E. Siegmund, *Rapid Commun. Mass Spectrom.* **7**, 971–988 (1993).

26. F. J. Schenck, R. Wagner, M. K. Hennessy, and J. L. Okrasinski, Jr, *J. AOAC Intl.* **77**(1), 102–106 (1994).
27. A. M. Gillespie, S. L. Daly, D. M. Gilvydis, F. Schneider, and S. M. Walters, *J. AOAC Intl.* **78**(2), 431–436 (1995).
28. F. J. Schenck, and M. K. Hennessy, *J. Liq. Chromatogr.* **16**(3), 755–766 (1993).
29. J. I. Mora, M. A. Goicolea, R. J. Barrio, and Z. Gomez de Balugera. *J. Liq. Chromatogr.* **18**(16), 3243–3256 (1995).
30. M. Hiemstra, J. A. Joosten, and A. de Kok, *J. AOAC Intl.* **78**(5), 1267–1274 (1995).
31. S. D. Regis-Rolle, and G. M. Bauville, *Pest. Sci.* **37**, 273–282 (1993).
32. J. R. Pardue, *J. AOAC Intl.* **78**(3), 856–862 (1995).
33. J. A. Cobin, and N. A. Johnson, Liquid chromatographic method for rapid determination of total avermectin B_1 and 8,9-*Z*-Avermectin B_1 Residues in apples: *J. AOAC Intl.* **78**(2), 419–423 (1995).
34. D. F. Hagen, C. G. Markell, and G. A. Schmitt, *Analytica Chimica Acta* **236**, 157–164 (1990).
35. L. Davi, M. Baldi, L. Penazzi, and M. Liboni, *Pest. Sci.* **35**, 63–67 (1992).
36. H. A. Moye, and W. B. Moore, A remote water sampler using solid phase extraction disks, *Proc. EPA 7th Annual Waste Testing and Quality Assurance Symp.* 1991, pp. 245–261.
37. T. McDonnell, and J. Rosenfeld, *J. Chromatogr.* **629**, 41–53 (1993).
38. S. A. Senseman, T. L. Lavy, J. D. Mattice, B. M. Myers and B. W. Skulman, *Environ. Sci. Technol.* **27**, 516–519 (1993).
39. G. Pavoni, Evolution of the extraction techniques in the research of pesticide residues in fruit and vegetable products, *Advanced Analytical Techniques in the Environmental Field*, Conference, Pontaccchio Marconi, Italy, Nov. 26–27, 1992.
40. J. A. Casanova, 1996. *J. AOAC Intl.* **79**(4), 936–940 (1996).
41. L. T. Taylor, *Supercritical Fluid Extraction*, J. Wiley, New York, 1996.
42. K. Zosel, German Patent 1,493,190 (1964).
43. D. R. J. Laws, N. S. Bath, and A. G. Wheldon, U.S. Patent 4,218,491 (1980).
44. V. J. Krukonis, *ACS Symposium Series*, Vol. 366, 1988, p. 30.
45. J. W. King, and M. L. Hopper, *J. AOAC Intl.* **75**(3), 375–378 (1992).
46. S. B. Hawthorne, *Anal. Chem.* **62**(11), 633A–642A (1990).
47. P. C. Sadek, P. W. Carr, R. M. Doherty, M. J. Kamlet, R. W. Taft, and M. H. Abraham, *Anal. Chem.* **57**(14), 2971–2978 (1985).
48. E. Stahl, and K. W. Quirin, *Fluid Phase Equilibrium* **10**, 269–275 (1983).
49. N. Alexandrou, and J. Pawliszyn, *Anal. Chem.* **61**(24), 2770–2776 (1989).
50. USDA *Pesticide Analytical Manual* Vol. I: *Multiresidue Methods*, 3rd ed., U.S. Food and Drug Administration, Washington, DC, 1994.
51. A. Di Muccio, S. Girolimetti, D. Attard Barbini, T. Generali, P. Pelosi, A. Ausili, F. Vergori, G. De Merulis, A. Santillo, and I. Camoni. *1st Eur. Pesticide Residue*

Workshop, Alkmaar (Egmond aan Zee), The Netherlands, June 10–12, 1996, abstract P-118.

52. S. J. Lehotay, N. Aharonson, E. Pfeil, and M. A. Ibrahim, *J. AOAC Intl.* **78**(3), 831–840 (1995).
53. K. D. Parker, M. L. Dewey, D. M. DeVries, and S. C. Lau, *Toxicol. Appl. Pharmacol.* **7**, 719–726 (1965).
54. J. E. Benville, and R. C. Tindle, *J. Agric. Food Chem.* **18**, 948–949 (1970).
55. E. A. Bunch, D. M. Altwein, L. E. Johnson, J. R. Farley, and A. A. Hammersmith, *J. AOAC Intl.* **78**, 883–887 (1995).
56. L. J. Mulcahey, and L. T. Taylor, *Anal. Chem.* **64**, 2352–2358 (1992).
57. R. J. Argauer, K. I. Keller, M. A. Ibrahim, and R. T. Brown, *J. Agric. Food Chem.* **43**, 2774–2778 (1995).
58. S. Bengtsson, T. Berglöf, S. Granat, and G. Jonsäll, *Pest. Sci.* **41**, 55–60 (1994).
59. M. F. Wolfe, D. E. Hinton, and J. N. Seiber, *Environ. Toxicol. Chem.* **14**, 1001–1009 (1995).
60. T. M. Fahmy, M. E. Paulaitis, D. M. Johnson, and M. E. P. McNally, *Anal. Chem.* **65**, 1462–1469 (1993).
61. S. J. Lehotay, *J. Chromatogr. A* **785**, 289–312 (1997).
62. S. J. Lehotay, and C.-H. Lee, *J. Chromatogr. A* **785**, 313–327 (1997).
63. I. A. Stuart, J. MacLachlan, and A. McNaughton, *Analyst.* **121**, 11R–28R (1996).
64. S. J. Lehotay, N. Aharonson, and K. I. Eller, unpublished results, 1996.
65. N. Aharonson, S. J. Lehotay, and M. A. Ibrahim, *J. Agric. Food Chem.* **42**, 2817–2823 (1994).
66. S. U. Khan, *J. Agric. Food Chem.* **43**, 1718–1723 (1995).
67. S. J. Lehotay, and A. Valverde-García, 1997. *J. Chromatogr. A* (in press).
68. T. Görner, J. Dellacherie, and M. Perrut, *J. Chromatogr.* **514**, 309–314 (1990).
69. F. Wach, *Anal. Chem.* **66**(6), 369A–372A (1994).
70. M. L. Hopper, J. W. King, J. H. Johnson, A. A. Serino, and R. J. Butler, *J. AOAC Intl.* **78**(4), 1072–1079 (1995).
71. J. W. King, J. H. Johnson, S. L. Taylor, W. L. Orton, and M. L. Hopper, *J. Supercrit. Fluids.* **8**, 167–175 (1995).
72. J. W. King, M. L. Hopper, R. G. Luchtefeld, S. L. Taylor, and W. G. Orton, *J. AOAC Intl.* **76**, 857–864 (1993).
73. S. J. Lehotay, and M. A. Ibrahim, *J. AOAC Intl.* **78**(2), 445–452 (1995).
74. S. J. Lehotay, and K. I. Eller, *J. AOAC Intl.* **78**(3), 821–830 (1995).
75. A. Valverde-García, A. R. Fernández-Alba, A. Agüera, and M. Contreras, *J. AOAC Intl.* **78**(3), 867–873 (1995).
76. A. Valverde-García, A. R. Fernandez-Alba, M. Contreras, and A. Agüera, *J. Agric. Food Chem.* **44**(7), 1780–1784 (1996).
77. H. A. Moye, *The 110th AOAC International Annual Meeting and Exposition*, Abstracts, Paper 10-A-001 Orlando, FL, Sept. 8–12, 1996.

CHAPTER

6

ENZYME-LINKED IMMUNOSORBENT ASSAY (ELISA)

H. ANSON MOYE

INTRODUCTION

About the time chromatographic methods were being developed for the analysis of pesticide residues in foods during the 1960s, immunoassays were catching on for the analysis of a wide array of analytes in the clinical chemistry laboratory, including assays for the assessment of thyroid (T_4, T_3), reproductive, oncological, cardiovascular, and infectious disease conditions in humans, with the assays for hormonal chemicals leading the way. Such chromatographic methods were largely being generated at the U.S. Food and Drug Administration (FDA) laboratories in Washington, DC, and about that time took on the character of what has come to be known as "multiresidue methods" (MRMs). From their inception, these methods were concerned only with those pesticides that were considered to be a threat to human health, the insecticides, which at that time were predominantly the organochlorines (1).

Because of their hydrophobicity, these early pesticides were extracted by water-immiscible organic solvents, such as petroleum ether, when foods were of a high fat content (2). However, when foods possessed less than 2% fat, they could be extracted with acetonitrile, and subsequently partitioned into petroleum ether (3,4). Subsequent cleanup was accomplished by preparative chromatography on columns packed with Florisil. Quantitations were performed on packed gas–liquid chromatography (GC) columns with either the electron capture detector or with the flame photometric detector when the newly developed organophosphate pesticides arrived. As the overall polarities of the pesticides grew, with the introduction of new types of organophosphates, and ultimately the *N*-methylcarbamate insecticides, the usefulness of using acetone became apparent (5,6). Removal or exchange of such extracting solvents has always placed constraints on the pesticide

Pesticide Residues in Foods: Methods, Techniques, and Regulations, by W. G. Fong, H. A. Moye, J. N. Seiber, and J. P. Toth. Chemical Analysis Series, Vol. 151
ISBN 0-471-57400-7

residue chemist, creating exposure and fire hazards in the analytical laboratory, adding significantly to analysis time, increasing analysis costs due to waste disposal, and increasing the opportunity for compromising sample integrity by the incorporation of unwanted interferences always present in less expensive solvents.

Also placing constraints on the pesticide analytical chemist in the early days were the ways in which a typical chromatographic analytical method was constructed, or laid out in time. Each pesticide analysis conducted using the conventional MRM analytical method has always been lengthy, tedious, and labor-intensive, largely because of the weight of sample required for a statistically significant analysis (100–200 g), the overall large volumes of solvent required for extraction and sample manipulation (400–800 mL), the need to replace or exchange that solvent as described above, the physical sizes of the glassware used in such operations, and finally the time required for a single analytical run on a typical gas chromatograph. Because of the expense and sizes of typical gas chromatographs, most laboratories employed only a few. Consequently, considering all these factors, only a few MRM analyses could be processed in a single day, and it was common to see samples waiting for the gas chromatographic determinative step, while in other laboratories samples could not be processed rapidly enough so as to be made ready for the gas chromatograph. Employing high-performance liquid chromatography (HPLC) for at least two of their specialty MRMs—those for the urea herbicides and the *N*-methylcarbamates—the FDA did not improve their capability for sample throughput, since the same constraints apply for HPLC-based methods as for the GC-based ones (6). It was clear that other approaches needed to be explored in order to manage the various monitoring programs underway in the United States at that time.

In the late 1980s a growing concern by the U.S. Congress about the purported small number of food samples analyzed for pesticide residues annually by the FDA prompted a study by the Office of Technology Assessment, the information-gathering arm of the U.S. Congress, to determine whether the development and implementation of newer pesticide residue analytical technologies would rectify the perceived problem (7). During that study the issue of further developing immunological techniques for pesticide residue monitoring was addressed, along with other technologies, such as supercritical fluid extractions (SFE), robotics, and biosensors. That study prompted the FDA to undertake a moderately extensive grants program to explore the opportunities there for development of immunoassay based methods, which also stimulated activity in the industrial sector, increasing sources for complete assay kits, antibodies, and associated analytical hardware, such as spectrophotometers, plate readers, reagent delivery systems, and other items that accelerated the technology (8,9). That agency's effort to propagate the use of ELISA for the analysis of pesticide residues in foods, unfortunately for the residue chemist, has not been as successful as efforts by other agencies charged primarily with assuring a clean environment, probably because of the many more challenges facing the method development chemist in dealing with foods rather than usually more simple environmental matrices, especially water. More successful has been the U.S. Environmental Protection Agency (EPA), which has stimulated the growth of ELISA kits, some of which have found wide usage on a regular basis for environmental monitoring (10–12).

Analytical methods have been categorized in a Codex document produced by the *Codex Committee on Residues of Veterinary Drugs in Foods*, using concepts that easily could be adapted by the pesticide residue scientific community and applied to making decisions as to when ELISA methods ought to be employed (13,14). Such methods were categorized as Type I, Type II, or Type III, depending on a number of attributes of the methods, including concentration present, confirmation of analyte identity, partial structural identification, and presence of groups or classes of compounds above a certain designated level of interest. Type I methods are the most rigorous of the three, providing both high precision and positive identification of the analyte in a single unified process, and typically require chromatographic separations coupled with mass spectrometry. Consequently, few regulatory metthods, in both the United States and Europe, qualify for Type I. Type II methods are not as rigorous, and do not unequivocally identify the analyte at the concentrations of quantitation, but they do provide some structural information as to what types of functional groups might be present or what types of atoms there are in the molecule. Such methods are generally reliable enough for the purposes of regulatory agencies, and are usually listed as reference methods by such organizations as the EPA, the USDA, and the FDA, and perform quantitation functions as well as confirming the presence of the analyte. Type III methods are much less specific, but generate information that can be very useful for the purposes to which they are put. They typically are used to perform the function of what can be called a *screening* method, one that determines whether an analyte is present, at some particular designated level of interest. The presence of the analyte above this designated concentration level would identify the sample as being a *positive* one, and one that does not elicit a response would be a *negative* one. If this Type III method has an acceptable level of false negatives, which is to say that it does not often miss samples having analyte concentrations above the designated level, then for many regulatory purposes no further analyses may be required. However, samples giving a *positive* response would require a reanalysis by a Type I or II method. Consequently, if Type III methods can be performed in a timely and cost-effective way, they can play a major role in any pesticide residue analytical program. ELISA-based analytical methods for pesticide residues definitely fall into the Type III category.

Probably the first analytical chemist to recognize the impact ELISA techniques might have on the pesticide residue analytical community was the late C. D. Ercegovich of Pennsylvania State University, who summarized various pieces of work in other fields that might be used to spin off pesticide analytical methods (15). Picking up the pesticide–ELISA torch, R. O. Mumma, also of Pennsylvania State University, and B. D. Hammock, of the University of California, Davis, began a long association by coauthoring a chapter in a monograph edited by the late Jack Zweig (16). The basis for this early work was a Ph.D. thesis at Penn State by Al-Rubae (17), other early publications from that laboratory (18,19), and early work at the Food Research Division, Bureau of Chemical Safety, Health and Welfare, Canada, by W. H. Newsome (20–22). This early work focused on assays for broad classes of pesticides (21) that included organophosphates, *N*-methylcarbamates, triazines, chlorinated hydrocarbons, chlorophenoxyacetic acids, pyrethroids, chitinase inhibitors, and biorational insecticides. It also supplied methods for specific compounds, benomyl, iprodione, maleic hydrazide, metalaxyl, and triadimefon.

Those attributes of ELISA that have propelled it into such prominence in North America and Europe—cost-effectiveness, speed, high selectivity, high sensitivity, and moderate simplicity—should make it attractive to those countries who are struggling to establish contemporary analytical chemistry laboratories for monitoring foods and environmental samples. Pesticide residue laboratories that have difficulty maintaining a sustained level of financial, personnel, space, and instrumental support by their sponsoring agencies may be able to accomplish a great deal using an ever-growing array of ELISA technologies. It is very important to understand that well-chosen ELISA technology is not second-rate to any of the instrumental technologies available today, including mass spectrometry (see Chapter 4), the chromatographies (see Chapter 3), and the newer extraction technologies (see Chapter 2). While it is the experience of this author that such technologies are difficult to implement in developing countries, the task is greatly simplified and facilitated for use with ELISA techniques. Whereas the pesticide analytical laboratory anchored in chromatographic techniques frequently requires detectors, glassware, and reagents that are unique to the measurement of pesticide residues, the equipment, laboratory support items, and the chemist's training on pesticide-based ELISA techniques can be extended to other trace analytical applications, such as drugs, microbial species, and disease conditions. Additionally, skills that a chemist acquires applying ELISA to pesticide residue analyses can also be used to monitor the quality of various products produced in a particular country, so that the food safety and environmental quality aspects of a laboratory can spill over into laboratories that are more product-quality-control oriented. As ELISA analytical devices continue to become more respected in the importing countries of the developed world, their trust and official acceptance of the products exported by developing countires relying on ELISA technques will rise.

In the past there have been widespread misconceptions that have slowed the acceptance of ELISA by the pesticide residue analytical community, which, fortunately for analysts everywhere, have significantly dissipated. These misconceptions are briefly reviewed here, primarily for the benefit of those considering the technique for the first time, so that we may all benefit from the efforts of those who struggled to propagate the technique, and whose efforts are paying great dividends today:

1. ELISA is not a *bioassay*; rather, it is a *chemical assay* in every way. It does not depend on the unpredictable behavior of a living animal to make it function properly. The components of the assay, as supplied by various vendors, may have been produced by a particular biological response of a mouse or rabbit, but they will have been carefully assayed to guarantee their composition and therefore their performance. If its limitations are clearly understood (see section on limitations below), ELISA can be used with the same degree of trust and certainty as in any conventional chromatographic technique.

2. ELISA does not require any biological "black magic" that is unattainable or incomprehensible to the analytical chemist. It is a physicochemical technique that just happens to use chemicals that are generated, to some degree, by biological organisms. A good analytical chemist can be trained to perform reproducible ELISAs in days, which may not be so for an immunochemist who would be asked to become an analytical or environmental chemist in that time period. Those concerns about proper glassware cleanliness, reagent manipulations, and sample preparations that the pesticide analytical chemist has always had, when applied to ELISA, are more than sufficient.

3. The terminology used by immunochemists, although baffling at first glance, is consistent from laboratory to laboratory, and quickly becomes a part of the pesticide residue chemist's vocabulary (see box following this paragraph for a glossary of terms). Some types of terminology are specific to immunoassay techniques other than ELISA, and will not be required for the readers of this chapter or those going on to master it in the laboratory (23).

KEY TERMS AND DEFINITIONS*

Antibody. A protein that selectively recognizes and binds to a target analyte or group of related analytes.
Antigen. A hapten–protein conjugate or other large molecule that can bind to a selective antibody.
Conjugate. A compound, usually a protein, covalently bound to a hapten or a label, such as an enzyme.
Hapten. A small molecule that can induce antibody production when covalently bound to a carrier molecule. A hapten can react with the specific antibodies produced in response to the hapten-carrier conjugate.
Immunoassay. A physical assay based on the reversible interaction of a specific antibody with a target analyte(s).
Immunogen. A hapten-carrier conjugate or other large molecule used for the production of selective antibodies.
Monoclonal antibody. A homogeneous antibody population derived from one specific antibody-producing cell.
Polyclonal antiserum. A heterogeneous population of antibodies varying in selectivity and affinity for a target analyte that is derived from several antibody-producing cells.

ELISA techniques have lately become more popular than any other immunoassay technique for the analysis of pesticide residues as well as most other environmental chemicals, and they have replaced its historic predecessor, *radioimmunonassay* (RIA), as the technique of choice, due largely to several very practical reasons. Newly developed ELISAs have overtaken and passed RIA assays when lower limits of detection (LLDs) are examined later in this chapter. For example, when RIA was compared to competitive direct ELISA (cdELISA) for the analysis of aflatoxins, lowest limits of detection were observed for the cdELISA (0.1 ng / mL), as compared to the RIA (7.8 ng / mL). It is interesting to note that when the cdELISA was performed with MAbs, the limit of detection was 2.5 ng / L, compared to the 0.1 ng / mL observed for the PAb, polyclonal antibody (24). Difficulties are encountered when analyses are conducted using the RIA technique that are associated with gaining approval from safety officials for use and storage of reagents. Also, radioactive particle counting devices are generally much more expensive than the spectrophotometers used for the ELISAs, and the spectrophotometers provide just as reliable results.

* *Source*: J. M. Van Emon and C. L. Gerlach, *Environ. Sci. Technol.* **29**, 312A–317A (1995).

HOW THE TECHNIQUE WORKS

Antibody Production

All higher animals have an immune system that, when challenged by various types of pathogens, responds by producing immunoglobins (proteins) that combine with that particular pathogen to eliminate its normal pathogenic action (25,26). A pathogen eliciting such a response is called an *antigen*, and does so by causing B lymphocyte cells to clone themselves, with each cell producing a single type of immunoglobin with a unique specificity for the antigen, complementing the activity of other cells in the immune system, including thymus-derived T lymphocytes and others. Such an immunoglobin is called an *antibody,* and is the molecular basis on which immuoassays are built. Small molecules (less than ~1000 daltons), called *haptens,* seldom produce such a response, and are not *immunogenic*, but become so when covalently bonded to a carrier protein. When an animal is immunized by injection with such a molecule, it typically forms antibodies to both the small molecule antigen and to the protein-complexed form as well. This dual behavior is fortuitous, because it allows for some clever antigen–antibody interactions to be manipulated so as to produce a highly sensitive and specific immunoassay. Such interactions are very specific ligand-binding events, which do not depend on any biological functions, such as enzyme production, cell growth, or cell division, and hence can be brought into play under a variety of ordinary laboratory conditions. Thus, the affinity of a particular antibody for a particular antigen can be expressed in ordinary chemical equilibrium terms:

$$[\mathrm{Ab}] + [\mathrm{Ag}] \underset{k_{\mathrm{back}}}{\overset{k_{\mathrm{forward}}}{\rightleftharpoons}} [\mathrm{AbAg}]$$

$$K_{\mathrm{a}} = \frac{k_{\mathrm{forward}}}{k_{\mathrm{back}}} = \frac{[\mathrm{AbAg}]}{[\mathrm{Ab}]\,[\mathrm{Ag}]}$$

As is implicit in this above equation, the antibody–antigen binding is a reversible one, one in which the interactions are proton donor–proton acceptor, dispersive, and van der Waals, and resemble closely the interactions that occur between an enzyme and its substrate, which has a very large steric component, and one in which the k_{forward} is much larger than the k_{back}.

Antibody Affinity, Specificity, and Cross-Reactivity

How tightly the antibody is bound to the analyte, in this case the pesticide, is critical, is termed the *affinity,* and is a measure of the relative concentrations

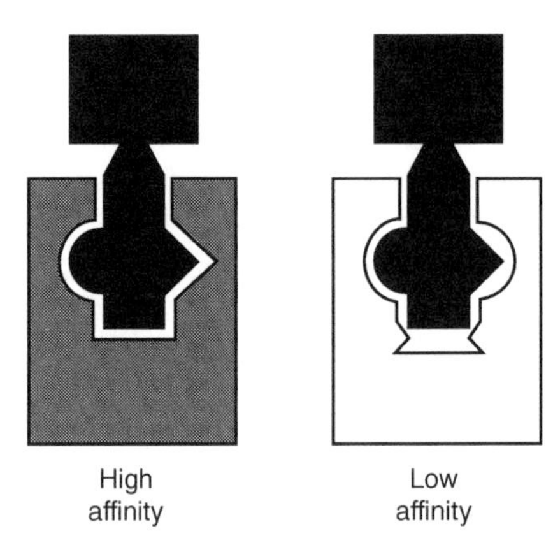

Figure 6.1. Diagrammatic representation of high affinity of a pesticidal chemical for an antibody (left) and of low affinity of a pesticidal chemical for an antibody (right). (From Ref. 7.)

of free pesticide and bound pesticide. As stated above, many types of interaction are at work in the antibody–antigen complex, and all are affected by molecular fit, which can be seen in Figure. 6.1, where multiple fits come into play for a high-affinity complex, on the left, and limited fits come into play on the right. When there is a high affinity for the pesticide, the ELISA technique usually provides high sensitivity, since the $k_{forward}$ is large. If multiple chemical structures can fit into the antibody, but with varying degrees of fit, then we can have varying degrees of *specificity* for the pesticide–antibody interaction. This ability of the antibody to bind with multiple chemical structures gives rise to what is termed *cross-reactivity.* To employ a highly selective ELISA method, one that is specific for only one pesticide in a whole family, the analyst would choose antibodies that have limited cross-reactivity. To assay for a group of pesticides in a semiquantitive manner, the analyst would choose antibodies that have extensive cross-reactivity. For example, to quantitate simazine in the presence of atrazine and propazine, all triazine herbicides, the analyst would choose antibodies that have no or little cross-reactivity to atrazine or propazine.

As was seen in the Introduction of this chapter, antibodies can be produced in one of two very distinctly different ways. The earliest, and still very well used, are the polyclonal antibodies. They are termed *polyclonal* because the animal that is injected with an antigen produces large numbers of B lymphocyte cells, groups of which respond somewhat differently than other groups to the antigen, producing an array of antibodies, differing widely in the mechanism in which they respond to the antigen. Some B lymphocyte cells respond only to the hapten, some respond to the linkage chemistry that connects the hapten to the carrier, and some respond only to the carrier molecule. Thus the serum of the animal containing the antibodies, called the *antiserum*, is a complex mixture of antibody molecules, some of which bind very tightly with the hapten, and some of which don't. Making practical use of such polyclonal antibodies to be used in immunoassays presents some practical problems that need to be overcome, and that are not insurmountable: (1) the nature and quantity of polyclonal antibodies found in animal sera are dynamic, varying with individual "bleeds" from an animal, and certainly varying from animal to animal, and even more so from species to species; and (2) since there are so many antibody types in some sera, the analytical utility can be lost by competing reactions. A phenomenon discovered in 1975 by two British scientists produced a technique for generating antibodies of more uniform characteristics that eliminates some of these problems (27).

In response to the difficulties associated with polyclonal antibodies, the *monoclonal antibody* (*MAb*) was given birth (27). This came about with the realization that an animal could be immunized with an antigen and B lymphocytes could be harvested from the animal's spleen and then fused with myeloma cells from tissue culture, producing hybrid cells that are heterokaryons, carrying the genetic information of both cells. Consequently, such cells generate antibodies characteristic of the B lymphocyte cells, but they also have unlimited growth potential in mass cultures. In addition, they can be implanted in the peritoneal cavities of mice, where they produce very high levels of antibodies in the ascites fluid (Fig. 6.2). Such hybridomas can be frozen in liquid nitrogen, and the monoclonal antibody in them is stable indefinitely under mild conditions of varied types. Hybridomas thus produced will be highly consistent from batch to batch, producing antibodies that do not change characteristics with individual animals, bleedings, and so on, unlike varied batches of antisera, which will give varied cross-reactivities and affinities (see text below).

An unfortunate criticism of monoclonal antibodies is that they frequently do not have as strong an affinity for the analyte as do polyclonal antibodies. Manufacturers of assay kits and reagents frequently have to do additional screening of the MAbs in order to select those with higher affinities. Such an operation is expensive and somewhat offsets gains in time and costs during their production. Consequently, most of the commercial assays available today for pesticides use polyclonal antibodies.

Success in generating antibodies has greatly depended on the design of *haptens*, small molecules mimicking the structure of the analyte, which are then covalently bonded to the protein-carrier molecule. Part of the hapten structure, the *spacer arm,* is the mechanical link between the hapten and the carrier molecule, and its chemical formula, location on the hapten, and

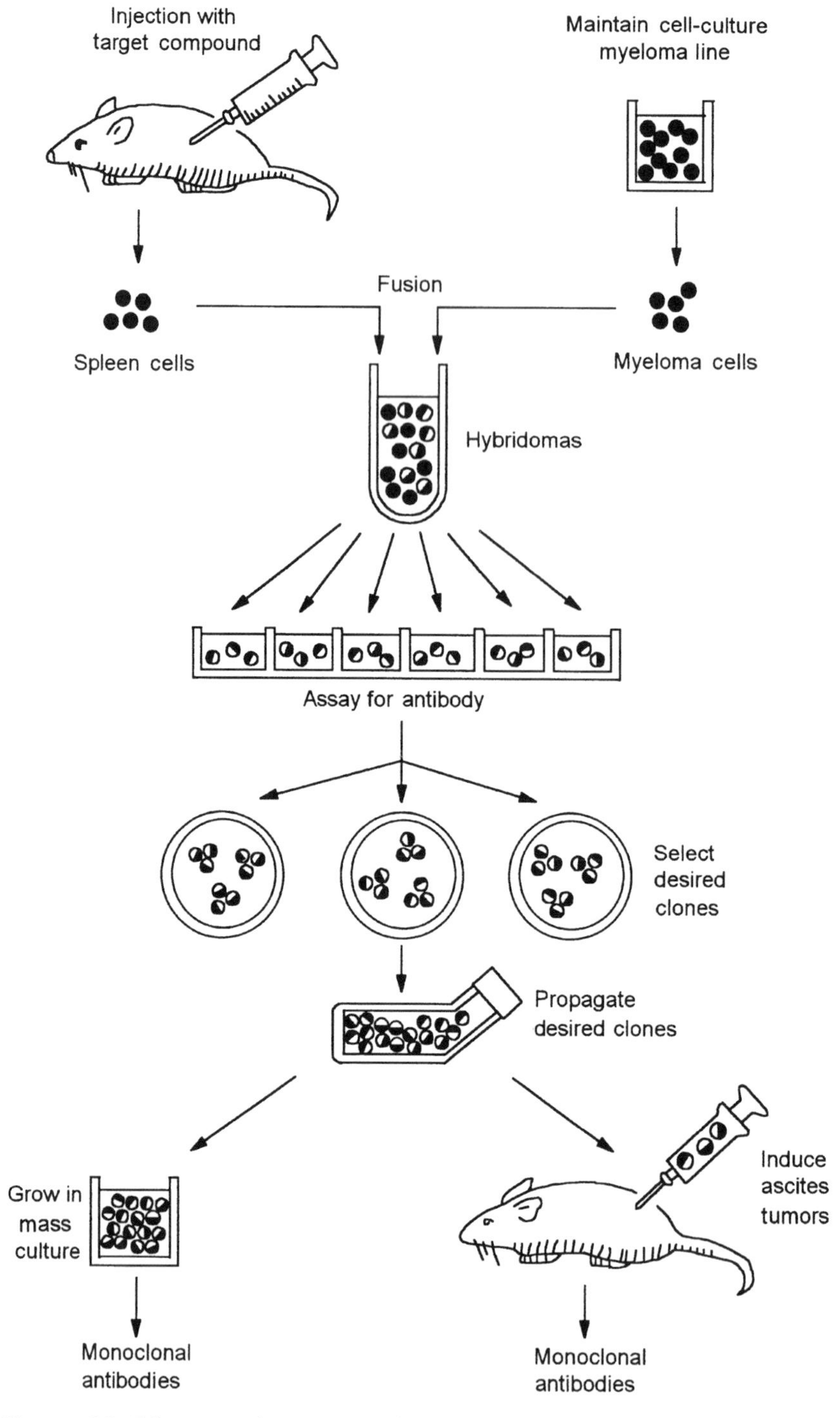

Figure 6.2. Diagrammatic representation of monoclonal antibody preparation. (From Ref. 7.)

length affect the numbers and characteristics of the antibodies that are produced, including the antibody's affinity for the analyte and its *cross-reactivity* (28). It is highly desirable that the animal's immune system not recognize the spacer arm at all, and respond only to the analyte part of the hapten. Consequently, it has been shown that alkyl chains usually perform best as spacer arms, and that arms with heteroatoms or aromatic rings usually produce antibodies that have an affinity for the arms themselves, reducing the antibodies' usefulness for ELISA purposes (29,30). Where the spacer arm is located on the molecule is of prime importance in retaining recognition of the analyte by the antibody (31). Usually, the spacer arm is positioned away from distinguishing atoms and functional groups of the hapten molecule. For example, in the lengthy development of an ELISA for the pesticide diflubenzuron, those haptens that were attached to the carrier molecule distal to the 2,6-difluorobenzoylurea portion elicited antibodies that were highly specific for this type of structure. Another study showed vast differences in antibody specificity for the insecticide ivermectin, depending on whether the spacer arm was located at the oxygen atom on carbon 5 or on the 4″ hydroxy group, at the other end of the molecule (32). When the spacer arm (succinoyl) was located at carbon 5, all antibodies examined bound more strongly than when the spacer arm was located at the 4″ hydroxyl group, giving I_{50} values as low as 2.0 ppb. More recently, MAbs were generated for the insecticide chlorpyrifos by beginning with the immunization of mice with conjugated haptens of several types (33). In one of them, designated AR3, the spacer arm was located at the position adjacent to the pyridyl nitrogen, in place of one of the chlorine atoms (Fig. 6.3). In another, designated TR2, the spacer arm was located in place of the thiophosphonic acid group of chlorpyrifos. As might be expected, the TR2 hapten elicited much more specific antibodies, giving rise to an I_{50} of 9 nM, where the AR3 hapten gave an I_{50} of 85 nM. Direct and indirect coating ELISA formats were evaluated, with comparable results.

When more than one pesticidal structure is recognized by a particular antibody or array of antibodies, the ELISA assay is said to be *cross-reactive.*

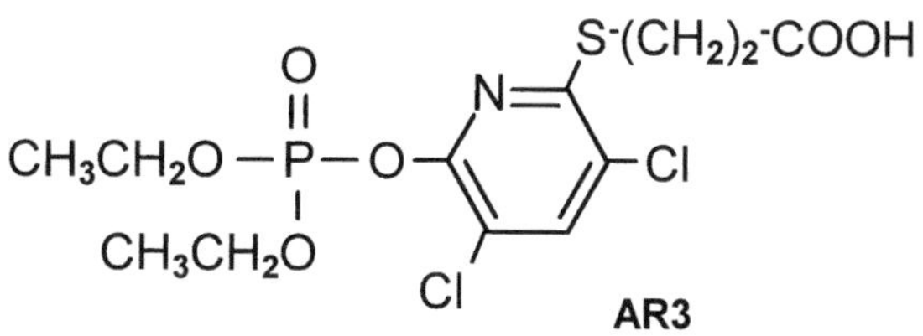

Figure 6.3. Structure of a hapten developed for chlorpyrifos insecticide, showing the sulfide-linked spacer arm. (From Ref. 33.)

Such a phenomenon may or may not be desirable, depending on the purposes of the assay. For example, if a particular assay is to be used for screening purposes, to determine whether one or more pesticides of a particular class are present, it may be attractive to design a large amount of cross-reactivity into the assay. This is usually done by proper design of the hapten, which is used to elicit an immune response. However, when there is a multiplicity of cross-reactive pesticides, interpretation of the data becomes very difficult, especially when applied to food residues, since in the United States tolerances on foods are written for individual pesticides, and not by groups. If cross-reactivity occurs, there are techniques available in the literature to at least partially deal with it, but they are usually beyond the resources of the analytical chemists. Such techniques include that of multianalyte ELISA (MELISA) (49), and principal-component analysis (PCA) (50), and are beyond the scope of this volume. Most ELISA kit and reagent suppliers spend a great deal of resources to produce antibodies that are *not* cross-reactive.

ELISA Formats

By far, the most popular ELISA format used for the analysis of small molecules, such as pesticides, is that of the competitive type, the cELISA, in either direct mode (cdELISA) or indirect mode (ciELISA) (34,35). Either plastic plates with wells in them or plastic tubes are used for manipulating the assays. As seen in Figure. 6.4, for the cdELISA, the pesticide specific antibody is immobilized to the surface of the well or tube, then the sample containing the pesticide to be quantified and a fixed amount of enzyme conjugated pesticide is added to the container. Time is allowed for the free unknown pesticide and the standard conjugated pesticide to compete for the antibodies on the tube; such an incubation can take from only a few minutes to up to one hour, depending on the affinities to the antibody. Next, the unbound pesticide and pesticide–enzyme conjugate is washed away, leaving behind portions of both that are now bound to the surface of the container. Finally, a substrate is added, which is catalyzed by the enzyme on the container, giving rise to a colored product. The density of the color is inversely proportional to the amount of unknown pesticide that was in the beginning solution.

In the ciELISA technique (Fig. 6.5) (36) the well or tube is coated with the protein-conjugated pesticide and the unknown pesticide is added along with a fixed amount of pesticide specific antibody. Incubation is again allowed to proceed, after which the container is emptied and rinsed to remove reactants not bound to the pesticide–protein conjugate. This is followed by the addition of a second antibody that is conjugated to an

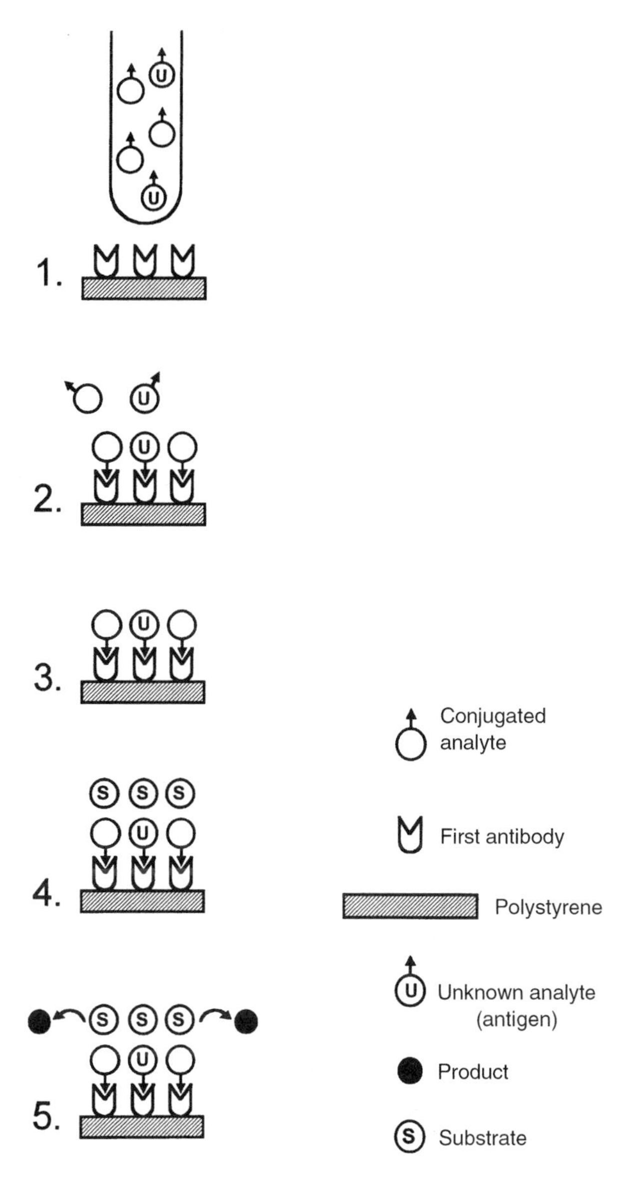

Figure 6.4. Representation of a competitive direct ELISA assay (cdELISA). (From Ref. 34.)

enzyme, and which binds to the first antibody. Again, the container is emptied and rinsed, leaving behind the bound second antibody–enzyme conjugate. Finally, a substrate is added that is converted to a colored product,

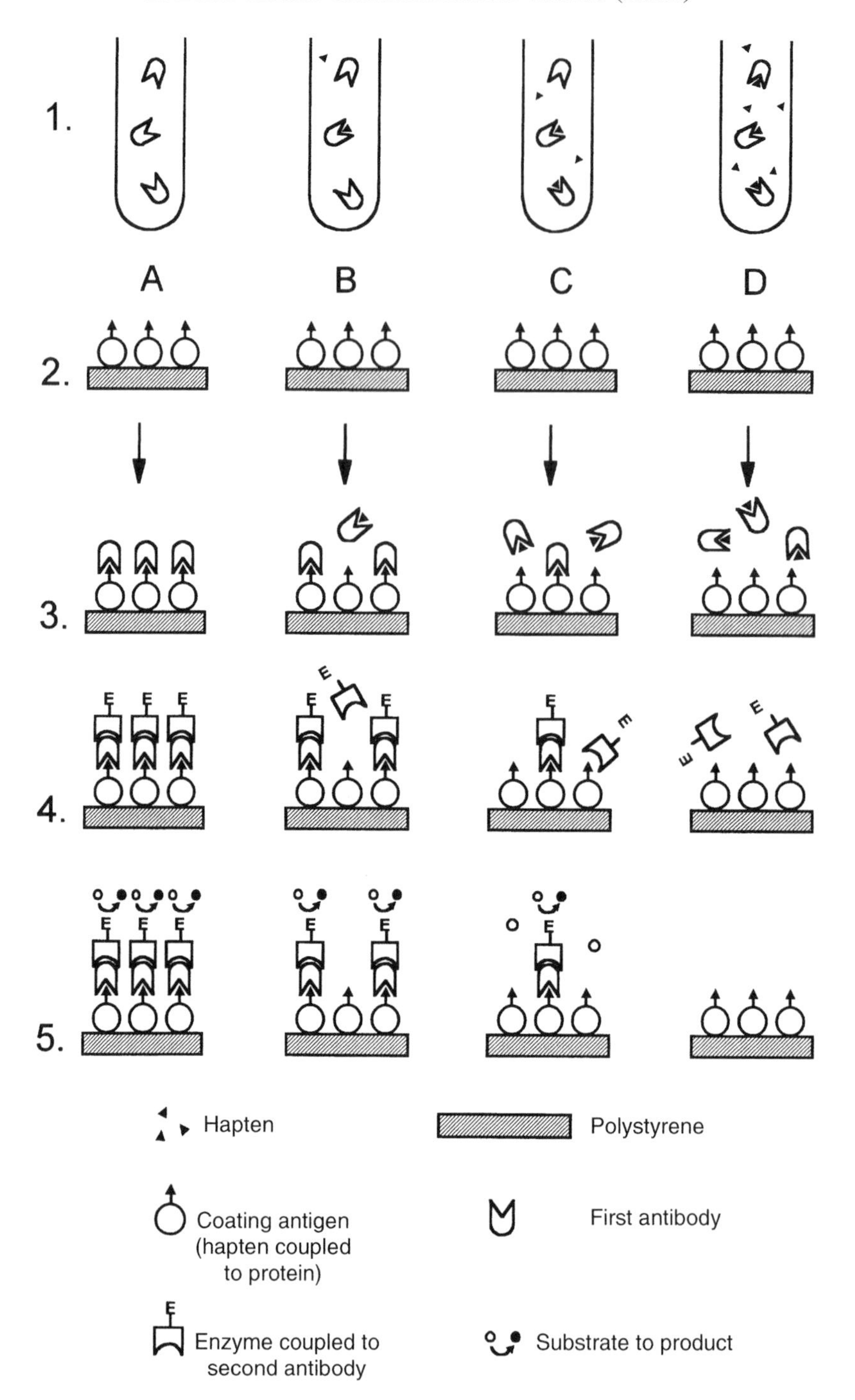

Figure 6.5. Representation of a competitive indirect ELISA assay (ciELISA). (From Ref. 35.)

and the density of the color is inversely proportional to the amount of unknown pesticide present in the first addition.

One alternative to adsorbing either the antibody or the pesticide–protein conjugate to the surfaces of plastic tubes or wells is to use suspended magnetic particles to which the antibody is covalently bonded (36). The tubes rest in a plastic rack that can be placed above another magnetic rack during washing steps, thus immobilizing the particles to the walls of the tube, preventing them from being washed away. While incubations are done, the particles remain suspended, and thus the antibody is made much more accessible to reagents, thereby accelerating reaction kinetics and shortening incubation times. The authors also claim the elimination of well-to-well variability, which can be a significant source of error in conventional techniques.

Although color can be correlated to pesticide amounts or concentrations in many ways, one of the more common ways is to plot *absorbance* versus *analyte concentration* (Fig. 6.6) (33). Or the data can be converted and expressed as *% inhibition of control*, giving a point on the curve that is the IC_{50}, that concentration of pesticide that produces 50% inhibition of the enzyme reaction (Fig. 6.7). Many other mathematical relationships relating analyte concentration to an observed response have been used, and will not be listed here. An excellent discussion of them has been prepared by Brady

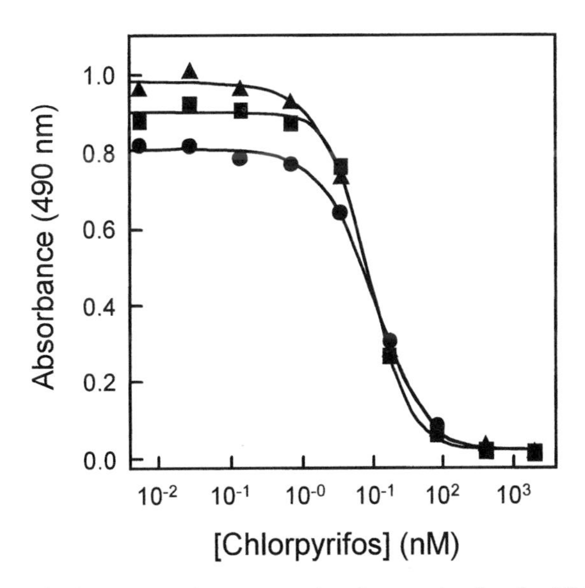

Figure 6.6. Typical concentration versus absorbance plot for the ELISA assay of chlorpyrifos insecticide. (From Ref. 33.)

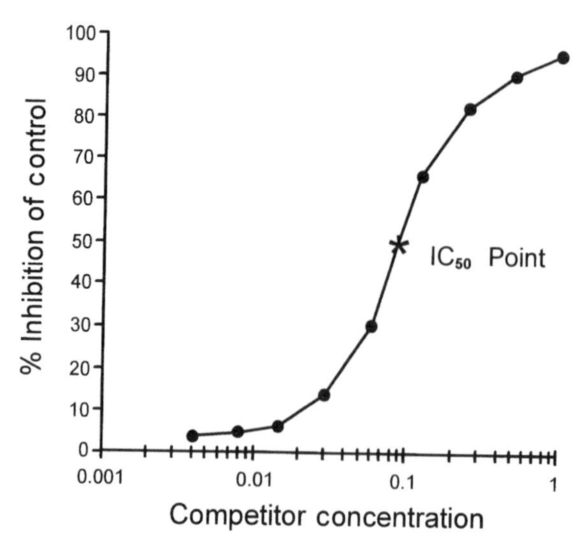

Figure 6.7. Typical IC_{50} plot for an ELISA assay.

(37). Of those presented in that discussion the *logit* transform merits mentioning (38). The *logit* relationship, given by the equation

$$\text{logit}\,(y) = \ln\left(\frac{y}{1-y}\right)$$

flattens out the sigmoidal plot, seen in Figure 6.8, by giving additional weight to the tail regions of that plot, where y is the absorbance reading. In using this relationship, the analyst must recognize that imprecision at the extreme ends of the plot goes up, still limiting the utility of the curve to only the central regions. However, most chemists feel more comfortable working with a linear plot, and will find the relationship useful.

Sample Preparation

Probably what has impeded the use of ELISA assays for pesticides in foods more than anything is the incompatibility of conventional sample preparation techniques with the antibodies and other chemicals associated with an ELISA-based method. Interferences can arise during sample preparation from two very different sources: (1) the solvents used for the extraction and (2) the sample matrix itself. Organic solvents always interfere with the pesticide–antibody interaction, but to varying degrees, depending on the antibody chosen and the solvent used (39). And such interferences always

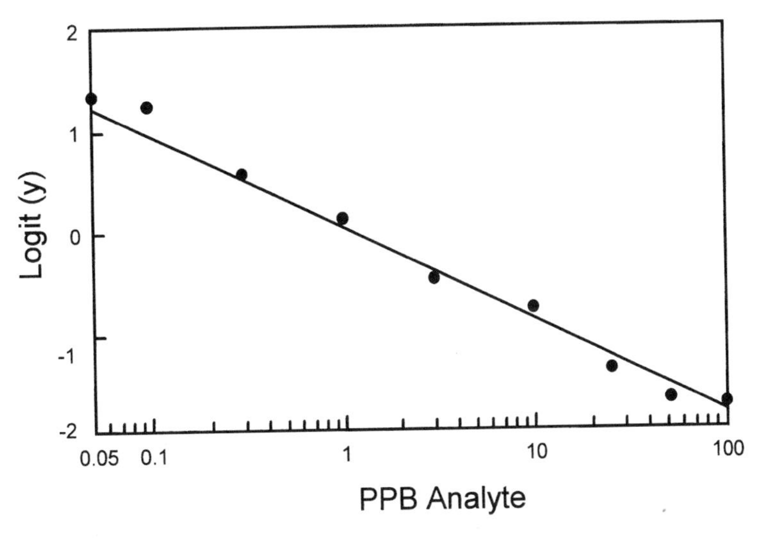

Figure 6.8. A *logit plot* for an ELISA assay. (From Ref. 37.)

increase greatest as pesticide concentrations decrease, as was seen when various concentrations of the popular multiresidue extraction solvents acetone and acetonitrile were compared to the less commonly used methanol (40). Methanol inhibited the cELISA analysis of atrazine in water much less than acetone, and somewhat less than acetonitrile. The analytical curve for atrazine in 50% methanol was usable all the way down to 0.1 μg/L, whereas in the presence of 50% acetonitrile, it was almost horizontal, having very little slope, and therefore was almost unusable. Other studies have confirmed the utility of methanol as an extraction solvent for pesticides in foods on the basis of not only its relative lack of interference as a solvent to the pesticide–antibody interaction but also its extraction efficiency.

Another aspect of solvent selection in the extraction of pesticides from foods is whether coextractives might not only inhibit the pesticide–antibody interaction but would also give false positives, meaning that a food co-extractive would elicit an enzyme response. For example, when apples, pears, peaches, and apricots were analyzed by ELISA for the fungicide tetraconazole, false positives were found for all control samples, with peach as the highest of the group (41). Methanol was used for extraction, and the samples were heated at reflux for 30 min according to a procedure reported by Newsome (42). Significant reductions in the control sample concentrations were observed, however, when the extract was washed with a 9 : 1 hexane–ethyl ether mix, particularly for the high-reading peach samples, with lesser reductions observed for the other three fruits. To account for this control matrix effect, they employed an *index of matrix interference*, defined by the following relationship:

$$\frac{OD_{\text{blank B}} - OD_{\text{blank A}}}{OD_{\text{blank B}}} \times 100 = I_{\text{m}}$$

where blank A and blank B are the unfortified fruit extracts and the simple buffer, respectively. Using this relationship they calculated the adjusted concentrations of tetraconazole using the following relationship:

$$C_x = C_{\text{determined}} \frac{100 - I_{\text{m}}}{100}$$

where C_x is the actual concentration of the fungicide in the fruit and $C_{\text{determined}}$ is the concentration read off the analytical curve; I_{m} is as defined above. Using this approach, Newsome et al. conducted 12 replicate experiments ($n = 12$) to determine recovery efficiencies at concentrations ranging from 0.025 μg/g to 10 μg/g, a total of nine concentrations. At the lowest fortification level, recoveries averaged 98% for all fruits; at the highest fortification level, recoveries also averaged 98%. Coefficients of variation were only slightly higher at the 0.025 μg/g level, 2.2%, compared to those at the 10 μg/g level, which averaged 2.0%, certainly more than comparable to typical multiresidue chromatographic methods.

One approach to dealing with false positives due to food coextractives has recently been published by workers at the California Department of Food and Agriculture (43). They used a commercially available magnetic particle ELISA kit for 2,4-dichlorophenoxyacetic acid (2,4-D), and applied it to the analysis of apples, grapes, white potatoes, and oranges. Samples were extracted with acetonitrile and cleaned up by elution through a C_{18} SPE cartridge. Split samples were analyzed by a GC procedure that required methylation with diazomethane. To test whether there might be false positives, each sample was analyzed with and without "standard added" 2,4-D; in the "standard added" tube, the pesticide was added at the 0.005-μg/g level. The optical density of the "standard added" tube was divided by the optical density of the unfortified tube, and if the ratio was equal to or greater than .80, the sample was considered positive, and the sample was analyzed by the GC procedure. In addition, if the ratio was less than .80, but the optical density was less than .20 for the unfortified tube, the sample was considered positive and was also analyzed by GC. Using these criteria, a study was reported which analyzed 91 apple, 13 peach, 56 orange, 25 grape, and 38 potato samples. Of these sample types, potatoes gave the highest percentage of false positives, 18.4%, while apples gave the lowest, 2.2%.

However, grapefruit, whose numbers weren't reported in the study, was an enigma, giving approximately 50% false positives, and was identified as a food that needed additional cleanup development. Grapefruit notwithstanding, the costs of the ELISA, considering reagents and analyst's time, were low enough to encourage that laboratory to employ it in their food-monitoring program. Apparently false negatives were not a problem.

One approach to cleanup for ELISA is to make more extensive use of solid-phase extraction devices, as illustrated in Chapter 5 of this volume. As seen in that chapter, several newly developed SPE particles are more tolerant to water than previous ones, and may perform mixed-mode-type separations with large percentages of water as the eluant. Among these, the Oasis-HLB material made by Waters, which is a proprietary *N*-vinylpyrrolidone-divinylbenzene copolymer, may offer some real advantages. Another proprietary material has been used successfully used for the cleanup of fruit and vegetable extracts in the batch mode after acetone extraction (40). In the analysis of carbendazim, the major breakdown product of benomyl fungicide, an 0.083 μg/g false positive for strawberries and an 0.025-μg/g one for grapes were both removed completely, giving a 0.025-μg/g detection limit for both of those fruits. Such material is offered by the Strategic Diagnostics Company of Newark, Delaware, as part of an array of pesticide kits employing the magnetic particle technique.

Another emerging technology that has already proved itself as being useful when applied to ELISA analyses is supercritical fluid extraction (SFE; see Chapter 5 of this volume). This technique, which is more amenable to the less polar pesticides, has the advantage that very little or no organic solvent is used, and the analyte can be collected in water or buffer as it is carried away from the substrate by supercritical CO_2. For difficult to extract, usually more polar pesticides, a small amount of organic solvent modifier, such as methanol or acetonitrile, may need to be added to the CO_2. Such small amounts of organic solvent seldom affect the ELISA assay, however. Recent work performed at the California laboratories of the Midwest Research Institute showed that limited success could be had with SFE extraction of total diet study (TDS) samples, samples consisting of composited foods, and can be either fatty or nonfatty (44). The fatty food TDS sample used in this work fortified with nine pesticides—alachlor, aldicarb, atrazine, carbaryl, carbendazim, carbofuran, cyanazine, 2,4-D, and metolachlor—all at the 0.002-μg/g level. Best recoveries were seen for carbofuran (98%) and atrazine (118%); only 2,4-D was not recovered at all, as might be expected from its polar/ionic nature. The organophosphate insecticide chlorpyrifos was also investigated in separate experiments at the 0.001-, 0.01-, and 1.0-μg/g levels. Apparently there was a slight false-positive response at the lowest level, since recoveries averaged 134% at that level, with an RSD of

14%. Less variable and recoveries averaging 94% were observed for the higher levels, with RSDs averaging 9%, quite respectable for a magnetic particle-based ELISA.

Fatty food samples present problems for ELISA that can be remedied, at least to some extent, by a proper selection of cleanup techniques. In an earlier study reported by workers at the USDA Agricultural Research Service laboratories in Illinois, an array of pesticides were used to fortify beef liver, ground beef muscle, and lard (45). Extractions were conducted in both the dynamic and static modes using CO_2 without organic modifiers, and the pesticides included alachlor, atrazine, benomyl, carbofuran, and 2,4-D. Fortified water was used as a reference matrix. Tissue fortifications were done at the 0.02-μg/g level for alachlor and atrazine, 0.05 μg/g for carbofuran, 0.10 μg/g for benomyl, and 0.02 μg/g for 2,4-D. While static extraction minimized interferences from the meat fat, extremely variable results were observed; the dynamic mode gave much more reproducible results, but large interferences were observed for alachlor and carbofuran. Employment of a simple filtration step with a 0.5-μm Millex-LCR membrane filter (Millipore Corp., Bedford, MA) after extractant capture with aqueous solutions of methanol greatly alleviated the fat interferences. Speculation was that microscopic fat globules were filtered out while leaving the pesticides in solution, although losses were significant for alachlor and atrazine. Such a simple filtration step as well as a conventional SPE cleanup with a C_{18} cartridge were performed. All analyses were performed with the magnetic particle ELISA kits now supplied by Strategic Diagnostics. Refinement of this approach gave excellent recoveries of all pesticides studied, with method detection limits well below those set in the USDA's residue program for meats (Table 6.1) (46), along with a considerable savings in time when compared to the standard USDA method for carbamate pesticides, which

Table 6.1. SFE/EIA Results for Dynamic Extraction of Pesticide-Fortified Meat Products

	SFE/EIA		FSIS Residue Program	
Compound	Recovery ± SD (%)	MDL (ppb)	LDL Limit (ppb)	Residue (ppb)
Alachlor	118 ± 13	1	—	20
Carbofuran	93 ± 10	3	5	50
Atrazine	98 ± 2	1	5	20
Benomyl	101 ± 7	5	50	100
2,4-D	140 ± 35	14	200	200

employs methylene chloride extraction, gel-permeation chromatography, SPE cleanup, membrane filtration, and HPLC with postcolumn fluorogenic labeling for detection.

Only recently explored for ELISA analyses is the use of *microwave-assisted extraction* (MAE) using water as the extracting solvent (47). This technique, first reported by Ganzler et al. in 1986, relies on the strong absorption of microwave energy by the water molecule to displace adsorbed analytes in a vigorous and rapid manner, which are then dissolved by the hot water and made available for analysis (48). Although it has not been applied to foods yet, the reader should be aware of its potential, since it has been shown to be applicable for the extraction of Aroclors from soils, with subsequent analysis by magnetic particle ELISA kits, which, when compared to analyses done by conventional gas chromatgraphic techniques, gave correlation coefficients of 0.999 (see Ref. 47; see the following "Future Directions" section for a further discussion of this technique).

Advantages of the Technique

Immunoassay methods have been used long enough in the clinical laboratories such that there has been developed an array of EIA analyzers that, through automation, eliminate most, if not all, manual sample manipulations. Operations such as sample dispensing, reagent addition, shaking, washing, incubating, reading, and data reduction can now be fully automated (see below under "Available Hardware"). Such hardware availability makes one of the objections to ELISA—that it still requires too lengthy procedures—now easier to deal with. And initial costs of this type of analyzer are not prohibitive, as they are comparable to those for a dual-detector gas chromatograph equipped with an autosampler and data system. Analysis time, when calculated per sample, can actually be significantly reduced when compared to conventional chromatographic methods. As with chromatographic methods, total analysis time is still determined by sample extraction requirements, which can be extensive for foods, when compared to such samples as environmental water.

Solvent costs, human exposure both in and out of the laboratory, solvent disposal issues, and hazards related to storage of large volumes of solvents, as is typical in the conventional pesticide residue analytical laboratory, all make ELISA methodologies appear attractive. And in the United States there has been the impact of Executive Order 12856, which requires the reduction of 17 USEPA priority chemicals in an effort to reduce the potential for environmental pollution by the government sector (51). ELISA methods for pesticides use almost immeasurable amounts of organic solvents, making their cost and disposal a nonissue. Consumption of energy is also an issue in the United States, and although no objective comparsions have yet been made, ELISA methods consume far less electrical energy than do chromatographic methods, especially gas chromatographic ones, although once again, much needs to be done in reducing energy consumption during sample extraction and cleanup.

Once an ELISA kit has been produced and validated, it can be acquired and used at far less the cost than a typical chromatography-based method, particularly if analyst time is factored in, and this has been clearly demonstrated by the California Department of Food and Agriculture (43). Their assessment of reagents and student assistant time for the analysis of 2,4-D herbicide came to $13.21 per sample (apples, grapes, potatoes, oranges), whereas the conventional chromatographic analyses cost $200–$300 and took almost 2 days for a single analysis. Earlier work with ELISA on the rice herbicide molinate compared the cost of that technique per sample with simultaneous analyses done by gas chromatography. Gas chromatographic analyses performed in house were estimated to cost $50 and for those done at contracting laboratories, $130–$200, whereas the in-house ELISA analyses were estimated to cost $5–$6 per sample. The ELISA was performed in a 96-well plate and had a detection limit of 0.020 μg / mL (52). Another cost analysis was done on the use of ELISA for the triazine herbicides atrazine, simazine, and propazine in water (53). This assay had a detection limit of 0.2–2.0 μg/L, and cost about $15 per sample, with no false positives seen on confirmation of residues with conventional means.

Limitations of the Technique

There are two dominant limitations of the technique, one inherent in its conceptual design, and the other part of the makeup of the vast array of sample types we call "foods." Immunoassay-based techniques, when properly designed and implemented, get their strengths from the high specificity of a properly generated antibody for binding to the analyte pesticide. It is desirable, therefore, to minimize cross-reactivity to gain this high degree of selectivity. Consequently, ELISA-based methods, by nature, are "one pesticide at a time" methods. Since there is no other dimension available for resolution of a multiplicity of pesticides in a single sample, as there is in chromatography-based methods that employ the dimension of *time* to quantitate more than one pesticide in a sample, there must be a library of antibodies to sequentially perform pesticide residue analyses in separate analyses. Such sequential analysis becomes problematic as the numbers of pesticides in a particular sample become more numerous, leading to increasing costs in both reagents and time. Another major limitation for the analysis of foods is the relative chemical complexity of the sample type, which frequently gives rise to competing reactions between the pesticide and its intended antibody. This gives rise to *false positives* as have been identified in the previous sections, and which require confirmation by other approaches. Such false positives appear much less frequently in environmentally related sample types, especially water. False positives have been observed in nearly every type of food sample analyzed, and must be dealt with by implementation of additonal cleanup steps, standard additions, alternate extraction solvents, and other approaches. We have already seen the importance in proper solvent selection for extractions, and how the solvent itself can affect the pesticide–antibody interaction.

Typical well- or plate-type devices are well known for the inhomogeneity of their sorptive surfaces, such that binding of proteins to the plastic surface becomes variable from plate to plate, frequently giving rise to large amounts of variation from sample to sample. This may well be one of the very real handicaps well-type ELISA techniques have to suffer, and applies to a limited extent to tube-type assays.

One major limitation of the ELISA technique in the monitoring of pesticides in foods is that it cannot give data on a particular pesticide that is not anticipated before the method is put into action. For example, the chromatographic techniques, especially when they are coupled to mass spectrometry, frequently unearth pesticide residue data of great value on pesticides and other chemicals that are totally unanticipated. If the chromatograms are searched with care, unknown peaks sometimes reveal the presence of pesticides that are totally unexpected. Such serendipity cannot be seen for ELISA-based techniques, because of their inherent nature.

How ELISA Fits in with Other Methods

As pointed out by Hammock (12), those pesticides best suited for ELISA method development are those that are large in size, hydrophilic, stable, nonvolatile, and foreign in structure compared to matrix-containing compounds. With the successes recently documented for such pesticides as the triazines, carbamates, pyrethroids, organophosphates, and organochlorines (54), it appears as if these compounds may not have been the best candidates for ELISA method development. From water solubility characteristics alone, even better results might have been realized if some attention had been given to such pesticides as the imidazolinones, such as imazaquin, imazapyr, and imazilil, which are involatile, and which have limited absorbing chromophores. Residue methods for the imidazolinones are either on based capillary electrophoresis or HPLC, since they do not hold up under gas chromatographic temperatures. Consequently, ELISA is a good complement

Table 6.2. Information to be Included in a Written Analytical Method

A. Summary of method
  1. Performance claims
  2. Matrix
  3. Intended use
B. Principle of method
C. Analytical procedure
  1. Materials
    a. Equipment
    b. Supplies
    c. Reagents / stability of reagents
    d. Analytical standards / calibrators
    e. Safety and health hazards of materials
  2. Method
    a. Reagent preparation (including standards if applicable)
    b. Instrument settings
    c. Stepwise procedure(s)
      i. Source / characterization of control samples
      ii. Sample preparation
      iii. Extraction
      iv. Fortification (if applicable)
      v. Cleanup (if applicable)
      vi. Instrument calibration
    d. Data interpretation (i.e., decision criteria for detection)
    e. Interferences
      i. Cross-reactivity
      ii. Matrix effects
      iii. Solvent and labware effects
    f. Valid nonimmunochemical confirmatory method (if applicable)
    g. Time required for analysis
    h. Modifications / potential problems (e.g., critical steps)
    i. Calculations
      i. Recovery
      ii. Conversion / dilution factor
      iii. Statistics (standard deviation, percent coefficient of variation)
    j. Representative raw data
    k. Other information needed to provide thorough description of the method
D. Results and discussion: summary of results of validation experiments
  1. Accuracy
  2. Precision
  3. Detection limit
  4. Limit of quantitation (quantitative range)
  5. Selectivity and specificity
  6. Correlation to non-immunological method (if applicable)
  7. Ruggedness testing (if performed)
  8. Limitations
E. Conclusions
F. Tables and figures
G. References

to GC methodologies, which are particularly suited to small, nonpolar, and volatile analytes.

An effort is being made in the United States to promulgate high technical and ethical standards when immunoassay-based products are developed and offered for sale; this activity is being led by the Analytical Environmental Immunochemical Consortium (55). This consortioum is made up of agrichemical and immunochemical companies that use or develop immunochemical methods and hardware for environmental and food analysis. Among their many activities is the establishment of guidelines for the validation of immunoassay-based methods, including ELISA (56). As part of their validation guidelines, they have developed an outline for what is to be included in the method validation report, to be prepared by the ELISA kit manufacturer (Table 6.2). Among the method performance criteria are those for *selectivity and specificity, limit of quantitation, accuracy*, and *precision*. For ELISA-based methods, much more emphasis is placed on demonstrating their specificity than for conventional chromatographic methods. The ELISA method must be critically evaluated for interferences from other pesticides, metabolites, or degradates. In addition, interferences from typical sample matrixes, solvents, labware surfaces, and other foreign materials must be evaluated. Correlation experiments should be performed when validating an ELISA method, which consists of fortifying control matrix with the pesticide of interest and then performing both ELISA and conventional chromatographic analyses on them. Statistical correlations between the two sets of data should then be made, and any false positives generated during the experiment should be noted.

RECENT ADVANCES

Recombinant Fab Antibodies

There has been rapid development since 1990 in obtaining and manipulating immunoglobin genes by recombinant antibody technologies and expressing them in a variety of hosts, including the popular *Escherichia coli* (57,58). Several advantages can be realized by employing recombinant fab (rFab) in ELISA methods, compared to MAbs and PcAbs. The rFabs are physically smaller then typical antibodies generated by immunization techniques, and usually have better physical and chemical properties. Affinities and specificities can be modified by generating various combinations of heavy and light chains or by employing mutagenisis. Antibodies having specificity for more than one ligand, called *diabodies*, can be made using rFab technologies. And a significant difference between the MAb and the rFab antibodies is that the latter is available for selection and modification of antibody properties without requiring reimmunization of the animal. Thus, the specificity and affinity of rFab antibodies can be designed to fit the requirements for food and environmental analyses.

Recent work at the University of California, Berkeley, used rFab technology to produce a water-soluble rFab consisting of the immunoglobulin V_H–C_{H1} (H) and V_L–C_L (L) sequences from hybridoma cell line 481 (59). They evaluated the performance of the various rFab antibodies that were generated with a previously used EIA method (60). Although the I_{50} values were approximately the same when diuron analyses were done using EIA with either MAbs or rFabs, at about 0.002 μg / mL in water, the rFab antibodies were much more sensitive to methanol in the binding step. Whereas the MAb could tolerate as much as 30% methanol without antibody inhibition taking place, the rFab was completely inhibitied at that methanol concentration. Cross-reactivities with other urea herbicides, monuron and linuron, were also considerably higher using the rFabs than when using the MAbs. No applications for the analysis of foods were reported, however. There has been an application that employs single-chain variable-fragment antibodies (scFv), a variant of the rFab approach, to the analysis of potato glycoalkaloids (61).

Polyclonal antibodies continue to hold the interest of many for the analysis of pesticide residues in foods (62). Wheat and barley products, including beer, barley cereal, wheat cereal, and wheat bread, were analyzed for the herbicide difenzoquat [1,2-dimethyl-3,5-diphenylpyrazolium methylsulfate], used for the control of wild oats in cereal crops. Pcabs from New Zealand white rabbits were generated by immunization with haptens having a 5-carbon spacer arm attached to one of the phenyl rings distal to the pyrazole center of the molecule. Advantage was taken of the water solubility of the cationic pyrazolium structure, so that 2 N hydrochloric acid could be used for extraction, which was subsequently neutralized before analysis by ELISA using the wells/plate approach. Of all pyridyl and bipyridyl herbicides examined for affinity and cross-reactivity, only cyperquat, a phenyl pyridyl, produced a response, but only at very high concentrations (2.1 μg / mL), compared to the IC_{50} of difenzoquat, 0.3 ng / mL. The developers of the method made clever use of a "matrix modifier," lager beer, for all the beer analyses. Water was used for the solid grain samples. Recoveries for samples of bread, the most difficult to analyze, ranging from 0.01 to 0.2 μg / g, were 72–98% for 12 different types of commercial breads. Fortified beers gave better recoveries, generaly in the 90% range. Precision was comparable to that for chromatographic techniques, with only a few exceptions.

Little attention has been given to luminescence techniques applied to ELISA methods for pesticides. This is somewhat surprising, given theoretical limits of detection of the approach, and given how it can be adapted quite easily to microtiter surface sample containment devices (63). Some attention has been given to molecular fluorescence techniques, and in particular polarization fluoroimmunoassay (PFIA), but little improvements were seen in method simplification, limits of detection, or specificity (64,65). Recently, a microformat imaging ELISA technique has been described that takes full advantage of microtiter plate technologies now available, contemporary luminescence measuring devices, and the power of enzyme linking (66). These authors chose to use a recently developed 1,2-dioxetene substrate, CSPD, to be acted on by the alkaline phosphatase enzyme, AP. In their scheme, antibody is preincubated with the antigen 2,4-D and the mixture is added to a flat-well support coated with 2,4-D–BSA conjugate, antimouse AP conjugate added. The 1,2-dioxetene substrate, CSPD, was added and the chemiluminsence measured with a Photometrix 200 CCD (charge-coupled device) camera, fitted with a 50 mm AF NIKKOR 1 : 1.8 objective; a 90-s exposure time was used at 542 nm. The Photometrix 200 CCD camera is a very low-noise device with cyrogenic cooling of the photomultiplier tube, which is capable of measuring very low light levels that are concurrent with small sample chemiluminescent measurements. This CCD camera-based ELISA was capable of reliably measuring 2,4-D herbicide at 0.006 ng / mL in water, almost 10,000 times more sensitive than a conventional colorimetric ELISA method. Another distinguishing feature of this approach was the good linearity of the standard calibration curve over 5 decades of concentration, even though the coefficient of variation, CV, began to climb steeply after the first 3 decades (Fig. 6.9). Also very significant about this method was the use of microscope slides with a proprietary hydrophobic layer, consisting of 2-mm^2 squares separated by 1 mm lines of hydrophobic surface. The hydrophobic areas where the sample was applied were only 100 μm thick.

The utility of the magnetic particle technique was recently demonstrated for the analysis of the insecticide methomyl in water and on grape leaves (67). For the grape leaves, only dislodgable foliear residues (DFR) were determined, and were removed from 1-in.-diameter organically grown leaf disks after the methomyl was added to each leaf in solution and allowed to dry overnight. A surfactant solution (0.04% Aerosol-OT) was used in 50 mL volumes to remove the residue by shaking. Of the 35 pesticides examined for cross-reactivity, only thiodicarb, a dimer of methomyl through the sulfur atom, showed any significant level (37%). Among those tested were aldicarb, aldicarb sulfone, and oxamyl. The major metabolite of methomyl, methomyl oxime, was less than 0.01% cross-reactive. Dilutions of the surfactant used for leaf extractions showed that it gave no matrix effects at a 50 : 1 dilution and lower. Recoveries of methomyl ranged from 86 to 109% at

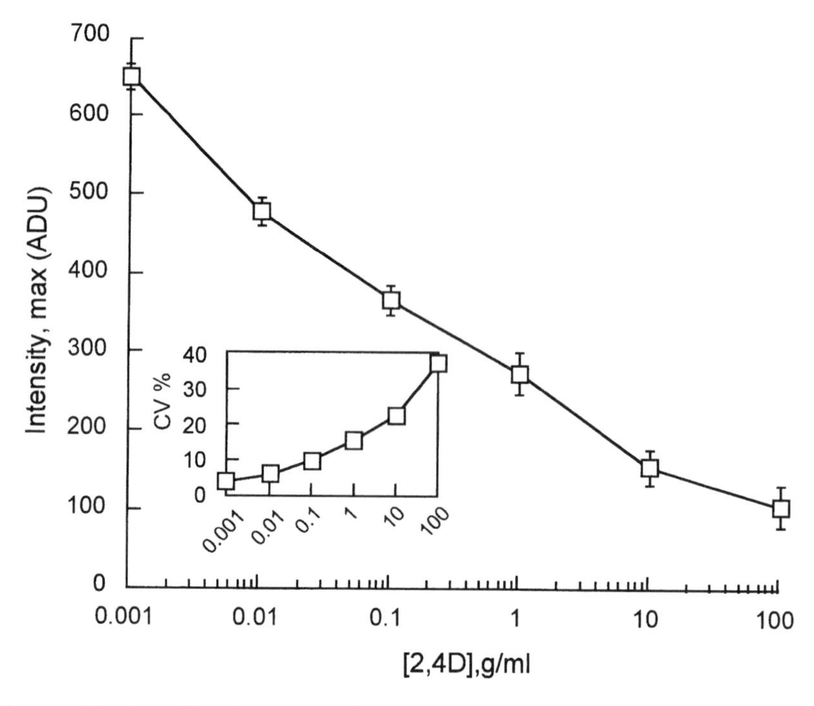

Figure 6.9. A calibration curve for the chemiluminescent analysis of 2,4-D herbicide illustrating the extended analytical range for this type of ELISA technique. (From Ref. 66.)

concentrations ranging from 0.025 to 0.10 μg / cm^2. Analysis time was less than one hour, comparing favorably with HPLC analysis, and can be used to measure methomyl residues well below regulatory levels following application to grape vines.

AVAILABLE HARDWARE

Pesticide ELISA kits and reagents available today are becoming increasingly more attractive for routine residue determinations, largely beause of the improvements in the hardware associated with reading microplates. Complete microplate analytical systems are now available, and at reasonable prices, which greatly extends the usefulness of microplate techniques, not only for pesticide residue analyses but also for many analyses performed in clinical chemistry laboratories, including tumor markers, infectious diseases, thyroid, fertility and steroid testing, and autoimmune screening. These systems are also finding use in veterinary, food chemistry, environmental, life science, and industrial laboratories. Some systems are truly "walkaway" in nature, which incorporate reagent addition, plate shaking, washing, incubating, reading, and data reduction steps. Systems having associated personal computer hardware and the capability of user programming complete methods, utilization of packaged quality control (QC) programs, creation of patient files, and the control of multiple assay methods simultaneously are also available. Systems can be assembled from modules or purchased as units.

One of the more versatile and sophisticated systems is the Bio-Rad CODA System (Hercules, CA), shown in Figure 6.10. This system is a unitized instrument that is PC-controlled and compatible with most ELISAs that are microplate-based. It can run as many as 270 or as few as 1 or 2 samples, and can manage up to 5 different assays simultaneously, with assay overlap for maximum efficiency. The steps in the program can be performed individually or in a programmed sequence. It comes equipped with a Coda Operation and Data Management software package on CD-ROM, which is upgraded by the

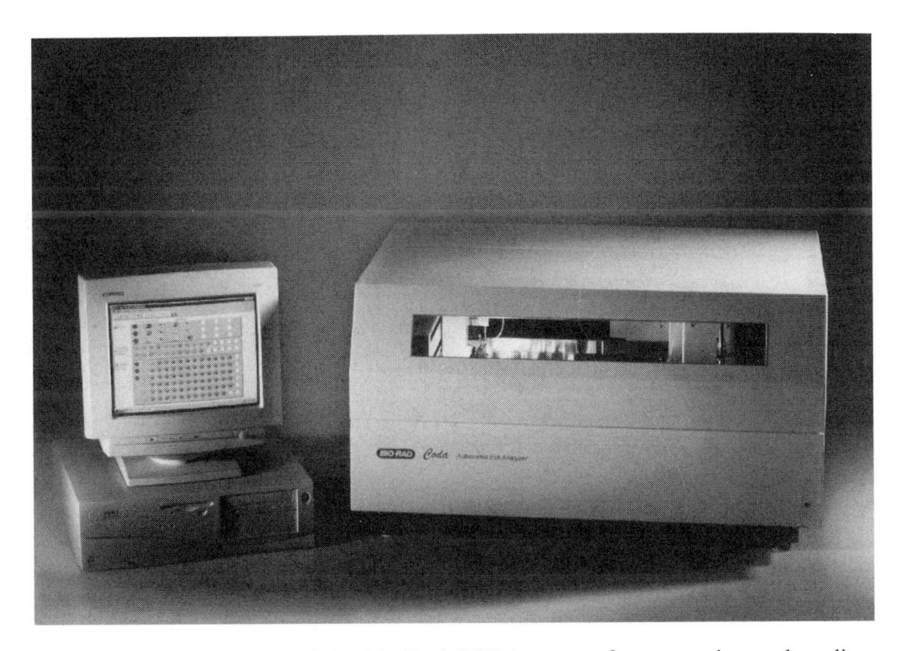

Figure 6.10. Photograph of the Bio-Rad CODA system for processing and reading ELISA microplates.

manufacturer periodically. This software operates in the Microsoft Windows 95 environment, and provides for setting up assays, monitoring of assay progress, generation of QC (quality-control) files, and generation of calibration curves. Calibration curves can be viewed, edited, overlaid, and saved to disk.

A more modular system, the Hyperion MicroReader 4/HyPrep Plus Workstation, is shown in Figure 6.11. Its deck allows for the configuration of multiple rack configurations for each assay, including tube-to-microplate, tube-to-tube-to-microplate, microplate-to-microplate, or tube-to-tube configuration. The use of a unique test tube and microplate location program allows the user to move the probe and center it over a location and enter its location by hitting ENTER on the PC. Its computer software, PLUS MULTITEST, allows for flexibility in choosing and entering assay procedures, including dilutions, sample transfers, reagent additions, incubation times, and plate washes. It can process up to four different assays at one time as well as process variable dilution factors for each assay's dispensing and aspirating routine. On-line help files with dialog boxes and the ProbeWalker feature provide for user friendliness. A final report of the samples including their well assignments can be printed or sent to a disk file. It is a walkaway

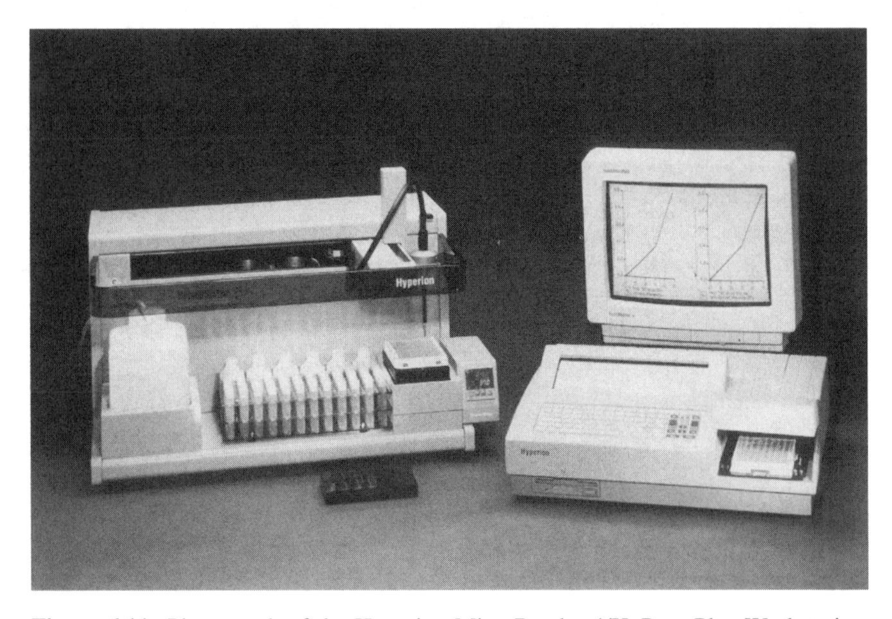

Figure 6.11. Photograph of the Hyperion MicroReader 4/HyPrep Plus Workstation for processing and reading ELISA microplates.

Table 6.3. Immunoassay Materials Commercially Available for Various Pesticides

Pesticide	Kit Type[a] Plate	Tube	Ticket[b]	Cassette	Antibody Only
2,4-D	1, 3, 5	1, 5	—	—	—
Acephate	—	—	4	—	—
Acetanilides, general	1, 3, 5	1	—	—	—
Acetochlor	1[c]	1[c]	—	—	—
Alachlor	1,[c] 5	1,[c] 5	—	—	—
Aldicarb	3, 5	5	4	—	—
Azinphos-methyl	—	—	4	—	—
Atrazine	1, 3	1, 5	—	—	2
Bendiocarb	—	—	4	—	—
Benomyl	1, 5	1, 5	—	—	—
Bioresmethrin	5	—	—	—	—
Captan	—	5	—	—	—
Carbaryl	—	5	4	—	—
Carbendazim	1	1	—	—	—
Carbofuran	—	5	4	—	2
Chlordane	—	5	—	—	—
Chlorothalonil	—	5	—	—	—
Chlorpyrifos-ethyl	—	—	4	—	—
Chlorpyrifos-methyl	—	5	4	—	—
Chlorpyrifos	1, 3, 5	1, 5	—	—	—
Chlorsulfuron	5	—	—	—	—
Cyanazine	3,[c] 5	5	—	—	—
Cyclodienes, general	1,[c] 5	1,[c] 5	—	—	—
DDT	3[b]	5	—	—	—
DDVP(dichlorvos)	—	—	4	—	—
Deet	3[c]	—	—	—	—
Demeton	—	—	4	—	—
Diazinon	1, 5	1	4	—	—
Dicrotophos	—	—	4	—	—
Dimethoate	—	—	4	—	—
Disulfoton	—	—	4	—	—
Endosulfan	5	—	—	—	—
EPN	—	—	4	—	—
Ethion	—	—	4	—	—
Ethoprop	—	—	4	—	—
Fenamiphos	—	—	4	—	—
Fenitorthion	5	5	4	—	—
Fenthion	—	—	4	—	—
Fluometuron	3	—	—	—	—

Table 6.3. *(contd.)*

Pesticide	Kit Type[a]				Antibody Only
	Plate	Tube	Ticket[b]	Cassette	
Fonofos	—	—	4	—	—
Formetanate	—	—	4	—	—
Hydrochloride	—	—	4	—	—
Imazapyr	5	—	—	—	—
Imidacloprid	3[c]	—	—	—	—
Isoproturon	3,[c] 5	—	—	—	—
Lindane	—	5	4	—	—
Malathion	—	—	4	—	—
Metalaxyl	3, 5	—		—	—
Methamidophos	—	—	4	—	—
Methidathion	—	—	4	—	—
Methiocarb	—	—	4	—	—
Methomyl	—	5	4	—	—
Methoprene	3,[c] 5	—	—	—	—
Methyl parathion	—	—	4	—	—
Metolachlor	1, 5	1, 5	—	—	—
Metribuzin	—	5	—	—	—
Metsulfuron	5	—	—	—	—
Mevinphos	—	—	4	—	—
Molinate	5	—	—	—	—
Monocrotophos	—	—	4	—	—
Naled	—	—	4	—	—
Oxamyl	—	—	4	—	—
Oxydemeton-methyl	—	—	4	—	—
Paraquat	3[c]	5	—	—	2
Parathion-ethyl	—	—	4	—	—
Parathion-methyl	—	—	4	—	—
Parathion	3,[c] 5	—	4	—	—
Phostebupirim	—	—	4	—	—
Phorate	—	—	4	—	—
Phosalone	—	—	4	—	—
Phosmet	—	—	4	—	—
Phosphamidon	—	—	4	—	—
Phosuchlor	—	—	4	—	—
Picloram	—	5	—	—	—
Pirimiphos-methyl	5	5	4	—	—
Procymidone	—	5	—	—	—
Propoxor	—	—	4	—	—
Prosulfuron	1	1	—	—	—
Silvex	—	5	—	—	—

Table 6.3. *(contd.)*

Pesticide	Kit Type[a] Plate	Tube	Ticket[b]	Cassette	Antibody Only
Simazine	—	5	—	—	2
Spinosad	—	5	—	—	—
Sulprfos	—	—	4	—	—
Terbufos	—	—	4	—	—
Thiabendazole	5	5	—	—	—
Triasulfuron	5	—	—	—	—
Triazines, general	1, 3, 5	1, 5	—	3	2
Trichlorfon	—	—	4	—	—
Trichloropyridinol	3	5	—	—	—
Triclopyr	—	5	—	—	—
Urea herbicides	5	—	—	—	—

[a] *Key*: (1) Beacon Analytical Systems, Inc., 4 Washington Avenue, Scarborough, ME 04074 (207/885-0440); (2) Biodesign International, 105 York Street, Kennebunk, ML 01013 (207/985-1944); (3) EnviroLogix Incorporated, 55 Industrial Way, Portland, ME 04103 (207/797-0300); (4) Neogen Corporation, 620 Lesher Place, Lansing, MI 48912 (517/234-5333); (5) Strategic Diagnostics Inc., 111 Pencader Drive, Newark, DE 19702 (800/544-8881).
[b] Test specific for cholinesterase inhibitors.
[c] In development.

system with temperature incubation, plate shaking, washing, and reagent addition all done without analyst attention.

The number of United States' based companies supplying ELISA kits and reagents for the analysis of pesticides have dwindled in the last 10 years, because of uncertainties in product demand resulting from progress in adoption of competing analytical techniques, government regulations, and a somewhat diminished emphasis on contaminant monitoring in general. Consequently, many companies who were active at that time now are no longer in business or have been bought out and merged with other companies. However, those that are still providing ELISA associated materials continue together to cover a comprehensive array of pesticides, as can be seen in Table 6.3. There are 90 different pesticides represented in this array, with an almost equal distribution between plate and tube kits. As can be seen, one company, Biodesign International, supplies only antibodies, and another, Neogen Corporation, supplies kits only in the "ticket" format, which provides only qualitative, or screening, information. Another company, EnviroLogix, manufactures a "cassette," which is designed to provide screening of water samples for an array of triazines. They are also developing a "lateral flow membrane" device, which will indicate human exposure to a pesticide, and which works much like a home pregnancy device, having a configuration similar to a piece of chewing gum.

Thus, the pesticide analytical chemist now has a powerful array of ELISA resources available from which to choose. Whether these resources continue to proliferate is open for speculation.

FUTURE DIRECTIONS AND CONCLUSIONS

Although still providing the pesticide residue analytical chemist with a significant challenge, ELISA techniques, methods, and devices have proved themselves to perform comparably to conventional chromatography-based methods. The quality of the residue data, as demonstrated by a growing number of food applications, has improved steadily, although there still remain problems with irreproducibility arising from adsorptive surface variations from plate to plate and from tube to tube. Many chemists, however, still feel the ELISA format, whether direct or indirect, is too involved, with too many places for error to occur. To them an HPLC or GC appears much simpler to understand and operate. Many immunochemists feel that a more direct route to immunoassays ought to be investigated than the ELISA approach, such as the magnetic particle/ruthenium/electron-transfer approach now offered by one of the immunoassay suppliers (68), or one of the other approaches at using biosensors. Clearly, the key lies in developing antibodies that have minimum cross-reactivity, can be produced inexpensively, and can be stored for long periods of time without degradation. Some immunochemists feel that these criteria point to MAbs, while others say that PAbs are the answer. Almost everyone believes that the short-chain antibodies produced by recombinant techniques are not stable enough, and will not pass the test of time.

Many of the author's colleagues believe that the fate of ELISAs and immunoaffinity tehniques in general for foods and environmental samples lie in the hands of those making the regulations, and will ultimately be driven by analytical method costs. Such considerations become more important when the analytical work load goes up, and when the private sector is required to perform more analyses to prove that their products are environmentally safe and pose no hazard to human health. Immunochemical methods are so popular in the clinical chemical field because the individuals place extreme value on knowing the condition of their health, and will pay for it through insurance premiums. How individuals feel about paying for food monitoring and environmental testing will ultimately determine the fate of ELISA and all immunochemical products for pesticide residue analyses.

REFERENCES

1. L. D. Sawyer, The development of analytical methods for pesticide residues, in *Pesticide Residues in Foods: Technologies for Detection*, Office of Technology Assessment, Congress of the United States, OTA-F-398, U.S. Government Printing Office, Washington, DC, Oct. 1988.
2. J. N. Seiber, Conventional pesticide analytical methods: can they be improved, in *Pesticide Residues in Foods: Technologies for Detection*, Office of Technology Assessment, Congress of the United States, OTA-F-398, U.S. Government Printing Office, Washington, DC, Oct. 1988.
3. P. A. Mills, J. H. Onley, and R. A. Gaither, *J. AOAC* **46**, 186–191 (1963).
4. M. A. Luke et al., *J. AOAC* **58**, 1020–1026 (1975).
5. M. A. Luke et al., *J. AOAC* **64**, 1187–1195 (1981).
6. B. M. McMahon and R. F. Wagner, eds., *Pesticide Analytical Manual*, Vol. I, *Multiresidue Methods*, 3rd ed. U.S. Food and Drug Administration, Washington, DC, 1993.
7. *Pesticide Residues in Foods: Technologies for Detection*, Office of Technology Assessment, Congress of the United States, OTA-F-398, Washington, DC: U.S. Government Printing Office, Oct. 1988.
8. B. M. Kaufman and M. C. Clower, Jr., *J. AOAC* **74**, 239–247 (1991).
9. B. M. Kaufman and M. C. Clower, Jr., *J. AOAC* **78**, 1079–1090 (1995).
10. J. M. Van Emon and Viorica Lopez-Avila, *Anal. Chem.* **64**, 79A–88A (1992).
11. R. G. M. Wang, C. A. Franklin, R. C. Honeycutt, and J. C. Reinert, eds., *Biological Monitoring for Pesticide Exposure*, ACS Symposium Series No. 382, American Chemical Society, Washington, DC, 1989.
12. B. D. Hammock, S. J. Gee, R. O. Harrison, F. Jung, M. H. Goodrow, Q. X. Li, A. D. Lucas, A. Szekacs, and K. M. S. Sundraam, Immunological technology in environmental analysis: addressing critical problems, in *Immunological Methods for Environmental Analysis*, J. M. Van Emon and R. O. Mumma, eds., ACS Symposium Series 442, American Chemical Society, Washington, DC, 1990, pp. 112–139.
13. R. L. Ellis, Rapid test methods for regulatory programs, in *Immunoassays for Residue Analysis: Food Safety*, R. C. Beier and L. H. Stanker, eds., ACS Symposium Series 621, American Chemical Society, Washington, DC, 1996, pp. 44–58.
14. *Residues of Veterinary Drugs in Food*, 2nd ed., Codex Alimentarius Commission; Joint FAO/WHO Food Standards Programme, Food and Agriculture Organization of the United Nations, Rome, Italy, 1993.
15. C. D. Ercegovich, in *Analysis of Pesticide Residues: Immunological Techniques*, American Chemical Society, Washington, DC, 1971, pp. 162–177.
16. B. D. Hammock and R. O. Mumma, in *Potential of Immunochemical Technology for Pesticide Analysis*, J. Harvey and G. Zweig, eds., American Chemical Society, Washington, DC, 1980, pp. 321–351.

17. A. Y. Al-Rubae, *The Enzyme-Linked Immunosorbent Assay, A New Method for the Analysis of Pesticide Residues*, Ph.D. thesis, Pennsylvania State Univ., 1978.
18. C. D. Ercegovich, in *Pesticide Identification at the Residue Level*, Advances in Chemistry Series 104, R. F. Gould, ed., American Chemical Society, Washington, DC, 1976.
19. C. D. Ercegovich, R. P. Vallejo, and R. R. Gettig, *J. Agric. Food Chem.* **29**, 559–563 (1981).
20. W. H. Newsome, *J. Agric. Food Chem.* **33**, 528–530 (1985).
21. W. H. Newsome, *Bull. Environ. Contam. Toxicol.* **36**, 9–14 (1986).
22. W. H. Newsome, in *Pesticide Science and Biotechnology*, R. Greenhalgh and T. R. Roberts, eds., Blackwell Scientific Publications, Oxford, U.K., 1987.
23. B. D. Hammock, in *Biotechnology for Crop Protection*, P. A. Hedin, J. J. Menn, and R. M. Hollingworth, eds., ACS Symposium Series 379, Washington, DC, 1988.
24. F. S. Chu, in *Analyzing Food for Nutrition Labeling and Hazardous Contaminants*, I. J. Jeon and W. G. Ikins, eds., Marcel Dekker, New York, 1995, pp. 283–332.
25. W. R. Clark. *The Experimental Foundations of Modern Immunology*, 4th ed., Wiley, New York, 1991.
26. I. M. Roitt, *Essential Immunology*, 7th ed., Blackwell, Oxford, U.K., 1991.
27. G. Kohler and C. Milstein, *Nature* **256**, 495–497 (1975).
28. F. Jung, S. J. Gee, R. O. Harrison, M. H. Goodrow, A. E. Karu, A. L. Braun, Q. X. Li, and B. D. Hammock, *Pest. Sci.* **26**, 303–317 (1989).
29. S. J. Gee, R. O. Harrison, M. H. Goodrow, A. L. Braun, and B. D. Hammock, in *Immunoassays for Trace Chemical Analysis*, M. Vanderlaan, L. H. Stanker, B. E. Watkins, and D. W. Roberts, eds., ACS Symposium Series 451, Washington, DC, 1991.
30. R. P. Vallejo, E. R. Bogus, and R. O. Mumma, *J. Agric. Food Chem.* **30**, 572–580 (1982).
31. S. J. Gee, T. Miyamoto, M. H. Goodrow, D. Buster, and B. D. Hammock, *J. Agric. Food Chem.* **36**, 863–870 (1988).
32. A. E. Karu, D. J. Schmidt, C. E. Clardson, J. W. Jacobs, T. A. Swanson, M. L. Egger, R. E. Carlson and J. M. Van Emon, in *Immunochemical Methods for Environmental Analysis*, J. M. Van Emon and R. O. Mumma, eds., ACS Symposium Series 442, American Chemical Society, Washington, DC, 1990, pp. 95–111.
33. J. J. Manclús, J. Primo, and A. Montoya, *J. Agric. Food Chem.* **44**, 4052–4062 (1996).
34. Z. Niewola, C. Hayward, B. A. Symington, and R. T. Robson, *Clin. Chim. Acta* **148**, 149–156 (1985).
35. M. Oellerich, *J. Clin. Chem. Clin. Biochem.* **18**, 197–208 (1980).
36. J. A. Itak, M. Y. Selisker, and D. P. Herzog, *Chemosphere* **24**, 11–21 (1992).

37. J. F. Brady, in *Immunoanalysis of Agrochemicals*, J. O. Nelson, A. E. Karu, and R. B. Wong, eds., ACS Symposium Series 586, Washington, DC, 1995, pp. 266–287.
38. D. Rodbard, in *Ligand Assay*, J. Langan and J. J. Clapp, eds., Masson Publishing, New York (1981), pp. 45–101.
39. J. H. Skerritt and B. E. Amita Rani, in *Immunoassays for Residue Analysis: Food Safety*, R. Beier and L. H. Stanker, eds., ACS Symposium Series 621, American Chemical Society, Washington, DC, 1996, pp. 29–43.
40. S. W. Jourdan, A. M. Scutellaro, M. C. Hayes, and D. P. Herzog, in *Immunoassays for Residue Analysis: Food Safety*, R. Beier, and L. H. Stanker, eds., ACS Symposium Series 621, American Chemical Society, Washington, DC, 1996, pp. 17–28.
41. S. Cairoli, A. Arnoldi, and S. Pagani, *J. Agric. Food Chem.* **44**, 3849–3854 (1996).
42. W. H. Newsome, J. M. Yeung, and P. G. Collins, *J. Assoc. Off. Anal. Chem.* **76**, 381–386 (1993).
43. S. J. Richman, S. Karthikeyan, D. A. Bennett, A. C. Chung, and S. M. Lee, *J. Agric. Food Chem.* **44**, 2924–2929 (1996).
44. V. Lopez-Avila, C. Charan, and J. Van Eamon, in *Immunoassays for Residue Analysis: Food Safety*, R. Beier and L. H. Stanker, eds., ACS Symposium Series 621, American Chemical Society, Washington, DC, 1996, pp. 439–449.
45. K. S. Nam and J. W. King, *J. Agric. Food Chem.* **42**, 1469–1474 (1994).
46. J. W. King and K. S. Nam, in *Immunoassays for Residue Analysis: Food Safety*, R. Beier and L. H. Stanker, eds., ACS Symposium Series 621, American Chemical Society, Washington, DC, 1996, pp. 422–438.
47. V. Lopez-Avila and J. Benedicto, *Trends Anal. Chem.* **15**, 334–340 (1996).
48. K. Ganzler, A. Salgo, and K. Valko, *J. Chromatogr.* **371**, 299–306 (1986).
49. G. Jones, M. Wortberg, S. B. Kreissig, D. S. Bunch, S. J. Gee, B. D. Hammock, and D. M. Rocke, *J. Immunol. Meth.* **177**, 1–17 (1994).
50. P. Y. K. Cheung, L. M. Kauvar, A. E. Engqvist-Goldstein, S. M. Ambler, A. E. Karu, and L. S. Ramos, *Anal. Chim. Acta* **282**, 181–190 (1995).
51. R. L. Ellis, in *Immunoassays for Residue Analysis: Food Safety*, R. Beier and L. H. Stanker, eds., ACS Symposium Series 621, American Chemical Society, Washington, DC, 1996, pp. 44–58.
52. R. O. Harrison, A. L. Braun, S. J. Gee, D. J. O'Brian, and B. D. Hammock, *Food Agric. Immunol.* **1**, 37–51 (1989).
53. E. M. Thurman, M. Meyer, M. Pomes, C. A. Perry, and A. P. Schwab, *Anal. Chem.* **62**, 2043–2048 (1990).
54. B. M. Kaufman and M. Clower, Jr., *J. Assoc. Off. Anal. Chem.* **78**, 1079–1090 (1995).
55. J. Rittenburg and J. Dautlick, in *Immunoanalysis of Agrochemicals*, J. O. Nelson, A. E. Karu, and R. B. Wong, eds., ACS Symposium Series 586, Washington, DC, 1995, pp. 301–307.

56. C. A. Mihaliak and S. A. Berberich, in *Immunoanalysis of Agrochemicals*, J. O. Nelson, A. E. Karu, and R. B. Wong, eds., ACS Symposium Series 586, Washington, DC, 1995, pp. 288–300.
57. J. McCafferty, A.D. Griffiths, G. Winter, and J. D. Chiswell, *Nature* **348**, 552–554 (1990).
58. C. F. Barbas and R. A. Lerner, *Methods: A Companion to Methods in Enzymology* **2**, 119–124 (1991).
59. K.-B. G. Scholthof, Guisheng Zhang, and A. E. Karu, *J. Agric. Food Chem.* **45**, 1509–1517 (1997).
60. A. E. Karu, M. H. Goodrow, D. J. Schmidt, B. D. Hammock, and M. W. Bigelow, *J. Agric. Food Chem.* **42**, 301–309 (1994).
61. C. Kamps-Holtzapple and L. H. Stanker, in *Immunoassays for Residue Analysis: Food Safety*, R. Beier and L. H. Stanker, eds., ACS Symposium Series 621, American Chemical Society, Washington, DC, 1996, pp. 485–499.
62. J. M. Yeung, R. D. Mortimer, and P. G. Collins, *J. Agric. Food Chem.* **44**, 376–380 (1996).
63. G. M. Whitesides and P. F. Laibinis, *Langmuir* **6**, 87–95 (1990).
64. S. A. Eremin, J. Landon, D. S. Smith, and R. Jackmann, in *Proc. of Food Safety and Quality Assurance: Applications of Immunoassay Systems*, Bowness-on-Windermer, Cumbria, U.K., March 19–22, 1991, Elsevier, London.
65. F. G. Sanchez, A. Navas, F. Alonso, and J. Lovillo, *J. Agric. Food Chem.* **41**, 2215–2219 (1993).
66. A. Dzgoev, M. Mecklenburg, P.-O. Larsson, and B. Danielsson, *Anal. Chem.* **68**, 3364–3369 (1996).
67. J. A. Itak, M. Y. Selisker, C. D. Root, and D. P. Herzog, *Bull. Env. Contam. Toxicol.* **57**, 270–277 (1996).
68. ORIGEN, IGEN International, Gaithersburg, MD, USA.

CHAPTER

7

REGULATORY ASPECTS: PESTICIDE REGISTRATION, RISK ASSESSMENT AND TOLERANCE, RESIDUE ANALYSIS, AND MONITORING

W. GEORGE FONG

INTRODUCTION

Advanced technology in food production has allowed for an adequate food supply for a rapidly increasing population in the world, particularly in developed countries. One of the many factors contributing to this success has been the judicious use of farm chemicals.

Farmers first began using synthetic pesticides on their crops in the mid-1800s. It wasn't until about 1910 that pesticide use in the United States entered the realm of public policy, and in 1920 health concerns arose about the effects of pesticides on the human population. This concern brought about pesticide residue analyses procedures for human food and food-producing animal feed products (1).

Significant developments in analytical methodology began in 1955. This coincided with the development and application of new classes of pesticides, specifically, from natural and inorganic chemicals to halogenated hydrocarbons, organic phosphates, and eventually to carbamate and synthetic pyrethroid pesticides. The introduction of gas chromatography (GC) in the early 1960s, along with the introduction of high-pressure liquid chromatography (HPLC), revolutionized multiresidue methodology (2).

Numerous biological pesticides were introduced along the way with synthetic chemical pesticides. They include biochemical and microbial products. Residues of biological pesticides are generally dissipated rapidly.

The U.S. Environmental Protection Agency (EPA) registers pesticides and establishes tolerances for pesticide residues in food and feed under the Federal Insecticide, Fungicide and Rodenticide Act (FIFRA) and the Federal

Pesticide Residues in Foods: Methods, Techniques, and Regulations, by W. G. Fong, H. A. Moye, J. N. Seiber, and J. P. Toth. Chemical Analysis Series, Vol. 151
ISBN 0-471-57400-7

Food, Drug, and Cosmetic Act (FFDCA) (3,4). The U.S. Food and Drug Administration (FDA), and to a lesser extent, the U.S. Department of Agriculture (USDA), enforce these tolerances for use in the United States. Pesticide registrants seeking an EPA registration must submit an analytical method suitable for enforcement of the tolerance, among other requirements.

Regulatory laboratories do not have pesticide application history for the majority of the samples analyzed; therefore, they must have the capability to analyze hundreds of pesticide chemicals that could possibly remain on the samples. In order to complete the analyses in a reasonable amount of time, the regulatory laboratories employ multiresidue methodology (5).

Until recently, method development for pesticide residue in foods was driven by laws and regulations that enforce established legal limits (tolerances). However, since the 1980s, pesticide residue method development has been driven by an additional factor, the consumer demand.

In this section on regulatory aspects, all regulatory facets of pesticides will be described, including registration data requirement, risk assessment, tolerance establishment and enforcement, analytical requirement quality assurance, pesticide residue database, and laboratory accreditation.

SECTION I. GOVERNMENT REGULATIONS

A. Pesticide Registration

1. Federal

All pesticides for sale or distribution must be registered under the Federal Insecticide, Fungicide, and Rodenticide Act (FIFRA). Section 3 of FIFRA sets forth the standard requirements for pesticide registrations. When Congress enacted the law, they recognized that pesticide uses could yield both risk and benefits. It directed the designated federal agency to consider both in deciding whether to permit particular uses of a pesticide. In granting the registration, it must be proved that the food production benefits outweigh the risks.

In Section 3 of the FIFRA, the registrant must demonstrate that the pesticide can accomplish its intended effect without causing unreasonable adverse effect (risk) to humans or the environment, and the agency, in making the registration decision, must take into account the economic, social, and environmental costs and benefits of the use of the pesticide.

Until the establishment of the Environmental Protection Agency (EPA) in December 1970, pesticide registration was the responsibility of the Pesticide Regulation Division of USDA, in conjunction with FDA reviewing more extensive toxicology and residue data. Currently, the Office of Pesticide Programs of the EPA is responsible for the registration of pesticides and the establishment of pesticide residue tolerances.

The procedures and requirements for registration and associated regulatory activities of a pesticide are detailed in Part 152, Chapter I of the *Code of Federal Regulations* (*CFR*) (6), Title 40, Protection of Environment. This part sets forth procedures, requirements, and criteria concerning the registration and reregistration of pesticide products under Section 3 of the Federal Insecticide, Fungicide and Rodenticide Act, and for associated regulatory activities affecting registration. These latter regulatory activities include data compensation and exclusive use and the classification of pesticide uses. This part also describes the requirements applicable to intrastate products that are not federally registered. Exemptions for pesticides of a character not requiring FIFRA registration are also included in this section.

Data requirements for registration are listed in Part 158. Table 7.1 lists the various guidelines of the pesticide program.

Data requirements under Residue Chemistry for Food Crop production by terrestrial, aquatic, or greenhouse are as follows:

Chemical identity
Direction for use
Nature of the residue: plants, livestock
Residue analytical method
Magnitude of the residue
  Crop field trials
  Processed food, feed
  Meat, milk, poultry, eggs
  Potable water
  Fish
  Irrigated crops
  Food handling
Reduction of residue
Proposed tolerance
Responsible grounds in support of petition
Submittal of analytical reference standards

FIFRA was passed in 1947 to replace the original Insecticide Act of 1910. In 1954, the Miller Pesticide Residue Amendment was added to the Federal Food, Drug and Cosmetic Act. A major amendment to FIFRA was passed in 1972, in which all uses of pesticides are to be classified as general or

Table 7.1. Pesticide Program[a]

Part	Title
152	Pesticide Registration and Classification Procedure
153	Registration Policies and Interpretations
154	Special Review Procedures
155	Registration Standards
156	Labeling Requirements and Pesticides and Devices
157	Packaging Requirements for Pesticides and Devices
158	Data Requirements for Registration
160	Good Laboratory Practice Standards
162	State Registration of Pesticide Products
163	Certification of Usefulness of Pesticide Products
164	Rule of Practice Governing Hearings under Federal Insecticide, Fungicide and Rodenticide Act
166	Exemptions of Federal and State Agencies for Use of Pesticides under Emergency Conditions
167	Registration of Pesticide and Active Ingredient Producing Establishments, Submission of Pesticide Reports
168	Statements of Enforcement Policies and Interpretation
169	Books and Records of Pesticide Production and Distribution
170	Worker Protection Standards
171	Certification of Pesticide Applicators
172	Experimental Use Permits
173	Procedures Governing the Rescission of State Primary Enforcement Responsibility for Pesticide Use Violations
177	Insurance of food additive regulations
178	Objections and Request for Hearing
179	Formal Evidentiary Public Hearing
180	Tolerances and Exemptions from Tolerances for Pesticide Chemicals in or on Raw Agricultural Commodities
185	Tolerances for Pesticides in Food
186	Pesticides in Animal Feed

[a] Subchapter E, 40 *CFR*, 7/97.

restricted according to the degree of hazard to users and/or the environment (Federal Environment Pesticide Control Act of 1972). Subpart I (Section 152.160–175) lists pesticide classifications.

A pesticide product may be unclassified, or it may be classified for restricted use or for general use. EPA does not normally classify products for general use; therefore, pesticides that are not restricted remain unclassified.

When a pesticide is classified for restricted use, it is to be used by a certified applicator, or under the supervision of a certified applicator.

Restricted use may also relate to the composition of the product, labeling, packaging, or distribution and sale, or to the status or quantitation of the user. Criteria for pesticides to be classified as restricted-use pesticides as hazardous to humans and to nontarget species are listed below (see also Table 7.2):

Hazards to Nontarget Species

1. Level of residue equals or exceeds one-fifth of the acute dietary LC_{50} or mammalian acute oral LD_{50}
   a. Level of residue equals or exceeds one-fifth of avian subacute dietary LC_{50}.
   b. Level of residue in water equals or exceeds one-tenth of the acute LC_{50} of nontarget aquatic organisms.
   c. The pesticide may cause discernible adverse effects on nontarget organisms, such as significant mortality or effects on the physiology growth, population levels, or reproductive rates of such organisms.

Two other minor amendments to FIFRA were passed. In 1975, an amendment to FIFRA (Public Law 94–140) was passed that eliminated registration of similar products solely on the basis of "established-use practices." The amendment requires standard procedures of submission of data or specific references of existing data in EPA files and provides

Table 7.2. Criteria for Pesticides to be Classified as Restricted-Use Pesticides Hazardous to Human

Toxicity Criteria	Residential Use	Non-Residential Use
Acute oral LD_{50}	<0.5 mg/kg	<50 mg/kg
Acute dermal LD_{50}	<2000 mg/kg	<200 mg/kg, <16 g/kg (diluted)
Acute inhalation LD_{50} 4-h	<0.5 mg/L	<0.05 mg/L
Corrosive to the eye	Irreversible destruction to ocular tissue	Corneal involvement or irritation persisting for 21 days
Corrosive to the skin	Tissue destruction into the dermis and/or scarring	Tissue destruction into the dermis and/or scarring
Subchronic, chronic, or delayed toxic effects	Significant causes	Significant cause

compensation by the registrant to the owner of the data. In 1978, the amendment of the Federal Pesticide Act provided the compensation rights of ownership of data. Requirements are listed in Sections 152.86 and 152.98, respectively.

The Environmental Protection Agency is also required by law to reregister existing pesticides that were originally registered years ago (prior to November 1, 1984) when the standards for government approval and test data requirements were less stringent than they are today. The comprehensive reevaluation of pesticide safety in light of modern standards is critical to protecting human health and the environment, and to maintaining public confidence in our food supply. In 1988, the FIFRA was amended to strengthen EPA's pesticide regulatory authority and responsibilities regarding the reregistration of pesticides (7). These amendments mandated an accelerated registration scheme to be carried out in the following five phases (FIFRA Section 4, [136B]):

Phase 1 EPA publishes list of pesticides

Phase 2 Registrant responds; identifies missing studies; agrees to do studies; pays fee

Phase 3 Registrant summarizes and reformats existing studies; certifies access to raw data; "flags" adverse effects; pays fee

Phase 4 EPA reviews Phase 2 and 3 submissions; identifies any other needed studies; publishes lists of missing studies; requires missing studies

Phase 5 After all studies are in, review in 1 year; product specific studies due 8 months later; review product specific studies in 3 months; reregister or take other action 6 months after receiving product specific studies.

The registration process is to be concluded in the late 1990s. The new law is known as FIFRA '88.

Food Quality Protection Act. On August 3, 1996, the Food Quality Protection Act of 1996 (FQPA) was signed into law, effective immediately. FQPA, an important new law that was agreed on by all public and private sectors, significantly alters the way in which EPA evaluates risks from pesticides. FQPA amends both the Federal Food Drug and Cosmetic Act (FFDAC) and the Federal Insecticide, Fungicide, and Rodenticide Act (FIFRA). Among other things, FQPA amended the FFDCA by establishing a new safety standard for setting tolerances. The FQPA did not, however, amend any of the existing reregistration deadlines set forth by FIFRA.

The new safety standards direct the EPA to consider information concerning the exposure to infants and children to pesticides in food, and available information concerning cumulative effects on infants and children of the pesticide residues and other substances that have a common mechanism of action. In addition, EPA must consider the cumulative effects of the pesticide and other substances that have a common mechanism of toxicity, and the aggregate exposure levels of these pesticides to the U.S. population and major subgroup of the population.

Specifically, in setting a tolerance for pesticide residues in food, FQPA directs EPA to consider the following:

- Use of an extra 10-fold safety factor to account for susceptibility of children
- The special susceptibility of children, including effects of in utero exposure
- Cumulative effects of exposure to the pesticide and substances having a common mode of action
- Aggregate exposure for all consumers (i.e., other routes, such as drinking water)
- Potential for endocrine-disrupting effects
- Establish residue tolerances for emergency and crisis pesticide registrations
- Reassess all pesticide residue tolerances (approximately 10,000) in 10 years

EPA uses the analogy of "risk cup" approach. EPA's decision logic is based on the concept that the total level of acceptable risk to a pesticide to represented by the pesticides reference dose (RfD). This is the level of exposure to a specific pesticide that a person could receive every day over a 70-year period without significant risk of a long-term or chronic noncancer health effect. The full cup represents the total RfD and each use of the pesticide contribute a specific amount of exposure that adds a finite amount of risk to the cup. As long as the cup is not full, meaning that the combined total of all estimated sources to the pesticide has not reached 100% of the RfD, EPA can consider registering additional uses and setting new tolerances. If it is shown that the risk cup is full, no new uses could be approved until the risk level is lowered. This explanation is focused on chronic noncancer risk, however, EPA will use a similar logic to assess acute risk and cancer risk.

EPA employ this strategy in assess aggregate exposure in to account how much of the "risk cup" should be set aside or reserved for sources of possible exposure for which EPA has limited or no actual data.

2. State

Numerous states with well-developed statutory authority regulate pesticides within each state. Two examples of state programs are described here.

California. California maintains its own pesticide registration system within the California Environmental Protection Agency (CAL EPA). California does not have a generic pesticide law analogous to the Federal Insecticide, Fungicide, and Rodenticide Act (FIFRA). Instead, California has enacted a series of laws over five decades that has established a comprehensive scheme of regulation. California maintains a complete pesticide evaluation and registration system, separate from that of EPA (8).

Registration of a product by EPA does not automatically ensure that it will also be registered by California. Data supporting all new pesticide active ingredients are reviewed by CAL EPA toxicologists. In addition to data on product chemistry, residue chemistry, efficacy, and environmental fate chemistry, registrants must submit adequate studies on general toxicity, reproductive toxicity, oncogenicity, mutagenicity, and neurotoxic effects.

Dietary exposure is assessed and a risk characterization is developed for those active ingredients where a possible adverse effect is indicated. In addition, all routes of exposure are addressed. Only products that do not pose significant worker, environmental, or dietary risk will be registered in California. In addition to the evaluation of data supporting new active ingredient registrations and new uses of previously registered active ingredients, CAL EPA operates ongoing programs for both chronic and acute data call-in and reevaluation of existing registrations. If adequate studies do not exist, they must be conducted.

Florida. Florida maintains its own pesticide registration system within the Florida Department of Agriculture and Consumer Services (FDACS) (9). The Pesticide Registration Section in the Bureau of Pesticides, Division of Agricultural and Environmental Services handles approximately 1500 new and amended registrations and approximately 12,000 product renewals each year.

Because of limited resources and the need to avoid the duplication of EPA efforts, the general objectives of FDACS will not include critical analysis of conclusions or decisions made by the EPA in connection with the acceptance of existing actions. However, clear defects, errors, or omissions that are identified should be evaluated and rectified if the circumstances warrant. Under certain conditions, Florida specific data are required.

Full data packages are not required for subregistrations of EPA-registered products. However, applications for registration of pesticide products with new active ingredients (NAI) or significant new uses (SNU) must be accompanied by additional data summaries.

Additional studies may be required with further review by a statutorily mandated Pesticide Review Council (PRC), which is composed of 11 scientific members representing state agricultural, environmental, and health agencies, as well as academia, the pesticide industry, and environmental organizations. The Council conducts special reviews of registered pesticides with potential adverse environmental or health effects and may recommend or conduct additional studies for any registered pesticide. The Council may make recommendations regarding continued sale or use of pesticides reviewed and also reviews biological and other alternative controls to replace or reduce pesticide use.

The following are examples of data to be required:

1. *New Active Ingredients*
   a. All products
      i. Acute mammalian toxicity
      ii. Residue chemistry data
   b. All "outdoor" uses
      i. Environmental fate
      ii. Wildlife and aquatic toxicity
   c. All food-use products
      i. Chronic toxicity
2. *Significant New Uses*
   a. Aquatic uses when all previous uses were terrestrial
      i. Environmental fate data for aquatic uses
   b. Terrestrial, greenhouse, forestry, or domestic outdoor uses when no previous uses were "outdoor"
      i. Requirements same as NAI for that use
   c. Food use when no previous use for food
      i. Requirements same as NAI for that use
   d. Data for other SNU categories will depend on the nature of the use pattern and potential for use expansion
3. *Experimental-Use Permits*
   a. New active ingredient
   b. Standard EUP data
   c. Analytical standard and analytical methodologies
   d. Active ingredient already registered
   e. Analytical methodology may be necessary depending on the request
4. *Special Local Need Registration.* Required only when the use is considered to be SNU. Data will be required according to the rules under the SNU section.

5. *Crisis and Emergency Exemptions* Four types of exemption are included; *specific*, *quarantine*, *public health*, and *crisis*. Application of exemption should include
   a. Identity of contact persons
   b. Description of the pesticides
   c. Description of the proposed use
   d. Alternative method of control
   e. Effectiveness of proposed use
   f. Discussion of residues for food uses
   g. Discussion of risk information
   h. Coordination with other affected state or federal agencies
   i. Notification of registrant or basic manufacturer
   j. Description of proposed enforcement program
   k. Interim report, summarizing results for repeated uses
6. *Products of Specific Concern.* Active ingredients may be of special concern for a period of time or for a particular reason. These active ingredients will be identified by PRC or PREC and a list of these products supplied to the Registration Section, FDACS. Data requests will be dependent on the nature of the registration request and the special concern. The Bureau of Pesticides maintains a Scientific Evaluation Section for data review consisting of 10 professionals with master's or doctoral degrees in a variety of disciplines. Representatives from the Pesticide Formulation Laboratory review the formulation method of the pesticide, and representatives from the Chemical Residue Laboratory review the residue method as well as the residue data of the food crop. In addition, significant registrations are reviewed by an advisory body, the Pesticide Registration Evaluation Committee (PREC). The PREC is composed of representatives from various state agencies, such as Florida Department of Environmental Protection, Florida Health and Rehabilitative Services, Florida Game and Freshwater Fish Commission and FDACS.

B. Pesticide Residue Tolerances

1. Establishment

A *tolerance* is the maximum concentration of a pesticide residue that is legally permitted to remain in or on a food. A tolerance regulation specifies each individual food or food group to which the limit applies. In general, each tolerance applies to the whole portion of a food commodity. No tolerance exists for a residue on a commodity unless the commodity itself, or the group to which it belongs, is specified. A tolerance regulation specifies the composition of pesticide residue for which the limit applies, that is, the parent form of active ingredient only, parent form plus one or more metabolites and/or degradation products, one or more metabolites and/or degradation products only, or some chemical moiety that can be measured analytically for calculating the pesticide residue. The tolerance is not expected to be exceeded if the pesticide's registered use directions are followed.

Under two federal statutes, The Federal Insecticide, Fungicide and Rodenticide Act (FIFRA) and The Federal Food, Drug and Cosmetic Act (FFDCA), pesticide residue tolerances are established by EPA and enforced by FDA and USDA on the federal level. FIFRA is administrated by EPA, and FFDCA is administrated by FDA, U.S. Department of Health and Human Services. Under FIFRA, registrants of pesticides intended to be used on food crops must submit petitions proposing tolerances or exemptions for pesticide residues in or on raw agricultural commodities (40 *CFR* 180.7). Setting and enforcing tolerances for pesticide residues in or on raw agricultural foods or feeds are under the provisions of Section 408 of the FFDCA, and for processed foods and feeds are regulated under Section 409 of the FFDCA.

In Section 408, FFDCA authorizes the establishment of tolerances for pesticide residues in or on raw agricultural commodities before they leave the farm gate. Raw agricultural commodities containing pesticide residues exceeding established tolerances or for which no tolerances have been established are deemed adulterated.

In Section 409, FFDCA authorizes FDA to regulate the purposeful addition of substances (food additives) to food. This provision includes the regulation of pesticide residues in processed food and animal feed. A tolerance on a raw agricultural commodity also applies to processed forms of the same commodity. In cases where processing may concentrate the residue, a food additive regulation may be issued in 40 *CFR* Part 185 to establish a higher tolerance on that processed commodity.

The Delaney Clause amendment, passed by Congress in 1958, was added to Section 409, FFDCA. The Delaney Clause states that "No additive should be deemed safe if it is found to induce cancer when ingested by man or animal or if it is found, after tests which are appropriate for the evaluation of the safety of food additives to induce cancer in man or animal." EPA has interpreted the Delaney Clause as subject to an exception for carcinogenic pesticides that post only negligible risk. This interpretation was adopted in 1988 on the recommendation of the National Academy of Science in 1992. However, the U.S. Court of Appeals for the Ninth Circuit (*Les v. Reilly*) overturned EPA's interpretation of the Delaney Clause, holding that the

Delaney Clause bars tolerances for carcinogen pesticides in processed foods without regard to the degree of risk (10). The Food Quality Protection ends the Delaney Clause (8).

In some cases, a specific residue may be present on a commodity for which no tolerance exists because of the environmental persistence of the pesticide rather than its direct application on the commodity. If in this type of case FDA considered the low level of such a residue to pose little risk to human health, FDA used to informally set regulatory residue levels called "action levels" at which FDA would take regulatory action and below which the food was not found to be violative [21 *CFR* Sections 109 and 509; FDA Compliance Policy Guides (11)]. The informal process by which these action levels have been set has been vacated by the Federal Appeals Court in the *District of Columbia Consumer Nutrition Institute v. Young*, 818 F.2d943 (D.C. Cir. 1987). FDA regulates these low level unavoidable pesticide residues on a case-by-case basis.

EPA uses its data collection authorities under FIFRA to require the pesticide manufacturer to produce residue chemistry and toxicity data to assess the validity of the proposed tolerances or exemption from the requirement of tolerance. In establishing pesticide residue tolerances, the following three areas must be addressed: (1) the identity of the chemical residue, (2) the concentration of the residue, and (3) the acceptability of the residue in the dietary level of exposure. The data must provide a reasonable assurance that under the prescribed conditions of the use of the pesticide, no unreasonable adverse effects will result in humans, including lifetime exposure. Product chemistry data requirements are listed under 40 *CFR* Part 158, Subpart C.

Product Chemistry Data. Data on the composition of pesticide products that includes (1) information on the manufacturing process, (2) chemical analysis to show the amount of the active ingredient and any associated impurities, (3) "certified limits" on the amounts of the ingredients in a product, and (4) analytical methods used to determine the composition of the pesticide.

In addition, data must include whether impurities could constitute a significant component of the residue in food or animal feed for evaluation in tolerance establishing processes.

Metabolism in Plants and Animals. These data are required to characterize the nature of the residue that occurs on crops intended for consumption as food or animal feed. The pesticide is labeled with a radioactive atom, usually ^{14}C-, and applied to the crop plant in accordance with proposed-use directions. The plants are harvested at maturity. The parent pesticide and one or more of the metabolites and degradation products remaining in the plants will stay to be radioactive. The ^{14}C-activity is separated into various fractions, and all chemicals associated with the activities are identified.

Plant metabolism studies are required for a minimum of three diverse crops: a fruit-bearing vegetable, a leafy vegetable, and a root crop. If the metabolism in each of these crops is similar, then the metabolism in other crops is assumed to be similar.

If a pesticide use will result in residues in animal feed or is intended for treatment of food-producing livestock, additional animal metabolism studies are required. Animal metabolism studies are generally carried out on ruminants (cows or goats) and poultry (chicken).

Animal metabolism studies also use radiolabeled pesticides. Animals are treated with the pesticide according to use directions. The radioactivities in meat or poultry products (muscle, liver, kidney, milk, and eggs) are analyzed and the chemicals associated with the activities identified.

Significant Metabolites and Tolerance Expression. Metabolites identified from plant and animal studies may or may not be included in the established tolerances of a pesticide. The decision is based on (1) the toxicity of the metabolites, (2) the percent and magnitude of its residue, and (3) whether a practical analytical methodology is available or can be developed to detect and measure the metabolite. Workable analytical methodology is mandatory for those metabolites that are toxicologically significant and occur at significant amounts. "Total toxic residue" is used to describe the sum of the parent pesticide and its degradation products, metabolites (free or bound), and impurities that are considered to be of toxicological significance, and therefore warrant regulation.

Residue Field Trial Data. Field experiments are carried out after the metabolism data indicate what chemicals to look for, and analytical methods are developed to measure the total toxic residue. In the studies, the pesticide is applied to crops at known application rates, in the same manner as the use directions that will eventually appear on the pesticide label. The field trial must reflect use conditions that could lead to the highest possible residues: highest permissible application rate, the maximum applications allowed, and the shortest pre-harvest interval permitted. Data are required for each crop or crop group for which a tolerance and registration is requested. Data are also required for each raw agricultural commodity derived from the crop, specifically corn residue analyses for both grain and feed items—forage, silage, and fodder.

Data should include all information necessary to provide a complete and accurate description of field trial treatment and procedures; sampling (harvesting), handling, shipping, and storage of the raw agricultural commodity; storage stability validation of the test chemical and metabolite(s) of specific concern; residue analysis of field samples for the "total toxic residue;" validation of the residue analytical methodology; reporting of the data and statistical analyses; and quality-control measure or precautions taken to ensure the fidelity of these operations (12). All components of the field trial study must be carried out in accordance with Good Laboratory Practice (GLP) protocols (43).

Analytical Methods. Tolerance petitioner must develop practical and workable analytical methods to detect and measure all components of the total toxic residue. A single method to cover all chemicals concerned is desirable; however, additional methods are acceptable if a single method is impossible. The petitioner must also test FDA multiresidue methods (MRM) published in the *Pesticide Analytical Manual* (*PAM*) for applicability to the components of the total toxic residue (13). All analytical procedures must meet GLP protocols.

Pesticide residue analytical methods are used for two purposes: (1) to obtain residue data on which dietary exposure assessments and tolerances are based and (2) to enforce the tolerance after it is established. EPA validates each new analytical method using a method trial to ensure that the procedures can actually be used for tolerance enforcement purposes by FDA, USDA, and state agencies.

Proposed Tolerance Level. A petitioner for a pesticide residue tolerance proposes a tolerance level based on residue field trial data that reflects the maximum residue that may occur under "worst-case" conditions as a result of the proposed use of the pesticide. The proposed tolerance level must include significant metabolites and must be high enough to cover all components of the total toxic residue. If one component of the residue is significantly more than other components, two levels should be included in the tolerance (40 *CFR*, Part 180.7).

Processing studies are required to determine whether residues in raw agricultural commodities may be expected to degrade or concentrate during food processing. If residues do concentrate in processing, one or more food or feed additive tolerances must be established. However, if residues do not concentrate in processed commodities, the tolerance for the parent raw commodity applies to all processed food or feed derived from it.

Whenever pesticide residues result in feed items, data on the transfer of residues to meat, milk, poultry, and eggs are required. These studies are also required if a pesticide is to be applied directly to food-producing animals. Data from these studies will provide the type of secondary residues that may result in meat, milk, poultry, and eggs.

The proposed pesticide residue tolerances are submitted to the Office of Pesticide Programs, EPA, for tolerance establishment evaluation. Traditionally, in making tolerance decisions, EPA has compared the reference dose with the Theoretical Maximum Residue Contribution (TMRC) of the pesticide to the daily diet. As a rule of thumb, where basic data requirements are satisfied, EPA has routinely established a proposed tolerance if the TMRC is less than the reference dose. However, as a routine practice before making tolerance decisions on a pesticide, EPA uses the Dietary Risk Assessment Evaluation System (DRES) to calculate TMRC and risk estimates for the general population and a number of subgroups, including the two most sensitive subgroups identified by the system, infants and children (15, 16).

2. Enforcement

A tolerance enforcement program provides a means of ascertaining that a pesticide was properly applied to a food commodity. Government pesticide residue tolerance enforcement programs generally consist of surveillance, compliance, and monitoring activities to investigate, control, and prevent illegal pesticide residues in food and animal feed supplies in the United States (17).

There are wide variations in pesticide residue tolerances. In order to enforce the tolerances effectively, one must be very proficient in understanding of the different tolerances. The many different established tolerances are listed under 40 *CFR*, Part 180:

Part 180
- Raw agricultural product tolerances — 40 *CFR*
- Subpart C—specific tolerances; Subpart D—exemptions from tolerances

Part 185
- Tolerances for pesticides in food (as food additives) — 40 *CFR*

Part 186
- Pesticides in animal feed (as feed additives) — 40 *CFR*

Certain pesticides are exempt from the requirement of a tolerance when the residues in all raw agricultural commodities for which it is useful, under conditions of use currently prevailing or proposed, will involve no hazard to the public health (40 *CFR*, Part 1001).

The following tolerances and administrative guidelines are not all listed in 40 *CFR*. They are published in the *Federal Register*:

Temporary Tolerance. Not approved for general use and limited by 40 *CFR* 180.31.

Proposed Tolerance. Does not have an assigned 40 *CFR* number or associated temporary tolerance and limited by 40 *CFR* 180.7.

Interim Tolerances or Time-limited Tolerance. Not approved for general use and limited by 40 *CFR* Part 319.

FIFRA Section 18. Under the New Food Quality Protection Act (FQPA), EPA must establish a formal tolerance even under emergency situations. The residues remaining in or on treated crops must meet the new safety standards.

Geographic Tolerance. Certain tolerances are established for specific pesticides that can be applied only to specified commodities in designated geographic locations of the nation. However, the treated commodities can be shipped to other parts of the nation for sale.

Administrative Guidelines. For some chemicals and products or legal action levels for which formal tolerances have not been promulgated. These guidelines are not tolerances. The criteria apply to objective samples. Legal action may be taken at residue levels lower than the administrative guideline if there is evidence of misuse of pesticides or whether other factors appear to warrant action.

Federal. EPA has no direct responsibility for enforcing pesticide residue tolerances in foods. FDA and USDA are responsible for enforcing tolerances and monitoring pesticide residues in food and animal feed. An advisory board of scientists from EPA, FDA, and USDA was established in 1985 to help keep pace with information on pesticide concerns.

Food and Drug Administration (FDA). FDA, under FFDCA, has the responsibility of enforcing the tolerances established by EPA in food (except meat and poultry, which are regulated by USDA) and animal feed, including processed products, in interstate commerce and for enforcing prohibition of a pesticide residue in food or feed for which no tolerance has been set or exemptions given (21 USC Sections 331–337, 1982 and Suppl. IV, 1986) (18,19).

To fulfill its regulatory responsibilities, FDA established a pesticide monitoring program that is designed to identify and quantify pesticide residues in food and animal feed. The two main objectives are to (1) monitor domestic and imported food and feed commodities for pesticide residues in support of regulatory actions against illegal residues and (2) obtain information on the incidence and levels of pesticide residues in food supply.

Monitoring of all pesticide–commodity combinations for all of these pesticides would far exceed the resources of FDA or any other regulatory agencies. FDA employs a selective monitoring approach that is based on (1) analytical method capability by using multiresidue methods (coverage is shown in Table 7.3) and (2) priorities in terms of propensity of health risk to consumers as the result of pesticide application. The risk assessment is based primarily on the FDA Surveillance Index (SI). Classifications of SI are listed in Table 7.4 (20).

The two major components of the FDA pesticide monitoring program are (1) general commodity monitoring and (2) Total Diet Study.

General Commodity Monitoring. This program is designed to enable the enforcement of tolerances established by EPA and to determine the incidents and levels of residues in domestic and imported raw agricultural commodities, processed food, and animal feed. It also serves to meet the objectives of (1) determining, on a geographic basis, pesticide residue levels of individual food commodities; (2) surveying, on a nationwide basis, pesticide residue levels of selected food commodities; (3) monitoring imported commodities and denying entry to those with illegal pesticide residues; and (4) identifying pesticide residues occurring in excessive levels as a basis for compliance and enforcement action.

Table 7.3. Number of Compounds Determined by FDA Multiresidue Methods, *Pesticide Analytical Manual* (as of May 1988)

Type of Chemical	Number of Chemicals in Database	Number of Chemicals Determined
Pesticides with tolerances	316	163
Pesticides with temporary or pending tolerances	74	10
Pesticides with no EPA tolerance	56	25
Metabolites, impurities, alteration products, and other pesticide-related chemicals	297	92

Table 7.4. Classifications of FDA Surveillance Index[a]

Class	Description
I	Pesticides posing high health hazards
II	Pesticides posing possible high risk
III	Pesticides posing moderate hazard
IV	Pesticides posing low hazard
V	Pesticides posing very little potential hazard

[a] As of May 1988, 205 pesticides were ranked.

The majority of samples are collected at random for monitoring purposes and are called *surveillance* samples. The remainder are compliance samples, collected after a violation has been found or there is evidence of a likely violation. FDA's ability to prevent violative food from reaching the consumer is constrained by the amount of time needed for sample transport and analyses. FDA can detain imported commodities until compliance analyses are completed, but does not have authority to detain domestic commodities.

The percentage of samples that violate EPA tolerances is called the *violation rate*. Annually, FDA collects and analyzes approximately 10,000 commodity samples with an overall violation rate of less than 2%. FDA believes that violation rates cannot be extrapolated to give the correct level of violation in the general food supply because of the biased nature of FDA sampling.

The Center for Food Safety and Applied Nutrition (CEFSAN), FDA, is responsible for much of the direction of the FDA-monitoring program, primarily through the development of its annual series of compliance program guidance manuals. Four types of sampling plans make up general commodity monitoring: core samples, special-emphasis surveys, headquarter-initiated surveys, and regional sampling plans (21).

Core samples, which must be analyzed by each district, are of commodities susceptible to environmental contamination and likely to bioaccumulat fat-soluble pesticides. Special-emphasis surveys permit each district to sample two domestic pesticide–commodity combinations and two imported pesticide–commodity–country of origin combinations subject to CEFSAN approval. These surveys focus on those pesticides neither adequately measured nor regularly analyzed by the routine multiresidue methods.

These pesticides may be selected for monitoring because of EPA requests, FDA investigatory reports, a high SI classification for a pesticide, or past violation problems. Headquarter-initiated assignments are those in which CEFSAN instructed a district to analyze a specific commodity. The regional sampling plan for domestic and imported food allows each region to determine what products it plans to sample on the basis of its knowledge of local crops, pesticide use, and coordination with state programs.

The Total Diet Study (TDS). In this study more than 200 foods items selected to represent the diet of the U.S. population are collected in retail markets 4 times annually, once from each of the four designated geographic areas of the United States: northeast, south, northcentral, and west. Sample collection consists of identical food from retail stores in three cities within each geographic area. Samples are sent to the FDA Total Diet Laboratory in Kansas City, MO, where the three samples of each food (from three different cities) are composited to form a single sample and analyzed using multi-residue methods (25).

The results of total diet studies are used to estimate dietary intake of selected pesticides by various U.S. age–sex groups. The design of TDS provides an estimate of public exposure to those pesticide residues detected by the analytical methods used in the study. FDA uses the data from TDS to make judgements about public health risk presented by pesticide exposure through food.

To further utilize available resources, FDA initiated the Partnership Agreement with states that have comprehensive pesticide monitoring and enforcement programs. In addition, the states must have high-quality analytical laboratory support. Since 1995, FDA and Florida Department of Agriculture and Consumer Services annually sign a Partnership Agreement to share the pesticide residue data generated by the Chemical Residue Laboratory in the Division of Food Safety, FDACS. With the FDACS data, FDA forgoes the monitoring of Florida grown agricultural products and exercise regulatory actions on violative interstate and imported products. (25).

U.S. Department of Agriculture Food Safety and Inspection Service (USDA FSIS). FSIS of USDA is responsible for enforcing pesticide residue tolerances in meat and poultry under authorities of the Federal Meat Inspection Act (21 *USC* Sections 601–695, 1982 and Suppl. IV, 1986) the Poultry Products Inspection Act (21 *USC* Sections 451–470, 1982 and Suppl. IV, 1986). Agricultural Marketing Service (AMS) of USDA is responsible for pesticide residue monitoring of raw egg products (dried, frozen, or liquid eggs) and tolerance enforcement at establishments having official USDA egg products inspection service under authority of the Egg Products Inspection Act (21 *USC* Sections 1031–1056). Under this act, FDA has jurisdiction over these products outside each establishment (20).

The USDA pesticide residue in food monitoring program is part of the National Residue Program (NRP), which addresses residue of pesticides, animal drugs, and environmental contaminants in meat, poultry, and raw egg products. The four components of the NRP are monitoring, surveillance, exploratory projects, and prevention, which are administered by FSIS. Sample analyses are performed by FSIS laboratories and by FSIS accredited laboratories. Samples with illegal pesticide residues are less than 2% of total samples analyzed.

Monitoring Program. This program involves random sampling of meat and poultry tissue from processing plants during routine inspection at slaughter of domestic animals and of imported products at the port of entry. These samples account for 80% of the total 5000 pesticide residue samples analyzed annually. The random sampling scheme used in this program is designed statistically to provide 95% assurance of detection over the course of a year with a violation rate of 1% or more in the national population. FSIS believes that monitoring data can be used to provide a good indication of violation rates in the general meat supply since the sampling in this program is random. In most cases, the monitoring program does not prevent violative products from reaching the consumers because of the time required to transport and analyze the samples. However, monitoring provides information on the occurrence of residue violations and helps to identify those producers who may be selected for surveillance sampling.

Surveillance Program. This program focuses on the investigation and control of movement of meat and poultry products that are suspected of contamination. The sampling in this program is directed specifically to those meat or poultry carcasses that have been implicated as sources of residues by either the monitoring program, investigation, or a prior history of violation by the supplier. Carcasses are held until the analysis is completed. Violative shipments are condemned, and the producer is prohibited from marketing animals until further samples show no illegal residues.

Exploratory Projects. These are surveys to determine whether a pesticide not currently detected should be included in the monitoring program. New methods that have not been validated by FSIS may be used in these surveys to detect the pesticides and to evaluate the value of the method.

Prevention Program. To complement the regulatory work, this program is based on producer testing and education. FSIS signs Memoranda of Understanding (MOU) with producers, who then pay for testing feed, feed additives, litter and some animals for chlorinated hydrocarbons. About seven FSIS accredited laboratories perform approximately 2000 analyses per month and provides FSIS the results. FSIS has also collaborated with USDA Cooperative Extension Services to produce educational materials for and provide counseling to producers on how to avoid chemical contamination of animals.

Raw Egg Products. AMS carries out a regulatory program for pesticide residues in raw egg products. At its laboratory in Gastonia, NC, approximately 400–500 samples are analyzed annually from the approximately 90 domestic, egg breaking and drying factories and imports using a multiresidue method can identify 50 pesticides. If violations are found in domestic products, AMS may analyze raw eggs to determine which producer is the source of the violative eggs. Violations in raw egg products are extremely low.

The Pesticide Data Program. The pesticide Data Program (PDP) in the Division of Science, Agricultural Marketing Service, U.S. Department of Agriculture is to collect comprehensive, objective, and statistically defendable data on pesticide residues in fresh fruits and vegetables as close to the consumer level as possible. This program began in 1991. PDP is a multifederal agency program with planning, policy, and procedural efforts coordinated among the U.S. Department of Agriculture, the Environmental Protection Agency, and the Food and Drug Administration. Sample collection and laboratory analytical work are performed by 10 states—California, Colorado, Florida, Maryland, Michigan, New York, Ohio, Texas, Washington—and Wisconsin—and two USDA laboratories. These states represent diverse geographic areas of the country, major agriculture-producing states, and more than 50% of the nation's population. PDP is now a critical component of the Food Quality and Protection Act of 1996. Title III of the Act directs the Secretary of Agriculture to provide improved pesticide residue data collection, including guidelines for the use of comparable analytical and standardized reporting methods and increased sampling for foods most likely consumed by infants and children.

PDP supports EPA's risk assessment process in the reregistration of pesticides vital for American agriculture to sustain a safe and abundant food supply. Other federal and state government agencies use the data to respond more quickly and effectively to food safety issues (27).

States. States also have pesticide residue provisions in individual state laws. State monitoring programs of pesticide residues in food vary widely in the number of samples they analyze and the purpose of their programs. For example, Florida's and Montana's programs are targeted to potential problem areas; Massachusetts has directed its program to dietary risks; where as California's program included random sampling as well as targeted (28). Table 7.5 illustrates the number of food monitoring samples analyzed for pesticide residues in 22 states, reported by FDA's FOODCONTAM in *Residue in Food*, 1992 (25). FOODCONTAM program terminated in Fiscal year 1998. Two of the state enforcement programs are explained in detail here.

California. California's pesticide residue monitoring program for fresh fruits and vegetables is administered by the state's Environmental Protection Agency's (CAL EPA) Department of Pesticide Regulation (DPR) and the California Department of Food and Agriculture (CDFA). The residue program is organized into five major components: marketplace surveillance, preharvest monitoring, priority pesticide monitoring, produce destined for processing monitoring, and pesticide registration evaluation. Each year the California program results in more than 14,000 samples. Processed foods are

Table 7.5. Summary of State Pesticide Residue Samples Reported to FOODCONTAM in 1992[a]

State	Total Sample Number	State	Total Sample Bumber
Arkansas	345	Minnesota	225
California	4364	New York	742
Connecticut	558	North Carolina	504
Delaware	12	Oregon	519
Florida	2982	Pennsylvania	455
Georgia	406	Rhode Island	118
Hawaii	336	Virginia	1139
Indiana	535	Washington	578
Kentucky	99	West Virginia	14
Maine	18	Wisconsin	585
Michigan	367	Wyoming	89

[a] Listed in FDA *Residue in Foods*, 1992.

monitored by Department of Health Services (DHS) Food and Drug Branch (29).

Marketplace Surveillance. The samples drawn by the DPR staff are submitted to one of the four pesticide residue enforcement laboratories operated by CDFA. The timely analysis of the marketplace surveillance samples is important to ensure violative samples (those with pesticide residues in excess of 40 *CFR* 180 tolerances or those with residues for which there are no established tolerances for the chemical–commodity combination) are quarantined and removed from channels of trade prior to consumer exposure. Samples are routinely completed and results reported within 6 h (six hours) of arrival, including confirmation analysis. This time requirement applies to both the multiresidue methods as well as the single analyte method.

Close cooperation between CAL EPA/DPR, DFA, and U.S. FDA is maintained to ensure a coordinated monitoring analysis and enforcement strategy for fresh fruits and vegetables grown and/or sold in California.

Preharvest Monitoring. This component consists of approximately 2500 samples taken from fields prior to harvest. These samples are analyzed by multiresidue screens with specific analyses "as needed." Early detection and deterrence of pesticide misuse is one of the major goals of this program.

PRIORITY PESTICIDE MONITORING. This component is a pesticide-based, rather than commodity-based, program. The total number of samples has been increased to 5000 recently. This program has a tremendous potential to generate real-world numbers for the conduct of dietary risk assessments. Each year, CAL EPA medical toxicologists identify pesticides of priority health concern, and commodities known to have been treated with those pesticides are sampled and analyzed for the specific pesticides. Dietary consumption patterns of various subpopulations are given consideration in choosing the commodities sampled.

PROCESSING FOOD MONITORING. This component consists of approximately 1500 samples of raw agricultural commodities destined for processing. Samples are taken in the field shortly before harvest or after harvest, at grading stations, and at processing plants prior to processing.

PESTICIDE REGISTRATION EVALUATION. This component consists of an in-laboratory working review of all analytical methods submitted in support of a pesticide registration for food crop use. This evaluation is, in addition to the review, conducted by U.S. EPA and focuses on method performance on the type of crops (minor crop) grown in California. Additionally, the methods must satisfy two (2) state analytical requirements, in addition to those required by EPA: (1) the method must be able to be completed within 24 h and, (2) the method must detect a moiety that is specific to the active ingredient that was applied to the crop. The results of the analytical method reviews are submitted to the DPR for consideration in the pesticide registration process.

California's extensive pesticide residue monitoring and enforcement program annually resulted in approximately 34,000 pesticide analyses conducted on more than 14,000 enforcement and monitoring samples. The results of the program can be summarized as follows:

Marketplace Surveillance	
No detectable residue	77.94%
Within tolerance	21.35%
Over tolerance	0.71%
(0.22% exceeded tolerance; 0.49% no tolerance)	
Priority pesticide	
Number of targeted pesticide	36
No detectable residues	90%
Within tolerance	9.9%
Over tolerance	1 sample

Produce destined for processing

No detected residues	91%
Within tolerance	9%
Over tolerance	0%

Florida. The Chemical Residue Section, Bureau of Food and Residue Laboratories in the Division of Food Safety, Florida Department of Agriculture and Consumer Services (FDACS), is responsible for the regulatory enforcement of federal pesticide residue tolerances adopted by the state, and monitors, pesticide residues in food produced or marketed in Florida, under the Florida Food Safety Act, Charter 500, F.S. (30). The mission is to ensure that food containing illegal pesticide residues do not enter channels of commerce and to ensure that all pesticides are applied according to label instructions.

The laboratory must assume that pesticides registered for use on crops could be present at harvest and must, therefore, have the capabilities to detect them. In order to complete the analyses within a reasonable amount of time and be able to detect literally hundreds of chemicals, multiresidue analytical procedures are used whenever applicable. In situations where the pesticide cannot be analyzed by multiresidue procedures, individual residue procedures are utilized when the need arises and resources are available.

Routine samples are collected from fields, packing houses, central warehouses, wholesalers, retailers, and import brokers. Fresh vegetable and fruit selection is based on a Surveillance Index (SI) procedure. The SI procedure consists of two primary parameters: (1) The propensity of the crop to accumulate significant pesticide residues, and (2) the particular characteristics of the pesticide applied to the crop. Consequently, it is a crop/pesticide index. The classification of crop groups from the standpoint of potential pesticide exposure of consumed plant parts is based on the book *Food and Feed Crops in the United States*, by Magness et al. and is listed in Table 7.6 (31).

Table 7.6. Surveillance Index

Crop Category Examples	Theoretical Residue Potential[a]
VI and VII (cabbage, celery, lettuce, etc.)	Highest
V (tomatoes, snap beans, peppers, summer squash, etc.)	High
III and IV (onion bulbs, sweet corn, peas, etc.)	Medium
I and II (carrots, potatoes, melons, etc.)	Low

[a] Based on surface preharvest applications.

Factors used to evaluate pesticides applied are listed as follows:

1. Acute oral toxicity
2. Persistency on the top
3. Toxic metabolite formes
4. Systemic property

Frequency of sampling category of commodities is based on several factors: propensity to accumulate residues, volume of production, and variety, as illustrated below:

Commodity	Pesticide	Combined SI
Potatoes	Methamidophos	H
	Aldicarb	H
	Malathion	L
Sweet corn	Methomyl	H
	Mevinphos	M
	Chlorothalonil	L
Tomatoes	Methomyl	H
	Methamidophos	H
	Permethrin	H
Lettuce	Methamidophos	H
	Methomyl	H
	Mevinphos	H

FDACS field inspectors survey vegetable fields and interview farmers to ascertain the types of pesticide used on fruit and vegetable crops and report the information to headquarters and the laboratories. The headquarters office also keeps in touch with EPA, FDA, USDA, and other state regulatory agencies to obtain pesticide usage information on food products shipped into Florida. The Florida regulatory office is constantly in touch with other state and federal agencies in exchange of pesticide residue data, monitoring, and enforcement strategies.

Approximately 200,000 analyses are performed on approximately 4000 samples annually. More than 65% of the food samples analyzed did not contain any detectable pesticide residues. The majority of the detected residues are below established tolerances or guidelines. The average violation rate during the past decade was around 2%. This violation rate is based on a biased sampling program described above and cannot be projected to the marketplace. FDACS continuously works with the various farm and grower organizations to reduce and prevent pesticide residue violations.

C. Pesticide Risk Assessment

There are four components to a complete risk assessment:

1. *Hazard Identification.* Gathering and evaluating data on the types of health injury or disease that may be produced by a chemical and the conditions of exposure under which injury or disease is produced. It may also involve characterization of the behavior of a chemical within the body and the interactions it undergoes with organs, cells, or even parts of cells. Hazard identification is not risk assessment. It is a scientific determination of whether observed toxic effects in one setting will occur in other settings.
2. *Dose–Response Evaluation.* Describing the quantitative relationship between the amount of exposure to a substance and the extent of toxic injury or disease. Data may be derived from animal studies or from studies in exposed human populations. Dose–response toxicity relationship for a substance varies under different exposure conditions. The risk of a substance cannot be ascertained with any degree of confidence unless dose–response relations are qualified.
3. *Human Exposure Evaluation.* Describing the nature and size of the population exposed to a substance and the magnitude and duration of their exposure. The evaluation could concern past or current exposures, or exposure anticipated in the future.
4. *Risk Characterization.* The integration of the data and analysis of the preceding three components to determine the likelihood that humans will experience any of the various forms of toxicity associated with a substance. When the exposure data are not available, hypothetical risk is characterized by the integration of hazard identification and dose–response evaluation data.

1. EPA Risk Assessment Procedure

The U.S. Environmental Protection Agency is the leading federal government agency conducting risk assessment on pesticides. EPA toxicologists use toxicology and residue data to assess possible health risks of pesticide's use on food and determine whether proposed tolerance levels would protect the public health within a practical certainty. The risk of pesticide residues depends on both the toxicity of the residues (i.e., their potential to cause adverse health effects) and potential human exposure to residues in the diet. EPA's risk assessment process for food-use pesticides consists of three steps: (1) determining the residue's toxicity, including a level of daily pesticide residue intake acceptable for humans (acceptable daily intake), (2) determining the maximum potential dietary exposure to pesticide residues (theoretical maximum residue contribution), and (3) comparing potential dietary exposure to the acceptable level of human intake. Risk assessments involve an element of uncertainty because results of animal tests must be extrapolated to humans. Uncertainty occurs because animals are biologically different from humans and because higher doses are used in animal tests than are expected for human exposure.

Determining the Pesticide Toxicity. Risk assessment of chemicals is based on data generated from animal toxicity testing and human epidemiological studies. In animal testing, there are two types of toxicity assessment: acute toxicity studies and chronicle toxicity studies. Most test models commonly employ species of mice and rats. Other species, such as chick embryos, rabbits, dogs, monkeys, and fish, are also used in toxicity testing. The majority of toxicology data are generated by pesticide registrants; however, EPA could use toxicological data from any credible source (32).

Human epidemiology is the science that attempts to determine the pattern of human diseases by studying the relationships between factors that are suspected to be the causative agents. It can identify the association between events that may be risk factors, but it rarely can determine cause and effect. It is the lack of precision, however, that makes it an excellent way to predict trends (33). There are no conclusive human epidemiology studies on pesticides to be used in risk assessments.

The results of testing animals when administrated with a chemical could range from no observable response or beneficial effects, to unfavorable results such as death, tumor formation, birth defects, cell mutation, and loss of weight or hair, depending on the type of the chemical and amount of the chemicals administrated.

Figure 7.1 is a simplistic illustration of animal testing results (34). When dosage is plotted against response, there could be no effect (Fig. 7.1 *a*) or an adverse effect with no threshold (Fig. 7.1 *b*) or an adverse effect with a threshold (Fig. 7.1 *c*) or beneficial effect at low dose and adverse effect at higher dose (Fig. 7.1 *d*).

Non-Carcinogenic Pesticides. In situations where there are thresholds as shown in Figure 7.1 *b–d* an "acceptable daily intake"(ADI) or "reference dose" (RfD) is established by applying a 100-fold safety factor to the "no-

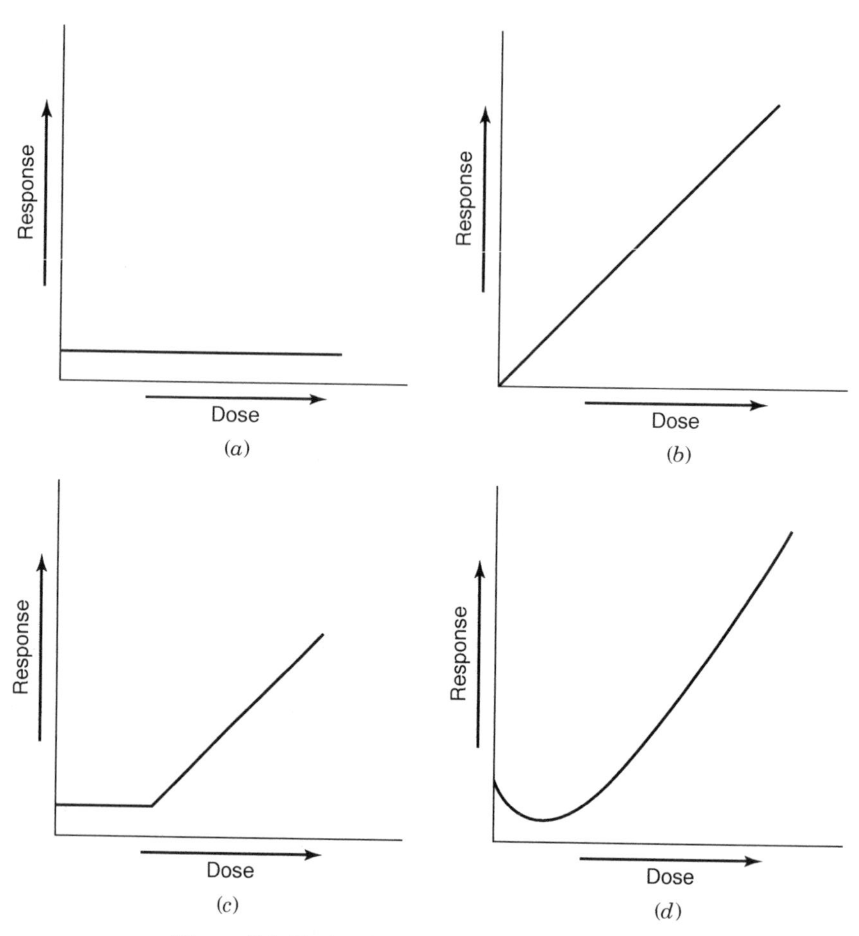

Figure 7.1. Toxic chemical dose–response curves.

observed-adverse effect level" (NOAEL) or the no-observed-effect level (NOEL) from the animal testing data (35,36). In determining the ADI, the first step is to determine the response above the base level. The next step is to determine the maximum daily dose above the control at which the response is zero or the same as the control. This is labeled as the "NOAEL" threshold. The NOAEL is then divided by a safety factor to obtain the RfD. The RfD is usually expressed as milligrams of the test substance per kilogram body weight per day. The safety factor to calculate the RfD from the NOAEL is usually understood to be 100, which is the combination of uncertainty factor (UF) of 10 and a modifying factor (MF) of 10. The rationale for using 100 is that humans may be 10 times more sensitive than rats, and humans themselves may vary in sensitivity by as much as 10 times ($10 \times 10 = 100$). Figure 7.2 illustrates logarithmic plot of doses versus. response and the safety factor (32).

The Food Quality Protection Act (FQPA) directs EPA to include an additional order of magnitude in calculating the RfDs for infants and children (Fig. 7.3). Thus, the safety factor of RfDs for infants and children becomes 1000 (37,38).

The "reference dose" is defined as "an estimate (with an uncertainty of one order of magnitude or more) of a lifetime daily dose of a chemical that is likely to be without significant risk to human population:

$$\text{Reference dose (RfD)} = \frac{\text{NOAEL}}{\text{UF} \times \text{MF}}$$

This concept of risk assessment for food safety has worldwide acceptance; however, it is not applicable for carcinogen pesticides.

Carcinogenic Pesticides. For carcinogenic pesticides, animal testings are subject to maximum tolerated doses (MTDs). MTD is the maximum amount of a chemical that can be administrated to an experimental animal without incurring extreme health consequences, such as death, while continuing to produce some measurable toxic effects. Current regulatory

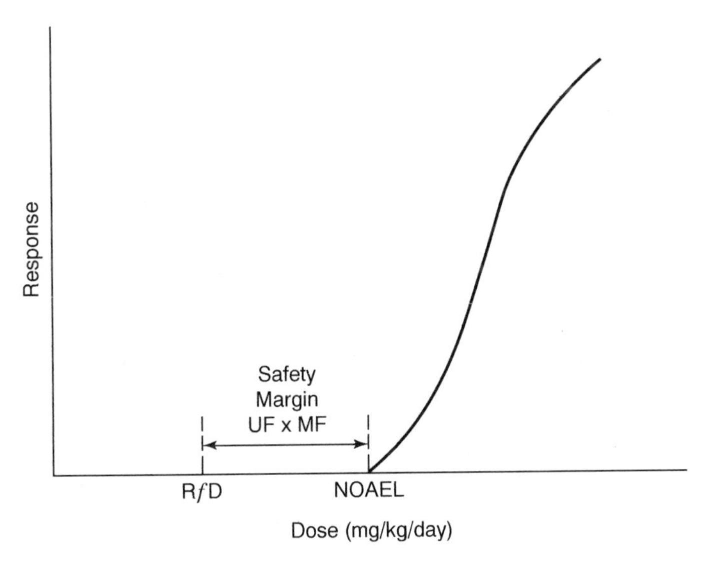

Figure 7.2. Establishing RfD.

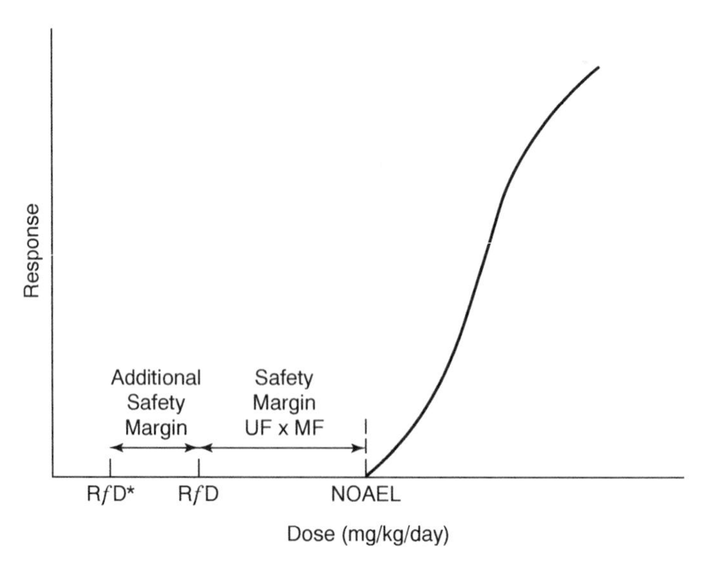

Figure 7.3. Additional safety margin RfD [RfD—recommended by NAS Report, Pesticides in the Diets of Infants and Children; dose in mg/(kg·day)].

theory holds that carcinogen effects do not have threshold and cannot be related to RfDs.

A pesticide's carcinogen potency is expressed quantitatively as a "Q-star" or Q^*. The Q^* represents the slope of the dose–response curve from animal studies yielding a positive oncogenic response, expressed in the units of excess tumor incidence per milligram of pesticide exposure/kg of body weight per day. Mathematical extrapolation are used to predict the human carcinogenic potency of a pesticide from tumor incidents resulting from very high doses administrated in animal tests to very low doses in human diets. The potency factor does not consider the type, site, or diversity of tumors observed. EPA often combines malignant and benign tumors for the potency factor. A high Q^* indicates a strong carcinogen response to the administrated dose; consequently, a low Q^* indicates a low response. The values of Q^* values represent the 95% upper-bound confidence limit of tumor induction likely to occur from a given dose and is considered to be a highly conservative model for quantifying carcinogen potency. Carcinogen risks below 10^{-6} (one in a million) are considered to be negligible risks (39,40).

Prior to FQPA, the Delaney Clause was problematic for EPA, especially since the decision of U.S. Court of Appeals for the Ninth District in *Les v. Reilly* in 1992. New and more sophisticated test procedures of both toxicological and analytical testing reveal new tumor-inducing potency of a currently registered pesticide and/or low levels of pesticide residues that previous analytical methods could not detect. These registered pesticides were found to meet the risk/benefit requirements of FIFRA and the negligible risk recommendation of NAS, but not the Delaney standards (41,42). These pesticides are now being re-evaluated under FQPA (8).

Endocrine System Disruption. The endocrine system may be disrupted by certain substances mimicking or blocking the effects of estrogen, a hormone generated by the ovaries, testes, and adrenal gland that plays many roles in the body, such as ovulation, blood clotting, bone growth, and modulating the immune system. In the fetus, estrogen plays a major role in organ development, including a part of determining whether the fetus develops male or female sex organs. Thus endocrine disruption can cause disruption of various reproductive and development processes and the possibility of increasing certain carcinogenesis. Endocrine disrupters provide a totally different view of toxic substances and great challenges to toxicologists. Dose–response toxicity evaluations of traditional substances do not hold for endocrine disrupters. The greatest effects of endocrine disrupters are often observed at smallest doses. Current methodology is not adequate for separation of the effects caused by synthetic estrogen mimics from the effects caused by estrogen generated by the body.

The Food Quality Protection Act of 1996 mandates the U.S. Environmental Protection Agency (EPA) in 2 years to develop a comprehensive strategy to screen and test common chemicals for endocrine disrupter effects and to implement the program a year after (8,38).

EPA's Office of Prevention, Pesticide and Toxic Substances established the Endocrine Disruptor Screening and Testing Advisory Committee (EDSTAC). EDSTAC will provide advice and counsel to the Agency on a strategy to screen and test chemical and pesticides that may be the cause of endocrine disruption in humans, fish, and wildlife. This strategy will be aimed at reducing or mitigation risk to human health and the environment. EPA expects the EDSTAC to take consensus approach to reaching their findings and recommendations.

2. Pesticide Residue Exposure Monitoring

The amount of a pesticide that people are exposed to through their diet depends on two factors: (1) the amount of pesticide remaining on various foods when they ingest them, and (2) what proportion of the diet these foods represent. The dietary exposure of a particular pesticide is the sum of all exposures calculated from each individual food that people consume and the residues of pesticide in the food.

The amount of pesticide remaining on various foods that they eat (calculating residues in foods). During tolerance risk assessment, the method used by EPA is called the *theoretical maximum residue contribution* (TMRC). The TMRC is computed by multiplying the tolerance of the pesticide residue on the raw agricultural commodity (RAC) by the corresponding food consumption factor. Therefore, TMRS assumes that 100% of the crop is treated with the pesticide residues at tolerance levels.

EPA intends to use anticipated residue data for risk exposure assessment. The anticipated residue data concept is to take into consideration the pesticide dissipation under natural conditions, such as heat and moisture, pesticide reduction by routine commercial postharvest practices, and good farming management practices, including integrated pest management (IPM).

EPA is also using the pesticide residue data generated by the U.S. Department of Agriculture's Pesticide Data Program. The pesticide–commodity pairs in the PDP program are targeted by EPA to conduct dietary risk assessment, address pesticide registration issues, and complete the special reviews of specific pesticides.

The Proportion of the Diet that These Foods Represent (Calculating People's Diet). EPA's dietary consumption is calculated by a dietary risk evaluation system (DRES). DRES is based on the National Survey of Individual Food Consumption conducted by USDA between 1977 and 1978, based on 31,000 people surveyed on their 3-day consumption of grams of food that was converted to RAC, per kilograms of body weight. Participants include people from all age and ethnic groups and from different parts of the country. Therefore, in addition to the general population, the survey also included 22 different subgroups, such as children, Hispanics, pregnant or nursing women, and people in certain sections of the country (32,33).

EPA compares the dietary exposure estimate for each pesticide with each RfD or with the level associated with an estimated lifetime cancer risk in the range of one in a million. If the dietary intake is less than the RfD / ADI or poses a negligible lifetime cancer risk, the pesticide is usually considered to be safe for food use. In the event, that the dietary exposure exceeds the RfD / ADI or poses a life cancer risk that is greater than one in a million, EPA will require the registrant to find ways to reduce overall dietary exposure for already registered food use and deny new food-use registration.

EPA and USDA established a Tolerance Reassessment Advisory Committee (TRAC) to provide a forum for a diverse group of individuals representing a broad range of interests and background from across the country to consult with and make recommendations to the Administrator of EPA and the Secretary of Agriculture on the matters relating to an approach for reassessing tolerances, including those for organophosphate pesticides, as required by the Food Quality Protection Act.

TRAC is composed of approximately 45 members approved by the Deputy Administrator of EPA and the Deputy Secretary of Agriculture. Members are selected based on their relevant experience and diversity of perspectives on organophosphate pesticide/food safety issues from the following sectors: environment and public interest groups; pesticide industry and trade associations; user, grower, and commodity organizations; pediatric and public health organizations; federal agencies, tribal, state, and local governments; academia; and consumer groups. The Deputy Administrator of EPA and the Deputy Secretary of Agriculture serve as co-chairs.

SECTION II. METHODOLOGY REQUIREMENTS AND METHOD VALIDATION

A. Methods

Pesticide residue methodology standards involve three (3) factors: (1) sampling, (2) analysis, and (3) data reporting. It is understood that there are different program purposes, and consequently different criteria, for pesticide residue tolerance enforcement and monitoring for risk exposure assessment (43).

1. Sampling

The objective of sampling is to obtain representative samples from the lot from which it was sampled for laboratory analysis. Each sample must be accompanied by a collection report prepared by the sampler, to include the following information: nature, origin, owner, grower, and approximate size of the lot; the time, date, and place of sampling; the method of sampling, and any other pertinent information.

The sample must be placed in a clean and inert container that provides adequate protection from damaging and contamination of the sample. The sample must be sent to the laboratory as soon as possible, via the fastest transportation method.

The following list tabulates the minimum amount of sample to be collected from the wholesaler or the retailer for laboratory analysis (44):

Small or light products (berries, peas, parsley, etc.)	2–3 lb
Medium-sized products (apples, carrots, potatoes, cucumbers, squash, etc.)	5 lb
Large-sized products (cabbage, melons, etc.)	10 lb
Dairy products (milk, cheese, butter, etc.)	1–2 lb
Meat, poultry, fish	2–3 lb

Oils and fats (cottonseed oil, margarine)	1–2 lb
Cereal and cereal products	1–2 lb

When sampling crops from a farm field, an imaginary grid is superimposed on the field so that the field is divided into approximately 100 subareas. Ten (10) of these subareas are selected at random. A 1-lb (one-pound) subsample from each subarea is collected, and subsamples are composited to a bulk sample for laboratory analysis. Samples are prepared representative to local harvesting practices, prior to shipping (45). Identification of the field and the location where the samples were collected must be entered on the sample collection report.

Enforcement. The enforcement sampling plan is based on prior application knowledge, compliance history, and anticipated potential areas. The plan is nonstatistical and is not representative of the overall pesticide residue situation. Samples are collected as close to the point of production as possible in order to be able to intercept violative products before reaching the consumer.

Monitoring. The monitoring sample plan must be statistically defensible and totally random. Sampling number and site selection are based on the amount of produce represented, with no bias. Sample collection points are as near to the consumption as possible.

2. Analysis

Enforcement. Enforcement samples are analyzed without washing or peeling in accordance with pesticide registration protocol. Quick screening procedures are used in order to perform timely interceptions of products containing pesticide residues in excess of established tolerances or guidelines. Method sensitivity must be capable of detecting residues at or less than the tolerances or guidelines.

Monitoring. Monitoring samples are prepared emulating the practices of the average consumer to more closely represent actual pesticide exposure. Lengthy and ultrasensitive analytical procedures are used in order to detect, measure, and confirm residues. Extra precautionary steps must be taken in order to avoid low-level cross-contamination.

Analyses under both programs must be performed under strict QA/QC (quality-assurance/control) protocols. Performance efficiency samples supplied by a third party should be analyzed periodically, and there must be a periodical third-party review of laboratory performance.

3. Data Reporting

The database of enforcement programs and monitoring programs should be entered separately and should include all corresponding methodology, QA/QC performed, limits of detection and quantitation, and similar data.

National Pesticide Residue Database. The National Pesticide Residue Database (NPRD) was initiated in 1992. The planning committee consisted of FDA, USDA, EPA, and the states of California and Florida. Funding to implement NPRD was not available until fiscal year 1997.

The data entered into the NPRD are to be individually coded for either enforcement or monitoring to allow independent analysis and reporting. Each data entry must be accompanied with standardized fields allowing for independent manipulation. The database will contain raw and processed food products. The purpose of the NPRD is to maximize the utility of pesticide residue data generated by various laboratories (46).

The following data elements must be included with all data submitted to NPRD:

1. Sampling
   a. Coordinator ID
   b. Sample ID
   c. Episode ID
   d. Sample description
   e. Sampling date
   f. Sampling rational
      i. Program category
      ii. Sampling basis
   g. Sampling origin
      i. Domestic / import
      ii. State collected
      iii. Farm / retail / wholesale
      iv. Country of origin
   h. Sample composition
   i. Postharvest pesticides (if any on the label)
2. Analytical
   a. Analysis date
   b. Commodity portion
   c. Sample classification

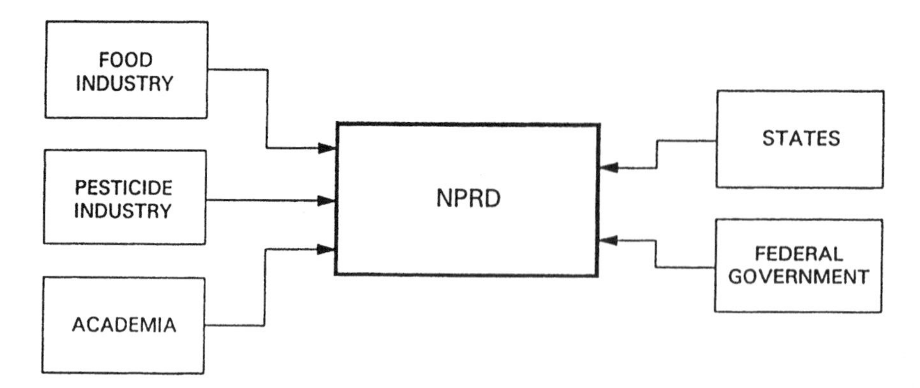

Figure 7.4. NPRD system potential contributors.

d. Sample preparation method
e. Analysis type
f. Extraction method
g. Determination method
h. Residue ID
i. Residue classification
j. Residue level
   i. Amount
   ii. Unit
k. Limit of Quantitation
   i. Amount
   ii. Unit
l. Limit of detection
   i. Amount
   ii. Unit
m. Spike level
   i. Amount
   ii. Unit

The benefits of NPRD are as follows:

- The availability of quality national pesticide residue monitoring data
- The ability of assessing realistic dietary exposure of pesticide residues
- The ability to better inform the public of the safety of the nation's food supply

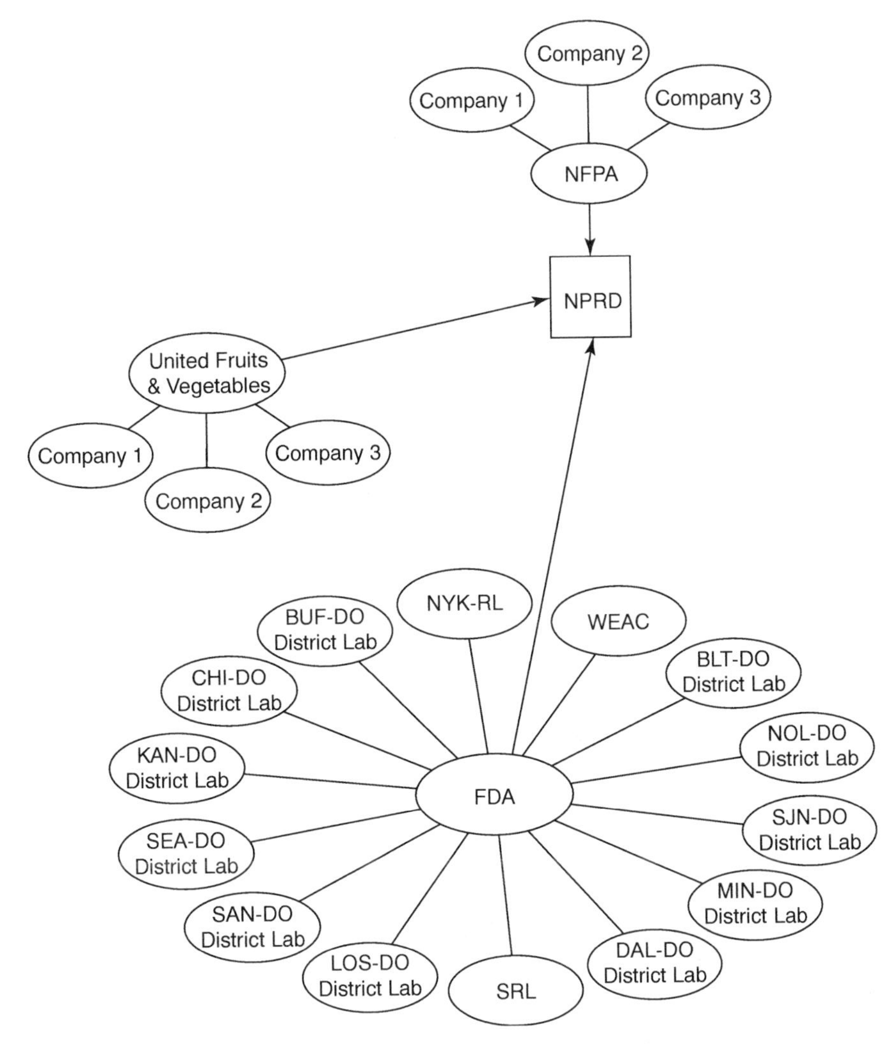

Figure 7.5. NPRD system general level.

- To provide the nation with safe, wholesome, and abundant food supply by making realistic assessment of the registration of currently "questionable" but "beneficial" pesticides and the reregistration of minor use pesticides

Figures 7.4–7.6 illustrate potential contributors to NPRD and database system levels.

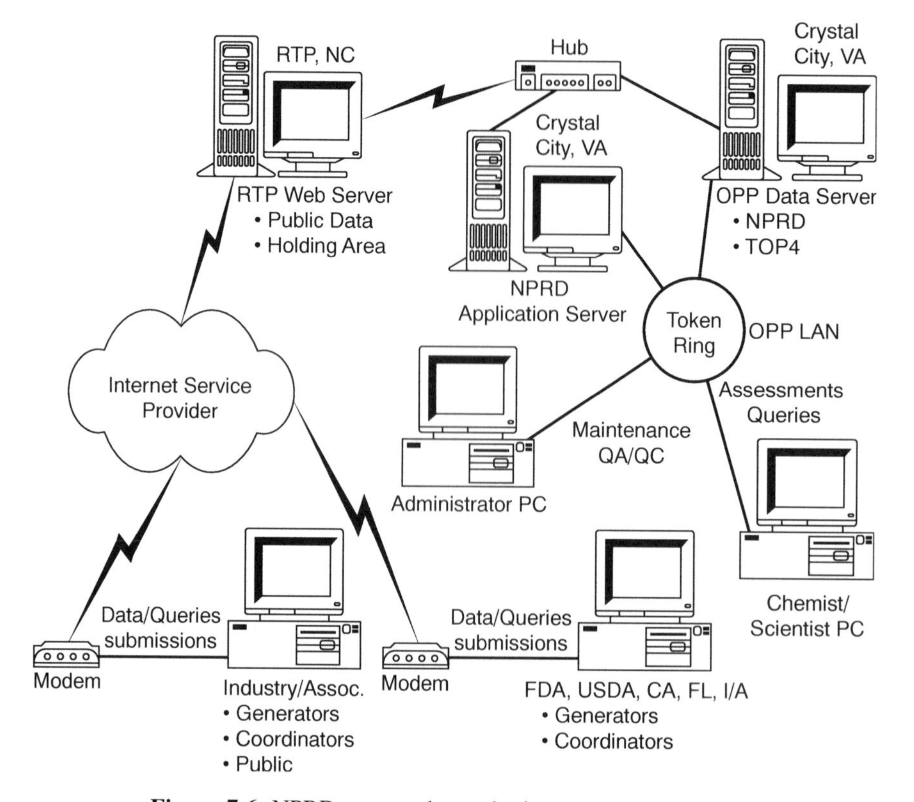

Figure 7.6. NPRD system data submitter coordinator level.

B. Method Validation

Association of Official Analytical Chemists (AOAC) Official Method Validation requires eight (8) laboratory collaborate studies and many years to complete. Often a method is out of date when it becomes an official AOAC method. Therefore, only a few pesticide residue methods have official AOAC status.

In many circumstances, laboratories must use available non-AOAC methods for analysis and also retain the credibility. Several published papers list the minimum criteria of QA / QC protocol for any analytical methods to be used in regulatory laboratories (47–50). The following parameters are extracted from the published papers that specify minimum analytical method validation requirements:

1. Accuracy
2. Recovery

3. Calibration curve
4. Linearity
5. Limit of detection
6. Limit of quantitation
7. Precision
8. Sensitivity
9. Specificity
10. Scope

Accuracy. *Accuracy* can be determined (average of a replicated set of trials) by use of certified reference materials, use of reference method of known uncertainty, or use of recovery from spiked samples.

Reference material and spiked samples should be carried through the entire procedure. The method of fortification of spiked samples should be described.

Recovery. *Recovery* can be determined by the amount of recovered added analyte over an appropriate range of concentrations. The number of replicated samples per study varies. Generally, the number of samples ranges from 3 to 12, depending on program protocol:

$$\text{Percent of recovery} = \frac{\text{analyte recovered}}{(\text{analyte added}) \times 100}$$

Calibration Curve. The *calibration curve* is defined as the responses of the method to a number of concentrations, minimum of 5 not including zero, of the analyte standards. Responses at various concentrations in pure or defined

Table 7.7. Inter- and Intralaboratory Coefficients of Variations

Analyte Concentration	CV(%)	
	Interlaboratory[a]	Intralaboratory[b]
10 ppm	11	7
1 ppm	16	11
100 ppm	23	15
10 ppb	32	21
1 ppb	45	30
0.1 ppb	64	43

[a] Taken from Horwitz curve.
[b] Two-thirds of Horwitz interlaboratory curves.

solvents and responses at various concentrations in matrix(ces) should be studied. For nonlinear curves, more concentrations of the standard, which are closer to the concentration of the analyte in the sample, are necessary.

Linearity. *Linearity* is determined by using external analyte standards at 4 nominal concentrations: $0x$, $1/2x$, $1x$, and $2x$, on 3 separate days (where x is the target concentration). A minimum linear correlation is .9995.

Limit of detection. There are several ways to define the *limit of detection* (LOD). Two examples are illustrated as follows:

1. The mean value of the matrix blank readings plus 3 standard deviations of the mean, expressed in analyte concentration
2. The amount, expressed in ppm or ppb (parts per million or billion), equivalent to 3 times the background signal contributed by the matrix blank

Limit of quantitation. There are several ways to define the *limit of quantitation* (LOQ). The values are established by repeated analysis of the appropriate samples, not by extrapolation. The examples are illustrated as follows:

1. The substrate blank plus 10 deviations
2. The amount, expressed in ppm or ppb, equivalent to 10 times the background signal contributed by the matrix blank

Precision. The *precision* of a method is assessed as the tightness of replicate fortifications measured by the relative standard deviation or coefficient of variation (CV). Interlaboratory CVs should be within the range estimated by Horwitz, et al., and intralaboratory CVs should be one-half to one-third of the value (Table 7.7) (48,50).

Sensitivity. *Sensitivity* is defined as the ability of the method to detect the analyte at the concentration of interest. To replace an existing method, the new method must be compared with the existing method. Ten comparative analyses are generally performed at the concentration level of interest.

Specificity. *Specificity* is defined as the ability of the method to actually determine the analyte, not interfering with the compounds. It is assessed by analyzing reagent blanks and sample matrix blanks. Chromatograms of reagent blanks and sample matrix blanks must be free of interfering peaks at the retention time(s) of interest. The method should be able to distinguish the analyte from all known interfering materials. The behavior of the analyte during analysis should be indistinguishable from the corresponding standard material in the appropriate matrix.

Scope. *Scope* refers to the number of different sample matrix to which the method can be successfully applied. To extend the scope of the method, additional method validation work must be performed on the sample matrix of interest.

SECTION III. LABORATORY MANAGEMENT

A. Laboratory Accreditation

Accreditation is defined as the formal recognition that an analytical laboratory is competent to carry out specific analyses or specific types of analysis. A scope of accreditation is presented to the laboratory when it receives its certificate to identify the specific analyses and types of analyses for which the laboratory is accredited (51).

There are numerous accreditation systems or bodies. Two environmental laboratory accreditation agencies are the National Environmental Laboratory Accreditation Conference (NELAC) and the American Association for Laboratory Accreditation (A2LA). General accreditation criteria are from the international standards, ISO/IEC Guide 25, 1990, "*General Requirements for the competence of Calibration and Testing Laboratories.*"

Elements of accreditation are based on the following:

Procedures	Documentation and adherence
Validation	Specificity, scope, accuracy, precision, range, limitation, and documentation
QA / QC	Standard operating procedures (SOPs), documentation, effectiveness, enforcement, etc.
Personnel	Qualification, troubleshooting ability, competence
Equipment	Appropriateness, maintenance, documentation

The following is the ISO Council Committee definition on Conformity Assessment:

Accreditation. Procedures by which an authorized body gives formal recognition that a body or person is competent to carry out specific tasks.

Certification. Procedures by which a third party gives written assurance (certificate of conformity) that a product, process, or a service conforms to specific requirements.

Registration. Procedures by which a body indicates relevant characteristics of a product, process, or service or particular of a body or person, in an appropriate publicly available list.

B. Good Laboratory Practice Standards

The Federal Insecticide, Fungicide and Rodenticide Act (FIFRA) Good Laboratory Practice (GLP) Standards were effective under 40 *CFR*, Part 160 (*1*) on October 16, 1989 and expanded regulations to include compliance for those studies routinely performed in laboratories, namely, residue chemistry, metabolism, environmental fate, and product chemistry studies. Each study should have an approved written protocol that clearly indicates the objectives and all methods for the conduct of the study (54,55).

EPA-FIFRA Good Laboratory Practice Standards protocol of pesticide studies includes the requirements of the following recordkeeping elements:

- Organization and personnel
- Facilities
- Equipment
- Testing facilities operation
- Test, control, and reference substances
- Protocol for conducting a study
- Records and reports

A typical GLP field audit procedure of an analytical chemistry laboratory includes a review of the following study components (56,57):

- Method development and validation
- Stability of test substance, metabolites, and degradation products in sample matrix
- Analyte reference standards for test substance, metabolites, and degradation products
- Reference standard solutions
- Test substance characterization, control, and handling
- Sample receipt and storage
- Sample preparation, extraction, and cleanup
- Sample analysis
- Quality control: replicates, controls, and reagent blanks

The details of these field audit study components are described in following paragraphs.

Method Development and Validation. The audit of an analytical phase of a study includes a review of the analytical methodology used to analyze specimen samples. Method development and validation are not required to be conducted according to the Good Laboratory Practice Standards regulations, except where the results of this work are submitted to the EPA as studies or parts of studies. However, method development and validation are vitally necessary to support any residue study; thus the associated raw data and records must be retained and available for review by the auditor.

The FIFRA Books and Records requirement as spelled out in 40 *CFR* 169.2 (k) also dictates retention of these records for the life of a pesticide registration.

The auditor will address the following issues relating to methodology:

- Were standard methods used [i.e., official (AOAC, EPA, FDA) or other recognized analytical methods]? If not, what was the source of the method? Was it appropriate for the intended purpose? Was it sufficiently sensitive, reproducible, and specific for the purpose of the study?
- Was the method validated by the analytical laboratory for actual analytical conditions? How was the validation conducted? Was the method validated by the same staff who conducted the sample analyses?
- Was the method validated by the analytical laboratory over the entire range of concentrations expected in the study?
- Was the method validated for all expected metabolites and/or degradation products expected in the study and required by the study protocol?
- Was the method validated for all sample matrices specified by the protocol?
- How were the limit of detection (LOD) and limit of quantification (LOQ) defined and determined? Were these limits appropriate and adequate? Were they properly calculated?
- Were all data generated during method validation retained by the laboratory and available for audit? Were there any discrepancies or deficiencies? Did the data support the final report?

Stability of Test Substance, Metabolites, and Degradation Products in Sample Matrix. Since there is generally a lapse of several days to weeks between the collection of a specimen sample and its analysis, the study personnel must be able to document that the pesticide and, where applicable, its metabolites and/or degradation products were stable in the sample under the conditions of storage for the maximum period that the sample was stored. The auditor will verify that the stability testing was done and that data and records are present and available for audit.

Stability testing may be conducted either prior to the study or concurrently. The auditor should verify the following:

- Was stability testing performed? What laboratory did the testing? Did the overall study protocol address stability testing, or was it a separate study with its own protocol?
- If stability testing was not performed, was there justification for this? Was stability testing conducted using a similar sample matrix rather than the identical sample matrix (carrots when the study was done on beets, corn fodder when the study was done on sorghum fodder, etc.)? Did this failure to conduct stability testing on the study sample matrix appear to affect the validity of the study test results?
- What fortification levels were used for stability testing? Were they similar to the expected residue levels in study samples?
- Did the timeframe of stability testing match or exceed the maximum storage period for the study samples?
- Were fortified samples analyzed using the same methodology employed for the study samples?
- Did the stability testing demonstrate that there was adequate storage stability for the conditions (temperature, humidity, etc.) of sample storage and analysis?
- Were all stability data available for audit and consistent with the reported findings?

Analytical Reference Standards. The analysis and characterization of analytical reference standards (reference substances), as well as the documentation of receipt and distribution, are Good Laboratory Practice (GLP) Standards issues.

Analytical reference standards may be obtained from a number of sources including the sponsor, commercial suppliers, custom synthesizers, and university and other research laboratories. In some instances, particularly with older materials predating the GLP Standards regulations, the source may not be known. The analysis and characterization of the reference standards may have been conducted by the supplier, who may furnish nothing more than a label statement of percent purity. In addition, the reliability of this analysis may not be known.

Under the new regulations, test facility management is responsible for assuring that the reference standard is appropriately tested for identity and purity, and must either obtain the raw data and records for this testing from the supplier or arrange to have the material independently analyzed. In either event, the analyses must be performed under GLP purview, and the raw data and records must be retained and available for audit. During the audit, the auditor will verify the source of the reference standards, review the analytical data supporting the identity and purity of these chemicals, and ascertain whether the data were generated according to GLP requirements. When the source of standard cannot be determined, appropriate additional documentation shall be obtained by the inspector, such as personal statements or copies of labels.

Analytical reference standards for pesticide metabolites and degradation products are often difficult to obtain in sufficient quantity to permit the type of detailed analysis and/or characterization that is possible for the parent pesticide and its more common metabolites. The auditor must exercise professional judgment in addressing the adequacy of any analysis, characterization, and archiving conducted (or not conducted) for these types of standards. The study records should address the source of the metabolite standard(s) and the method(s) of synthesis, if known. They should also contain as much information as possible concerning analysis for identity, purity, and stability and should specifically address any problems or restrictions in obtaining detailed analyses. The inspector and/or auditor will pay particular attention to reconstruction of the synthesis of radiolabeled compounds to assure that the label is in the specified position of the molecule.

In general, if any analyses of analytical reference standards were conducted at the audited facility, the auditor will conduct an audit of the raw data and other records. These analyses are also required to have been conducted under GLP Standards regulations since October 1989, and the underlying raw data and records must be retained.

Additional points to be considered when reviewing the identity and purity of analytical reference standards include the following:

- Has stability been demonstrated? How long ago were the reported chemical analyses conducted? Were they recent enough to preclude any subsequent significant changes in purity or composition, if no stability data were available?
- Were the specific activity and radiochemical purity determined for radiolabeled analytical standards? Was stability verified for these compounds, if appropriate?
- Were the reference standards stored so as to minimize degradation? Was labeling adequate to prevent mixup and to meet requirements of 40 *CFR* Sections 160.105 and 160.107?

Standard Reference Solutions and Instrument Calibration. Several additional issues concerning analytical reference standards will also normally be addressed by the auditor, including preparation and storage of analytical reference solutions, determination of detector response and detector linearity, and any other factors relating to the reliability and/or validity of the analytical reference standard solutions. When auditing the raw data and records relating to these issues, the auditor will consider the following:

- Were data retained that document the preparation of stock solutions and working dilutions of analytical reference standards? Who prepared the standard solutions? Was there an SOP in place and being followed for preparation and handling of stock and working standard solutions?
- How were standard solutions stored? What kind of containers were being used? Did changes in concentration caused by evaporation of the solvent appear to be minimal? Were standard solutions protected from degradation by light?
- Was proper labeling used to uniquely identify personnel, notebook reference, expiration date, and other data to preclude mixup and use of outdated standards?
- How often were fresh working standards prepared? Were practices consistent with applicable SOPs and/or protocols?
- Were standard solutions used exclusively for this study, or were they used for more than one study? Did more than one person appear to have used them? Was there evidence of possible contamination or other loss of integrity for the standard solutions?
- Did the instrument detector response to the analytical standards solution remain relatively stable, or did it change with time? If it changed, was there a steady progression to either greater or lesser response, or did the variation appear to be random? Were study personnel aware of any changes in detector response? Was there documentation of the cause and, if appropriate, any remedial action taken?
- Was quantification made from the calibration curve or from single standards? Was there documentation that the detector response was established over the entire range used for quantification of samples? If the response was not linear, how was this addressed in the calculations?
- How often was a new calibration curve prepared? How often, relative to analysis of samples, was the instrument recalibrated with analytical reference standard(s)?

Test Substance Characterization Control and Handling. Test substance analysis, characterization, stability/solubility, receipt, distribution, and handling are specific GLP Standards issues.

Since radiolabeled test substances are often used in chemical fate and metabolism studies, the auditor should be certain that the specific activity and radiochemical purity of the test substance are adequately determined and documented. It is also very important that the synthesis of a radiolabeled test substance be fully and adequately documented, since the position of the label in the molecule is usually germane to the study results and adherence to test guidelines.

Analysis of Test Substance Mixture. Preparation and analysis of mixtures of test substance with carrier and preparation and analysis of agricultural tank mixes are addressed separately in GLP Standards.

Sample Receipt and Storage. The study protocol or facility SOPs should address procedures to be followed for receipt of samples, maintenance of records to track the distribution and disposition of samples, and storage of samples prior to analysis. The auditor will verify that these procedures were adequate and were followed. Review of records and documents and interviews with study personnel should enable the auditor to answer the following questions:

- What records were maintained to document receipt of the sample(s)? Did the records document date and time of receipt and condition of the sample(s) receipt?
- Was there a logbook or other documentation for the storage location and for the distribution of the sample(s)? Where were the samples stored prior to analysis? If refrigerated or frozen, were there records of temperatures during storage?
- When were samples distributed for analysis? Who obtained them? How much was distributed, and on what occasions? How was surplus material disposed of?
- Was sample preparation (grinding, sieving, drying, etc.) adequately documented as to when, where, and who performed the operations, particularly if performed by a group separate from the analytical personnel?
- Were storage and handling procedures adequate to ensure the integrity of the sample(s)?

Sample Preparation, Extraction, and Cleanup. Analytical methodology should specify procedures to be used for sample preparation (homogenizing, mixing, grinding) and isolating the analyte(s) from the sample matrix, usually by extraction and some form of cleanup. The auditor will determine through review of available documentation that this method was followed, and that required procedures were followed in the event that deviations from the written and approved SOP or protocol occurred. The areas to be reviewed by the auditor include the following:

- Who prepared and extracted the sample?
- How were data recorded? Were worksheets and/or notebooks properly used, with dates and identification of technicians recorded?
- Were balances calibrated? How often and by whom? Were calibration procedures appropriate for their use in the analyses?
- When sample preparation and analysis could not be completed in 1 day, how and where were samples and/or extracts stored? Did the raw data adequately document storage conditions?
- How much time elapsed between receipt of samples and subsequent analysis? How much time elapsed between the beginning of sample preparation and final quantitative or qualitative measurement? If delays occurred, were reasonable explanations offered? Were delays the possible result of too few technicians or analysts?
- Were delays caused by instrument or facility problems? Were the quality-assurance unit and study director aware of unusual delays? Were stability data generated or available that covered the encountered delays?

Sample Analysis. Quantitative and/or qualitative analysis is normally the final technical phase of a residue study. The auditor will determine that this was conducted, as described in analytical methodology and facility SOPs. Initially, at least 10% of the analytical data should be audited. When discrepancies, data gaps, or other problems are encountered, additional data (up to 100%) should be audited, in order to determine the magnitude of any data problems. The auditor will also verify that all standards of good analytical practice are followed. At a minimum, the data audit should address the following:

- Were there any changes in procedure from that given in the protocol and/or SOPs? Were these properly documented?
- Were instruments always properly calibrated? How was this documented?
- Was automated data collection (ADC) used? Were hard copies of the data collected in this manner archived with other study data? Were tapes and/or disks also archived? Were there written SOPs for ADC procedures and data storage?

- Were SOPs or protocol directions available to define criteria for (1) reanalysis of samples, (2) number of significant figures to be reported, (3) rounding, and (4) reporting analytical values that were less than the limit of quantification and less than the limit of detection?
- Were analytical reference standards analyzed concurrently and at subsequent appropriate intervals? Were calculations made using linearity curves or from single reference points?
- Was it possible to reconstruct the study from the raw data and other records? Was it possible to follow manipulations of data, application of correction factors, averaging of results from replicate analyses, and other procedures used to produce derived or summarized data, tables, or figures?
- Did the quality-assurance unit review the analytical results? Was the review adequately documented and reported? Did the study report contain errors that should have been found during internal reviews by study personnel, the quality-assurance unit, or the study director?
- What percent of reported data and results were verified by the EPA auditor? Were all necessary data and records available for audit? Were any discrepancies noted between raw data and reported study results? Were any irregularities noted? Were they few and sporadic or significant in number and/or importance?
- Were all the analytical data that were generated during the study used in the final report? If some data were not used, what was given as the reason? Was the reason scientifically valid? Could the study conclusions have been affected by the exclusion of these results?

Quality Control: Replicates, Controls, and Reagent Blanks. Because of the potential for analytical problems in residue determinations, a certain amount of quality control (QC) should be conducted concurrently with the study sample analyses. This QC usually takes the form of some or all of the following: (1) periodic analysis of replicate samples to determine reproducibility of analytical results; (2) analysis of controls (samples from portions of the test system that were not treated with the test substance) to determine potential interference from the sample matrix, containers, sampling equipment, or other contamination introduced from soil, water, agricultural practices, or other sources during the conduct of the field portion of the study; (3) analysis of reagent or blanks to determine interference or contamination from the reagents or glassware used in the analyses; and (4) analysis of control samples that have been fortified with known quantities of the analyte(s) to determine the analytical efficiency of the methodology.

The study protocol and/or SOPs should define which of the above QC procedures are to be included with the study and how often the procedures are to be conducted. The auditor will verify that the requirements of the protocol and/or SOPs are met. If no or inadequate QC procedures were conducted in conjunction with the study, the auditor should determine the reasons and make an evaluation of the significance of this lapse. Failure to perform proper quality control could seriously compromise study results.

C. Good Automated Laboratory Practice

Good Automated Laboratory Practice (GALP) ensures data integrity in automated laboratory operations. The Environmental Protection Agency published a draft document, Good Automated Laboratory Practices, in December 1990 for laboratories that provides data to EPA through requirements of the Toxic Substances Control Act (TSCA) and the Federal Insecticide, Fungicide, and Rodenticide Act (FIFRA) (58). The document made recommendations in the following six (6) principles: data, formulas, audit, change, SOPs, and disaster, and distinguishes the following six (6) operational roles: laboratory management, responsible person (RP), quality-assurance (QA) unit, archivist, vendor, and users.

Principles

Data. The system must provide a method of assuring the integrity of all entered data. The demonstration of control necessitates the collection of evidence to prove that the system provides reasonable protection against data corruption.

Formulaes. The formulas and decision algorithms employed by the system must be accurate and appropriate. The formulas must be inspected and verified.

Audit. An audit trail that tracks data entry and modification to the responsible individual is a critical element in the process. The trail identifies the person(s) entering a data point and generates a protected file logging all unusual events.

Change. A consistent and appropriate change control procedure capable of tracking the system operation and application software is a critical element in the control process. All software changes should follow carefully planned procedures, including a preinstallation test protocol and appropriate documentation update.

Standard Operating Procedure (SOP). Control of even the most carefully designed and implemented system will be thwarted if appropriate user procedures are not followed. This principle implies the development of clear directions and standard operating procedures, the training of all users, and the availability of user support documentation.

Disaster. Consistent control of a system requires the development of alternative plans for system failure, disaster recovery, and unauthorized access. The principle of control must extend to planning for reasonable unusual events and system stresses.

Operational Roles

Laboratory Management. Laboratory management has ultimate responsibility for all GALP standards. The management shall designate the responsible person; arrange for QA oversight of the system; provide the necessary resources, facilities, and equipment that may be required; receive and respond to QA reports and audits; and provide all other laboratory personnel with the guidance, training, or supervision they require to perform successfully in their assigned roles.

Responsible Person (RP). The identification of the system RP eliminates the confusion of exactly who or what organization unit is ultimately responsible for a specific system. The RP is generally a professional with some computer background, in a position of authority related to the control and operation of the automated data system. The RP's responsibilities include training of users, implementing appropriate security measures, developing or reviewing SOPs for system use, enforcing change control procedures, and responding to emerging problems.

Quality-Assurance (QA) Unit. This unit must be independent from the RP of the automated data system in order to establish its legitimacy and credibility. It is responsible for reviewing system SOPs, inspection and audit of the system, review of final reports and data integrity, and review of archives.

Archivist. The archivist is responsible for the safe storage and retrieval of all records required by EPA statute or legal judgment to be retained.

Vendor. The vendor is the one who designs, codes, supports, licenses, and/or distributes automated systems that meet the requirements specified in the GALP. If the vendor is an outside source, the laboratory management is responsible for informing the vendor of the GALP requirements. If the vendor is within the in-house system, the RP is responsible for ensuring satisfactory compliance with GALP requirements.

Users. All system users are responsible for familiarity with and conformity to SOPs and are expected to comply with and support management policies. The GALP assumes laboratory professionals are personally motivated to follow the principles of their professions and that they will take every practical step to ensure the accuracy and reliability of the data and analyses produced by their laboratory. Compliance with GALP will assure the reliability of much of the data that EPA uses in reaching decisions on human health and the environment.

Conclusion. Although pesticide residue analysis methodologies are well established, there is still some need for more rapid, and cost-effective, and environmentally friendly procedures to satisfy the knowledge–quest government officials and consumers. Pesticide residue analysts have the responsibility to produce true scientifically based data to ensure that federal, state, and local government agencies, industries, academia, and consumers are able to share the responsibility of the distribution of accurate information in regard to the balance of food and environmental safety versus the need for affordable food supply for the exploding world population.

D. National Laboratory Accreditation

The 1990 Farm Bill (Public Law 101–624) authorized the creation of a National Laboratory Accreditation Program (NLAP) for laboratories that request accreditation and conduct pesticide residue testing of agricultural products or that make claims to the public or buyers of agricultural products concerning pesticide residue levels on agricultural products. The program is designed to ensure the safety of agricultural food products by ensuring that organizations that make claims to the public concerning pesticide residue levels based on laboratory analyses of agricultural food products meet minimum quality and reliability standards. The standards for the NLAP will be developed by the U.S. Food and Drug Administration (FDA) under the Department of Health and Human Services (DHHS) with the operation of the program under the Agricultural Marketing Services and the Food Safety Inspection Services under the U.S. Department of Agriculture (USDA) (51).

The NLAP is in the rule-making stage. However, the standards will include, but not be limited to, laboratory facilities, analytical methods, training and supervision of personnel, provisions for handling of samples, recordkeeping, supervisory education requirements, in-house quality-control mechanisms, and quality-assurance programs (proficiency check sample and on-site inspection by members of an accreditation body). The NLAP is voluntary and applies only to commercial laboratories (53).

The accreditation body consists of qualified state or nonprofit entities approved by the Secretaries of DHHS and USDA. The operating revenue of the NLAP will be generated by charging fees to laboratories requesting accreditation. All current authorities of the Secretary of DHHS under the Food, Drug, and Cosmetic Act (FFDAC) regarding pesticide residue in foods, labeling of foods, and any health-based claims about such foods, will not be substituted or affected by the NLAP.

DISCLAIMER

This chapter of the book has *not* been prepared on behalf of the Florida Department of Agriculture and Consumer Services, and therefore does not represent any official policy of that Department.

REFERENCES

1. *Pesticide and Public Policy*, National Agricultural Chemicals Association, Washington, DC, 1960.
2. W. G. Fong, "Pesticide residue methodology in foods—an overview of the past, current initiative and future directions," Paper No. 012, presented at the 1991 Pittsburgh Conference and Exhibitions.
3. *The Federal Insecticide, Fungicide and Rodenticide Act as Amended*, U.S. Environmental Protection Agency, revised 1988.
4. *The Federal Food, Drug and Cosmetic Act, as Amended and Related Laws*, U.S. Food and Drug Administration, revised 1989.
5. *Florida Pesticide Residue Monitoring Program*, Chemical Residue Laboratory, Division of Food Safety, Florida Department of Agriculture and Consumer Services, Tallahassee, FL 1998 revision.
6. *Code of Federal Regulations*, Title 40, Parts 150–189, The Office of the Federal Register National Archive and Records Administration, Washington, DC, 1997.
7. *Pesticide Reregistration*, U.S. Environmental Protection Agency, Office of Pesticide and Toxic Substances, Washington, DC, March 1991.
8. Food Quality Protection Act of 1996, including Title IV—Amendment to the Federal Food, Drug, and Cosmetic Act, 104th Congress, 2nd Session, Report 104–669, Part 2.
9. *Pesticide Regulation (PR) Notice 97–1*, U.S. Environmental Protection Agency, Jan. 31, 1997.

10. *Code of Federal Regulations*, Title 7, Part 205, The Office of the Federal Register National Archive and Records Administration, Washington, DC, 1993. [Docket Number TMD-94-00-2], RIN: 0581-AA40. "Compliance and Testing" Sections 205.430–205.433.
11. *California Pesticide Regulatory Program*, Department of Pesticide Regulation, California Environmental Protection Agency, Sacramento, California, Nov. 1991–Dec. 1993.
12. "The Florida Pesticide Law," *Florida Statutes*, Charter 487, Florida Department of Agriculture, Tallahassee, FL, 1997 revision.
13. "EPA and the Interpretation of the Delaney Clause," U.S. Environmental Protection News Release, Jan. 19 and Feb. 1993.
14. FDA Compliance Policy Guides, Office of Enforcement, Division of Compliance Policy, Associate Commissioner for Regulatory Affairs, Food and Drug Administration, Washington, DC.
15. "Magnitude of the Residues: Crop Field Trials—Hazard Evaluation Division Standard Evaluation Procedure," Office of Pesticide Programs, U.S. Environmental Protection Agency, Washington, DC, 1985.
16. *Pesticide Analytical Manual*, Vols. I and II, U.S. Food and Drug Administration, Washington, DC, 1994 Revision.
17. M. F. Kovacs, Jr. and C. L. Trichilo, *Development of Analytical Methods for Tolerance Enforcement—an EPA Overview of Current Initiatives and Future Directions*, Office of Pesticide Programs, U.S. Environmental Protection Agency, Washington, DC, 1989.
18. "Pesticide Tolerances," *Environmental Fact Sheet*, Office of Pesticides and Toxic Substances, U.S. Environmental Protection Agency, Jan. 1990.
19. *Pesticides—EPA's Formidable Task to Assess and Regulate Their Risks*, United States General Accounting Office Report to Congressional Requesters, April 1986.
20. *Pesticides—Need to Enhance FDA's Ability to Protect the Public from Illegal Residues*, United States General Accounting Office Report to Congressional Requesters, Oct. 1986.
21. *Requirements of Laws and Regulations Enforced by the U.S. Food and Drug Administration*, HHS Publication No. (FDA) 85–1115, U.S. Food and Drug Administration, Rockville, MD.
22. 21 U.S.C. Sections 301–391, Chapter 675, Public Law 75–717.52, Statute 1040 as amended.
23. *Pesticide Residues in Food—Technology for Detection*, Congress of the United States, Office of Technology Assessment, OTA-F-398, Washington, DC, Oct. 1988.
24. D. V. Reed, The FDA surveillance index for pesticides: establishing food monitoring priorities based on potential health risk, *J. Assoc. Off. Anal. Chem.* **68**(1), 121 (1985).

25. *Residue in Food*, 1987, 1988, 1989, 1990, 1991, 1992 and 1993, U.S. Food and Drug Administration, Washington, DC.
26. "Partnership Agreement," signed by U.S. Food and Drug Administration and Florida Department of Agriculture and Consumer Services.
27. "Pesticide Data Program, January–June, 1996 Report," U.S. Department of Agriculture, Agricultural Marketing Science, Washington, DC, July 1993.
28. J. W. Wells and W. G. Fong, State pesticide regulatory programs and the food safety controversy, in *Pesticide Residue and Food Safety—a Harvest of Viewpoints*, a special conference sponsored by the Division of Agrochemicals of the American Chemical Society, Jan. 1990.
29. Personal correspondence of B. Cusick, California Department of Food and Agriculture, Sacramento, CA, Dec. 1993.
30. "The Florida Food Safety Act," *Florida Statutes*, Chapter 500, Florida Department of Agriculture and Consumer Services, Tallahassee, FL, 1993.
31. J. R. Magness, G. M. Markle, and G. C. Compton, *Food and Feed Crops of the United States — A Descriptive List Classified According to Potential for Pesticide Residues*, New Jersey Agricultural Experimental Station, Rutgers University.
32. P. A. Fenner-Crisp, "Risk assessment methods for pesticides in food and drinking water," Office of Pesticide Programs, U.S. Environmental Protection Agency, Paper presented at the Florida Pesticide Review Council Meeting, July 7, 1989.
33. F. J. Francis, *Food Safety: The Interpretation of Risk*, Council for Agricultural Science and Technology, Ames, Iowa, No. CC-1992-1, April 1992.
34. F. J. Francis, Testing for toxicity, *J. Sci. Food Agric.* **4**(1), 10 (1986).
35. *A Proposed Food Safety Evaluation Process*, Final Report of the Board of Trustees, Nutrition Foundation, Food Safety Council, Washington, DC, 1982.
36. F. J. Francis, Testing for carcinogen, *J. Sci. Food Agric.* **4**(2), 10 (1986).
37. *Pesticides in the Diets of Infants and Children*, National Research Council, National Academy Press, Washington, DC, 1993.
38. B. Ballantyne and W. G. Fong, Toxicology, *Kirk-Othmer Encyclopedia of Chemical Technology,* in 4th ed., Vol. 24, Wiley, New York, 1997.
39. *Regulating Pesticides in Food—The Delaney Paradox*, National Research Council, National Academy Press, Washington, DC, 1987.
40. C. K. Winter, Pesticide tolerances and their relevance as safety standards, *Regul. Toxicol. Pharmacol.* **15**, 137–150 (1992).
41. "The Delaney Paradox and Negligible Risk," *Environmental Fact Sheet*, Office of Pesticides and Toxic Substances, Environmental Protection Agency, Washington, DC, Jan. 1990.
42. *Assessing Pesticide Risks*, United States General Accounting Office Report to Congressional Requesters, 1986.
43. W. G. Fong, Pesticides in foods, *Environ. Test. Anal.*, (Burbank, CA) **3**(2), 24, 26 (1994).

44. *Recommended Method of Sampling for Determination of Pesticide Residues, Guide to Codex Recommendations Concerning Pesticide Residues*, Part 5, Codex Alimentarius Commission, Joint FAO/WHO Food Standards Programme.
45. "Pesticide Sampling," *Inspection Operations Manual*, Food and Drug Administration.
46. "Data Element Requirements," National Pesticide Residue Database Meeting, Jan. 9, 1998, U.S. Environmental Agency Head Quarters, Crystal City, VA.
47. H. B. S. Conacher, Validation of methods used in crisis situations: task force report, *J. Assoc. Off. Anal. Chem.* **72**(2), 332–335 (1990).
48. G. A. Parker, Validation of methods used in the Florida Department of Agriculture and Consumer Services' Chemical Residue Laboratory, *J. Assoc. Off. Anal. Chem.* **74**(5), 868 (1991).
49. *AOAC Peer-Verified Methods Program—Manual on Policies and Procedures*, Association of Official Analytical Chemists International, Washington, DC, 1993.
50. W. Horwitz, Evaluation of analytical methods used for regulation of foods and drugs, *Anal. Chem.* **54**(1), 67A–76A, (Jan. 1982).
51. P. S. Unger, *ISO 90000 versus ISO/IEC Guide 25 for Laboratories*, American Association for Laboratory Association, Gaithersburg, MD, March–April 1996.
52. National Laboratory Accreditation Program, *Fed. Reg.* **57**(225), 54727–54728 (Nov. 20, 1992).
53. Draft document of "National Laboratory Accreditation Program," 7 *CFR* Part 98, RIN 0581-AA38, U.S. Department of Agriculture Agricultural Marketing Service and Food Safety Inspection Service and U.S. Food and Drug Administration, Washington, DC, Nov. 1991.
54. *Enforcement Response Policy for the Federal Insecticide, Fungicide and Rodenticide Act Good Laboratory Practices (GLP) Standards*, Office of Compliance Monitoring, Office of Pesticides and Toxic Substances, U.S. Environmental Protection Agency, Sep. 30, 1991.
55. G. Burnett, *Practical Protocols for the Laboratory*, Ciba-Geigy Corporation, Agricultural Division, Greensboro, NC, 1990.
56. *Glossary of GLP Terms*, GLP-S-05, Office of Compliance Monitoring, Office of Pesticides and Toxic Substances, U.S. Environmental Protection Agency, 1991.
57. *Auditing Filed Studies (Analytical Chemistry)*, GLP-DA-01, Office of Compliance Monitoring, Office of Pesticides and Toxic Substances, U.S. Environmental Protection Agency, 1991.
58. *Good Automated Laboratory Practices*, U.S. Environmental Protection Agency, Draft Document, Dec. 1990.

APPENDIX

PESTDATA

This table* serves as a quick reference for identification of residues analyzed by *Pesticide Analytical Manual* (1993) multiresidue methods. Standards should be chromatographed in the individual laboratory's GLC system for accurate comparisons. PESTDATA does not include all details; individual *PAM* tables that accompany each method should be consulted for more detailed information.

Explanation of PESTDATA

Name. Preferred name for each chemical. Asterisk (*) indicates chemicals with multiple GLC peaks.

RRT. Retention times relative to chlorpyrifos. Conditions under which RRTs were obtained are described in detail in *PAM*. Approximate operating conditions are as follows:

GLC Column	Percent Liquid-Phase Load	Target RRT for Marker Compound
OV-101	5	3.1 ± 0.06 (p,p'-DDT)
		2.56 ± 0.05 (ethion)
OV-17	3	3.5 ± 0.07 (p,p'-DDT)
OV-225	5	3.6 ± 0.06 (p,p'-DDT)
		3.9 ± 0.1 (ethion)
		0.69 ± 0.02 (lindane)
DEGS	5	—

*Adapted from more complete "Appendix I" in the *Pesticide Analytical Manual* (1993).

Pesticide Residues in Foods: Methods, Techniques, and Regulations, by W. G. Fong, H. A. Moye, J. N. Seiber, and J. P. Toth. Chemical Analysis Series, Vol. 151.
ISBN 0-471-57400-7

1. All columns are 1.8 m × 2 or 4 mm in diameter; liquid phase coated on 80/100-mesh Chromocarb WHP or equivalent.
2. Column and injection temperatures are 200°C. Carrier-gas flow is about 60 mL/min. All columns are preconditioned as per common GLC practice.
3. Note that most RRT data are for packed columns. Megabore open-tube columns are now preferred over packed columns. RRTs are not substantially altered by the substitution.

Detection. PESTDATA in *PAM* (1993) contains response information on the following detectors:

Flame photometric
^{63}Ni electron capture
Nitrogen/phosphorus
Electroconductivity
Microcoulometric

Recoveries. Data on recoveries of each compound through *PAM* multi-residue methods are listed. For method 304, approximate elution fractions are given. Recovery codes are as follows: C, complete (>80%); P, partial (50–80%); S, small (<50%); V, variable; R, recovered but no quantitative information available; NR, not recovered.

PESTDATA Chemicals in Order by Chemical Name

Name	Molecular Formula	RRT/c OV-101	RRT/c OV-225	RRT/c OV-17	RRT/c DEGS	Recoveries				
						302	303	304	Ethers	$MeCl_2$
1,1′-(2,2-Dichloroethylidene) bis(2-methoxybenzene)	$C_{16}H_{16}Cl_2O_2$	2.9	—	3.7	—	—	R	—	—	—
1,2,3,4-Tetrachlorobenzene	$C_6H_2Cl_4$	—	—	0.09	—	—	—	—	—	—
1,2,3,5-Tetrachlorobenzene	$C_6H_2Cl_4$	—	—	0.07	—	—	P#	—	6	1
1,2,3-Trichlorobenzene	$C_6H_3Cl_3$	0.08	—	—	—	—	C	P	6	1
1,2,4,5-Tetrachloro-3-(methylthio)benzene	$C_7H_4Cl_4S$	0.49	0.35	0.48	—	R	C	—	6	1
1,2,4,5-Tetrachlorobenzene	$C_6H_2Cl_4$	—	—	0.07	—	—	—	—	—	—
1,2,4-Triazole	$C_2H_3N_3$	0.2	—	0.27	0.37	V	NR	NR	6–15–50	1–2–3
1-Hydroxychlordene	$C_{10}H_6Cl_6O$	0.99	1.63	1.07	—	—	R	—	15	—
1-Methyl cyromazine	$C_7H_{13}N_6$	—	—	0.72	—	—	—	—	—	—
10,10-Dihydromirex	$C_{10}H_2Cl_{11}$	2.87	—	—	—	—	C	—	6	—
10-Monohydromirex	$C_{10}HCl_{11}$	4.26	—	—	—	—	C	—	6	—
2,3,5,6-Tetrachloroanisidine	$C_7H_5Cl_4NO$	0.59	0.73	0.66	—	—	C	—	6	2
2,3,5,6-Tetrachloroanisole	$C_7H_4Cl_4NO$	0.24	0.15	0.22	—	—	C	—	6	1
2,3,5,6-Tetrachloronitroanisole	$C_7H_3Cl_4NO_3$	0.56	0.63	0.56	—	—	C	—	6	1 + 2
2.3,5,6-Tetrafluoro-4-hydroxy-methylbenzoic acid	$C_8H_4OF_4$	—	—	—	0.46	—	—	—	—	—
2,3,5-trimethacarb	$C_{11}H_{15}NO_2$	0.35	0.60	0.38	0.91	C	S#	NR	50	1–2–3
2,3-Dihydro-3,3-methyl-2-oxo-5-benzofuranyl methyl sulfonate	$C_{11}H_{12}O_5S$	0.68	2.89	0.93	6.2	—	—	—	—	—
2,4,5-T BEP ester*	$C_{17}H_{23}C_{13}O_3$	0.16	0.14	—	—	—	—	—	—	—
		0.68	0.66	—	—	—	—	—	—	—

PESTDATA Chemicals in Order by Chemical Name *(contd.)*

Name	Molecular Formula	RRT/c OV-101	RRT/c OV-225	RRT/c OV-17	RRT/c DEGS	Recoveries 302	303	304	Ethers	$MeCl_2$
		2.85	1.19	—	—	—	—	—	—	—
		3.3	2.78	—	—	—	—	—	—	—
		5.3	3.28	—	—	—	—	—	—	—
		7	7.7	—	—	—	—	—	—	—
2,4,5-T butoxyethyl ester*	$C_{14}H_{17}C_{13}O_4$	—	2.66	—	—	—	—	—	—	—
		2.91	3.3	—	—	—	—	—	—	—
2,4,5-T butyl esters*	$C_{12}H_{13}Cl_3O_3$	—	—	0.86	—	—	—	—	—	—
				1.05						
2,4,5-T ethylhexyl ester	$C_{16}H_{21}Cl_3O_3$	3.38	—	2.62	—	—	—	—	—	—
2,4,5-T isobutyl ester	$C_{12}H_{13}Cl_3O_3$	0.94	—	—	—	—	—	—	—	—
2,4,5-T isoctyl ester*	$C_{16}H_{21}Cl_3O_3$	—	2.69	—	—	—	—	—	—	—
		2.56	3.1	—	—	—	—	—	—	—
		2.96	3.4	—	—	—	—	—	—	—
		3.25	3.8	—	—	—	—	—	—	—
2,4,5-T isopropyl ester	$C_{11}H_{11}Cl_3O_3$	0.67	0.65	—	—	—	—	—	—	—
2,4,5-T methyl ester	$C_9H_7Cl_3O_3$	0.49	0.63	0.47	—	—	—	—	—	—
2,4,5-T *n*-butyl ester	$C_{12}H_{13}Cl_3O_3$	1.1	—	—	—	—	—	—	—	—
2,4,5-T propylene glycol butyl ether esters	$C_{15}H_{19}Cl_3O_4$	2.37	—	—	—	—	—	—	—	—
2,4,5-Trichloro-alpha-methyl-benzenemethanol	$C_8H_7OCl_3$	0.34	—	0.25	—	R	R	—	15	—
2,4,-D BEP ester*	$C_{17}H_{24}Cl_2O_4$	—	0.08	—	—	—	—	—	—	—
		0.69	0.74	—	—	—	—	—	—	—
		1.66	1.18	—	—	—	—	—	—	—

		2	1.79	—	—	—	—	—	—	—
		3.22	2.09	—	—	—	—	—	—	—
		4.1	5.1	—	—	—	—	—	—	—
		10.2	13	—	—	—	—	—	—	—
2,4-D butoxyethyl ester*	$C_{14}H_{18}Cl_2O_4$	—	1.67	1.44	—	—	—	—	—	—
		1.82	2.08	1.79	—	—	—	—	—	—
2,4-D ethyl hexyl ester*	$C_{16}H_{22}Cl_2O_3$	—	1.51	—	—	—	—	—	—	—
		2.1	1.78	1.68	—	—	—	—	—	—
2,4-D isobutyl ester	$C_{12}H_{14}Cl_2O_3$	0.62	0.62	0.49	—	—	—	—	—	—
2,4-D isooctyl ester*	$C_{16}H_{22}Cl_2O_3$	—	—	1.48	—	—	—	—	—	—
		2.04	1.78	1.78	—	—	—	—	—	—
2.4-D isopropyl ester*	$C_{11}H_{12}Cl_2O_3$	—	0.62	—	—	—	—	—	—	—
		0.42	0.74	0.33	—	—	—	—	—	—
2,4-D methyl ester	$C_9H_8Cl_2O_3$	0.3	0.38	0.25	—	—	—	—	—	—
2,4-D *n*-butyl ester	$C_{12}H_{14}Cl_2O_3$	0.72	—	—	—	—	—	—	—	—
2,4-D propylene glycol butyl ether ester*	$C_{15}H_{20}Cl_2O_4$	—	1.42	—	—	—	—	—	—	—
		1.54	3.6	—	—	—	—	—	—	—
2,4-DB methyl ester	$C_{11}H_{12}Cl_2O_3$	0.62	0.72	—	—	—	—	—	—	—
2,4-Dichloro-6-nitro-benzenamine	$C_6H_4Cl_2N_2O_2$	0.3	—	—	—	—	R	—	15	2
2,6-Dichlorobenzamide	$C_7H_5NOC_{12}$	0.39	1.3	0.52	5.5	C	NR	NR	6–15–50	1–2–3
2,8-Dihydromirex	$C_{10}H_2CL_{10}$	2.41	—	—	—	—	C	—	6	—
2-Chloroethyl caprate	$C_8H_{15}ClO_2$	0.32	—	—	—	—	C	C	15	2
2-Chloroethyl laurate	$C_{14}H_{27}ClO_2$	0.59	—	—	—	—	C	C	15	2
2-Chloroethyl linoleate	$C_{20}H_{35}ClO_2$	4.1	—	—	—	—	V	P	15	2
2-Chloroethyl myristate	$C_{16}H_{31}ClO_2$	1.17	—	—	—	C	V	V	15	2
2-Chloroethyl palmitate	$C_{18}H_{35}ClO_2$	2.35	—	—	—	—	V	P	15	2
2-Hydroxy-2,3-dihydro-3,3-methyl-5-benzofuranyl methyl sulfonate	$C_{11}H_{14}O_5S$	1	6.6	1.46	30	—	—	—	—	—

PESTDATA Chemicals in Order by Chemical Name *(contd.)*

Name	Molecular Formula	RRT/c OV-101	RRT/c OV-225	RRT/c OV-17	RRT/c DEGS	Recoveries 302	303	304	Ethers	$MeCl_2$
2-Methoxy-3,5,6-trichloropyridine	$C_6H_4C_{13}NO$	0.19	0.08	0.1	—	CP#	—	C	6 + 15	1 + 2
3,4,5-Triimethacarb	$C_{11}H_{15}NO_2$	0.45	0.78	0.5	1.37	C	NR	NR	50	1–2–3
3,4-Dichloroaniline	$C_6H_5Cl_2N$	0.2	0.32	0.16	0.5	V	S	—	15	—
3,4-Dichlorophenylurea	$C_7H_6Cl_2N_2O$	0.22	0.14	0.1	0.48	—	NR	NR	6–15–50	—
3,5-Dichloroaniline	$C_6H_5Cl_2N$	0.18	0.27	0.14	0.56	S	S	S	6 + 15	1 + 2
3-(3,4-Dichlorophenyl)-*I*-methoxyurea	$C_8H_8Cl_2N_2O_2$	0.21	—	1.36	—	R	NR	NR	6–15–50	—
3-(3,4-dichlorophenyl)-*i*-methylurea	$C_8H_8Cl_2N_2O$	0.17	0.14	0.1	—	—	NR	NR	6–15–50	—
3-Aminophenol	C_6H_7NO	—	—	—	1.29	—	—	—	—	—
3-Chloro-5-methyl-4-nitro-1H-pyrazole	$C_4H_4ClN_3O_2$	1.07	—	—	—	C	—	—	—	—
3-Hydroxycarbofuran	$C_{12}H_{15}NO_4$	—	—	—	13.4	—	—	—	—	—
3-Hydroxymethyl-2,5-dimethylphenyl methylcarbamate	$ClIH_{15}NO_3$	0.8	—	1.03	-	—	NR	NR	6–15–50	1–2–3
3-Methyl-4-nitrophenol	$C_7H_7O_3N$	0.38	0.63	0.26	3.96	V	NR	NR	6–15–50	1–2–3
3-Methyl-4-nitrophenol methyl ether	$C_8H_9O_3N$	0.17	0.22	0.13	0.29	—	—	—	—	—
3-Phenoxybenzyl alcohol	$C_{13}H_{13}O_2$	1.28	—	1.6	—	—	—	—	—	—
3-*tert*-Butyl-5-chloro-6-hydroxymethyluracil	$C_{19}H_{13}ClN_2O_3$	1.35	2.27	2.55	5	—	NR	NR	6–15–50	1–2–3
4-4′-Dichlorobiphenyl	$C_{12}H_8Cl_2$	—	—	0.51	—	—	—	—	—	—
4-Chlorobiphenyl	$C_{12}H_9Cl$	—	—	0.2	—	—	—	—	—	—

4-Chlorophenoxyaniline*	$C_{12}H_{10}ClNO$	0.87	—	1.07	—	S	—	—	—	—
		1.28	—	1.31	—	—	—	—	—	—
4-Hydroxymethyl-3,5-dimethylphenyl	$C_{11}H_{15}NO_3$	0.18	—	0.22	—	—	NR	NR	15–50	1–2–3
methylcarbamate*		0.27	—	0.31	—	—	—	—	—	—
6-Chloro-2,3-dihydro-3,3,7-methyl-5*H*-oxazolo(3,2-a)-pyrimidin-5-one	$C_9H_{13}ClN_2O_2$	0.43	1.34	0.6	1.69	—	NR	NR	6–15–50	1–2–3
6-Chloro-2,3-dihydro-7-hydroxymethyl-3,3-methyl-5*H*-oxazolo (3,2-a) pyrimidin-5-one	$C_9H_{13}ClN_2O_3$	0.86	—	1.55	2.11	—	NR	NR	6–15–50	1–2–3
6-Chloronicotinic acid*	$C_6H_4NO_2Cl$	—	—	—	1.83	—	NR	NR	6–15–50	1–2–3
		—	—	—	2.7	—	NR	NR	—	—
8-Monohydromirex	$C_{10}HCl_{11}$	3.74	—	—	—	—	C	—	6	—
Acephate	$C_4H_{10}NO_3PS$	0.15	0.64	0.19	1.65	C	—	—	—	—
Acetochlor	$C_{14}H_2ONO_2Cl$	0.75	0.88	0.67	0.97	C	C#	P	50	3
Acifluorfen	$C_{14}H_7ClF_3NO_3$	1.05	1.47	0.88	2.29	—	NR	NR	6–15–50	1–2–3
Alachlor	$C_{14}H_2OClNO_2$	0.8	1	0.72	—	C	C	C#	50	3
Aldrin	$C_{12}H_8Cl_6$	1.05	0.58	0.76	0.47	C	C	C	6	1
Allethrin	$C_{19}H_{26}O_3$	1.36	1.22	—	—	—	C	C#	50	3
Allidochlor	$C_8H_{12}ClNO$	0.09	—	—	—	C	NR	—	6–15	1–2–3
α-Cypermethrin	$C_{22}H_{19}Cl_2O_3N$	14	—	—	—	C	C	—	—	2
Ametryn	$C_9H_{17}N_5S$	0.77	1.1	—	—	C	—	—	—	—
Aminocarb	$C_{11}H_{16}N_2O_2$	0.56	—	—	1.28	C	—	—	—	—
Amitraz	$C_{19}H_{23}N_3$	—	—	—	12.6	S	—	—	—	—
Anilazine	$C_9H_5Cl_3N_4$	1.24	1.88	1.47	—	V	S	P	15 + 50	2 + 3
Aramite*	$C_{15}H_{23}ClO_4S$	2	2.77	—	—	C	P	NR	15	—
		2.14	3.05	—	—	—	—	—	—	—

PESTDATA Chemicals in Order by Chemical Name *(contd.)*

Name	Molecular Formula	RRT/c OV-101	RRT/c OV-225	RRT/c OV-17	RRT/c DEGS	Recoveries 302	303	304	Ethers	$MeCl_2$
Atrazine	$C_8H_{14}ClN_5$	0.43	0.74	0.44	—	C	S#	NR	50	1–2–3
Azinphos-ethyl	$C_{12}H_{16}N_3O_3PS_2$	6.9	—	14.8	47.2	C	P	S	50	3
Azinphos-methyl	$C_{10}H_{12}N_3O_3PS_2$	5.2	—	11.8	54.6	C	NR	NR	6–15–50	1–2–3
Azinphos-methyl oxygen analog	$C_{10}H_{12}N_3O_4PS$	3.7	—	10.1	—	C	—	—	—	—
Benazolin methyl ester	$C_9H_6O_3SNCl$	0.99	—	—	—	—	—	—	—	—
Bendiocarb	$C_{11}H_{13}NO_4$	0.32	—	—	1.28	C	—	—	—	—
Benfluralin	$C_{13}N_{16}F_3N_3O_4$	0.37	0.28	0.18	—	C	C	C	6	2
Benodanil	$C_{13}H_{10}INO$	2.43	—	4.5	28	C	—	—	—	—
Benoxacor	$C_{11}H_{11}Cl_2NO_2$	0.64	1.06	0.7	1.55	C	P	C	15 + 50	2 + 3
Bensulide	$C_{14}H_{24}NO_4PS_3$	9.5	—	20.2	—	C	P	C	50	3
Benzoylprop-ethyl	$C_{18}H_{17}Cl_2NO_3$	4.3	8.4	6	—	P	NR	NR	6–15–50	1–2–3
BHC, α	$C_6H_6Cl_6$	0.4	0.48	0.35	—	C	C	C	6	1
BHC, β	$C_6H_6Cl_6$	0.43	1.62	0.56	—	C	C	C	6	1
BHC, δ	$C_6H_6Cl_6$	0.5	1.71	0.67	—	C	C	C	6 + 15	1
Bifenox	$C_{12}H_9C_{12}NO_5$	5	14.9	8.8	—	C	C	P	15 + 50	2 + 3
Bifenthrin	$C_{23}H_{22}ClF_3O_2$	4.9	3.8	4.5	6.2	V	C	—	6 + 15	2
Binapacryl	$C_{15}H_{18}N_2O_6$	2.19	4.2	2.38	5.23	C	P	P	15	—
Bis(2-ethylhexyl)phthalate	$C_{24}H_{38}O_4$	6.4	4.5	6.1	—	—	C	C	15 + 50	—
Bis(trichloromethyl)disulfide	$C_2Cl_6S_2$	0.19	—	—	—	—	R	—	6	—
Bitertanol*	$C_{20}H_{23}N_3O_2$	9.4	—	11.8	—	—	—	—	—	—
		9.7	—	12.5	—	—	—	—	—	—
Bromacil	$C_9H_{13}BrN_2O_2$	0.8	4.8	1.36	—	C	NR	NR	6–15–50	1–2–3

Bromofenoxim methyl ether	$C_{14}H_9Br_2O_6N_3$	0.3	—	—	—	—	—	—	—	—
Bromophos	$C_8H_8BrCl_2O_3PS$	1.11	1.29	1.16	1.52	C	C	C	6	—
Bromophos-ethyl	$C_{10}H_{12}BrCl_2O_3PS$	1.51	1.42	1.45	1.33	C	C	P	6	—
Bromopropylate	$C_{17}H_{16}Br_2O_3$	4.4	6.5	—	—	C	C#	C#	15 + 50	1–2–3
Bromoxynil butyrate	$C_{11}H_9Br_2NO_2$	0.78	—	—	1.53	—	V	—	15 + 50	2
Bromoxynil methyl ether	$C_8H_5Br_2NO_2$	0.3	—	—	—	—	—	—	—	—
Bromoxynil octanoate	$C_{15}H_{17}Br_2NO_2$	3.14	—	—	4.6	—	V#	—	15 + 50	2
BTS 27271-HCl	$C_{10}H_{14}N_2.HCl$	—	—	—	1.1	—	—	—	—	—
BTS 27919	$C_9H_{11}NO$	—	—	—	0.76	C	—	—	—	—
Bufencarb*	$C_{13}H_{19}NO_2$	—	—	—	0.6	—	—	—	—	—
		—	—	—	0.72	—	—	—	—	—
		—	—	—	0.81	—	—	—	—	—
		—	—	—	0.93	—	—	—	—	—
		—	—	—	1.14	—	—	—	—	—
Bupirimate	$C_{13}H_{24}N_4SO_3$	2	3.7	2.6	5.9	C	—	—	—	—
Butachlor	$C_{17}H_{26}ClNO_2$	1.73	1.83	1.46	—	C	C	—	50	—
Butralin	$C_{14}H_{21}N_3O_4$	1.15	1.22	0.93	1.13	V	C	—	6 + 15 + 50	—
Butyl benzyl phthalate	$C_{19}H_{20}O_4$	3.06	5.1	4.5	—	—	C	P	15 + 50	—
Butylate	$C_{11}H_{23}NOS$	0.22	—	—	—	V	—	—	—	—
Butylisodecyl phthalate	$C_{22}H_{34}O_4$	—	—	0.82	—	—	—	—	—	—
Cadusafos	$C_{10}H_{23}O_2PS_2$	0.37	0.27	0.29	0.24	C	NR	NR	6–15–50	1–2–3
Captafol	$C_{10}H_9Cl_4NO_2S$	3.11	—	5.4	—	C	P	—	50	3
Captan	$C_9H_8Cl_3NO_2S$	1.2	3.49	1.85	4.5	C	P	C	50	3
Carbaryl	$C_{12}H_{11}NO_2$	0.75	—	1.05	5.1	C	—	—	—	—
Carbetamide	$C_{12}H_{16}N_2O_3$	0.96	—	1.32	—	C	—	—	—	—
Carbofuran	$C_{12}H_{15}NO_3$	0.39	—	—	1.65	C	—	—	—	—
Carbofuran-3-keto-7-phenol	$C_{10}H_{10}O_3$	—	0.24	—	—	—	—	—	—	—
Carbofuran-7-phenol-DNP ether	$C_{16}H_{14}N_2O_6$	—	18.1	—	—	—	—	—	—	—

PESTDATA Chemicals in Order by Chemical Name *(contd.)*

Name	Molecular Formula	RRT/c OV-101	RRT/c OV-225	RRT/c OV-17	RRT/c DEGS	Recoveries				
						302	303	304	Ethers	$MeCl_2$
Carbophenothion	$C_{11}H_{16}ClO_2PS_3$	2.94	4..2	3.7	5.2	C	C	P	6	2
Carbophenothion oxygen analog	$C_{11}H_{16}ClO_3PS_2$	2.17	4.2	3.06	6.2	C	NR	NR	6–15–50	1–2–3
Carbophenothion oxygen analog sulfone	$C_{11}H_{16}ClO_3PS_2$	3.8	—	7.1	—	—	—	—	—	—
Carbophenothion oxygen analog sulfoxide	$C_{11}H_{16}ClO_4PS_2$	4.2	—	2.87	6	—	—	—	—	—
Carbophenothion sulfone	$C_{11}H_{16}ClO_4PS_3$	5.1	—	9.2	4.9	C	C	P	6	1
Carbophenothion sulfoxide	$C_{11}H_{16}ClO_3PS_3$	5.4	—	4	5	—	—	—	—	—
Carbosulfan	$C_{20}H_{32}N_2O_3S$	5.4	—	5.3	—	P	—	—	—	—
Carboxin	$C_{12}H_{13}NO_2S$	1.87	—	—	11.5	C	NR	NR	6–15–50	—
CGA 100255	$C_{15}H_{12}NO_5$	1.8	—	2.96	—	S	—	—	—	—
CGA 118244	$C_{15}H_{13}Cl_2N_3O_3$	7	—	11.4	—	V	NR	NR	6–15–50	1–2–3
CGA 120844	$C_8H_9NSO_3$	0.6	2.65	0.9	16	—	NR	NR	6–15–50	1–2–3
CGA 14128	$C_{12}H_{21}N_2O_4PS$	0.75	0.8	0.68	1.08	C	—	—	50	1–2–3
CGA 150829	$C_5H_{14}N_4O$	0.22	—	0.14	0.47	V	—	—	—	—
CGA 171683	$C_6H_5F_4N_3O_2$	0.06	0.08	0.04	0.28	C	—	—	15 + 50	3
CGA 189138	$C_{13}H_8O_3Cl_2$	1.39	1.89	1.54	—	—	—	—	—	—
CGA 205374	$C_{16}H_{11}N_3O_2Cl_2$	12	8.9	6.1	0.33	—	NR	NR	6–15–50	1–2–3
CGA 205375	$C_{16}H_{13}N_3O_2Cl_2$	6.7	—	1.59	—	—	—	—	—	—
CGA 236431	$C_8H_7F_3N_2O_2$	0.17	—	0.11	0.31	—	—	—	—	—
CGA 236432	$C_9H_9F_3N_2O_2$	0.26	—	0.13	0.28	—	—	—	—	—
CGA 27092	$C_8H_7F_3N_2O$	—	—	0.62	—	—	—	—	—	—

CGA 37734	$C_{10}H_{13}NO_2$	0.4	—	0.47	5.2	C	NR	NR	6–15–50	1–2–3
CGA 51702	$C_9H_9F_3N_2O$	0.46	—	0.49	6.1	—	—	—	—	—
CGA 72903	$C_7H_6F_3N$	0.22	—	0.14	0.8	—	—	—	—	—
CGA 91305	$C_{10}H_8Cl_2N_3O$	1.15	4.3	1.54	16.1	V	NR	NR	6–15–50	1–2–3
CGA 94689A	$C_{15}H_{21}NO_5$	1.53	6.5	2.41	12.6	V	NR	NR	6–15–50	1–2–3
CGA 94689B	$C_{15}H_{21}NO_5$	1.54	6.6	2.45	12.9	S	NR	NR	6–15–50	1–2–3
Chloramben methyl ester	$C_8H_7Cl_2NO_2$	0.44	1.03	—	—	—	—	—	—	—
Chlorbenzide	$C_{13}H_{10}Cl_2S$	1.39	1.62	1.54	—	C	S	P	6	1
Chlorbromuron	$C_9H_{10}BrClN_2O_2$	1.27	3.39	1.42	—	V	V	V	50	3
Chlorbufam	$C_{11}H_{10}ClNO_2$	0.42	0.75	0.45	—	C	—	—	15	2 + 3
Chlordane*	$C_{10}H_6Cl_8$	0.45	0.16	—	—	C	C	C	6	1
		0.63	0.5	—	—	—	—	—	—	—
		0.73	0.52	—	—	—	—	—	—	—
		0.81	0.85	0.23	—	—	—	—	—	—
		0.97	0.9	0.53	—	—	—	—	—	—
		1.16	1.45	0.61	—	—	—	—	—	—
		1.45	1.54	0.88	—	—	—	—	—	—
		1.62	2.69	1.33	—	—	—	—	—	—
		2.61	3.33	1.47	—	—	—	—	—	—
Chlordane, *cis-*	$C_{10}H_6Cl_8$	1.66	1.54	1.48	—	C	C	C	6	1
Chlordane, *trans-*	$C_{10}H_6Cl_8$	1.49	1.46	1.34	—	C	C	C	6	1
Chlordecone	$C_{10}H_8Cl_{10}O_5$	2.75	1.67	2.38	—	—	S#	P#	15 + 50	1–2–3
Chlordene	$C_{10}H_6Cl_6$	0.56	0.4	0.32	—	—	C	C	6	1
Chlordene epoxide	$C_{10}H_6Cl_6O$	0.84	0.65	—	—	—	C	—	15	—
Chlordene, α-	$C_{10}H_6Cl_6$	0.82	0.64	0.67	—	—	—	—	—	—
Chlordene, β-	$C_{10}H_6Cl_6$	0.98	0.84	0.89	—	—	—	—	—	—
Chlordene, γ	$C_{10}H_6Cl_6$	0.98	0.89	0.88	—	—	—	—	—	—
Chlorfenvinphos, α-	$C_{12}H_{14}Cl_3O_4P$	1.21	1.58	1.29	—	C	—	NR	6–15–50	—
Chlorfenvinphos, β-	$C_{12}H_{14}Cl_3O_4P$	1.29	2	1.52	2.57	C	S#	—	50	1–2–3

PESTDATA Chemicals in Order by Chemical Name *(contd.)*

Name	Molecular Formula	RRT/c OV-101	RRT/c OV-225	RRT/c OV-17	RRT/c DEGS	Recoveries				
						302	303	304	Ethers	$MeCl_2$
Chlorflurecol methyl ester	$C_{15}H_{11}ClO_3$	1.73	—	1.88	—	C	—	—	—	—
Chlorimuron ethyl ester	$C_{15}H_{15}ClN_4O_6S$	0.13	0.15	0.1	0.3	P	NR	—	—	—
Chlormephos	$C_5H_{12}ClO_2PS_2$	—	—	0.11	0.02	C	—	—	—	—
Chlornitrofen	$C_{12}H_6Cl_3NO_3$	2.85	4.7	—	—	C	C	C	6 + 15	2
Chlorobenzilate	$C_{16}H_{14}Cl_2O_3$	2.31	3.26	2.61	—	C	C#	P#	15 + 50	3
Chloroneb	$C_8H_8Cl_2O_2$	0.19	0.19	—	—	C	C	—	6	2
Chloropropylate	$C_{17}H_{16}Cl_2O_3$	2.33	2.9	2.41	—	P	C	C	15 + 50	3
Chlorothalonil	$C_8Cl_4N_2$	0.55	1.44	0.74	1.95	S	C#	C#	6–15–50	2 + 3
Chlorothalonil trichloro impurity	$C_8HCl_3N_2$	0.32	—	—	—	R	R#	NR	6–15–50	2 + 3
Chloroxuron	$C_{15}H_{15}ClN_2O_2$	0.81	0.85	—	—	C	NR	NR	6–15–50	1–2–3
Chlorpropham	$C_{10}H_{12}ClNO_2$	0.32	0.43	0.25	0.75	C	C	C	15	2
Chlorpyrifos	$C_9H_{11}Cl_3NO_3PS$	1	1	1	1	C	C	P	6	2
Chlorpyrifos oxygen analog	$C_9H_{11}Cl_3NO_4P$	0.95	1.51	1.08	1.98	C	NR	—	6–15–50	—
Chlorpyrifos-methyl	$C_7H_7Cl_3NO_3PS$	0.72	0.86	0.79	0.91	C	C	—	6	2
Chlorsulfuron	$C_{12}H_{12}ClN_5O_4S$	1.3	8.9	—	0.41	—	NR	NR	6–15–50	—
Chlorthiamid	$C_7H_5Cl_2NS$	0.69	—	—	—	—	—	—	—	—
Chlorthiophos oxygen analog	$C_{11}H_{15}Cl_2O_4PS$	2.22	4.1	2.99	5.4	C	NR	NR	6–15–50	1 + 2 + 3
Chlorthiophos sulfone	$C_{11}H_{15}Cl_2O_5PS_2$	5.3	18.8	9.1	34	C	C	—	50	3
Chlorthiophos sulfoxide	$C_{11}H_{15}Cl_2O_4PS_2$	4.7	10.3	6.9	14.5	C	NR	NR	6–15–50	1–2–3
Chlorthiophos*	$C_{11}H_{15}Cl_2O_3PS_2$	2.24	—	2.58	3.12	C	C	C	6	2
		2.36	—	2.77	3.37	—	—	—	—	—
		2.56	—	3.16	4.1	—	—	—	—	—

CL 202,347	$C_{13}H_{19}N_3O_5$	2.96	11.5	4.1	—	—	—	—	—	—
Clofentezine	$C_{14}H_8Cl_2N_4$	5.9	—	9.8	—	R	S	—	15	2
Clomazone	$C_{12}H_{14}ClNO_2$	0.45	0.59	0.46	—	C	—	—	50	3
Clopyralid methyl ester	$C_7H_4Cl_2NO_2$	0.18	—	—	—	—	—	—	50	—
Compound K*	$C_{10}H_6Cl_8$	0.83	—	—	—	—	C	—	—	1
		2.53	2.66	—	—	—	C	—	—	1
Coumaphos	$C_{14}H_{16}ClO_5PS$	9	40	18	73	C	NR	C#	6–15–50	3
Coumaphos oxygen analog	$C_{14}H_{16}ClO_6P$	8	45	16	100	C	NR	NR	6–15–50	1–2–3
CP 108064, methylated	$C_{15}H_{21}NO_4$	0.73	—	0.67	0.86	—	—	—	—	—
CP 51214	$C_{14}H_{21}NO_3$	0.7	—	0.58	1.14	C	NR	NR	6–15–50	1–2–3
Crotoxyphos	$C_{14}H_{19}O_6P$	1.37	2.85	1.9	5.6	C	NR	NR	6–15–50	1–2–3
Crufomate	$C_{12}H_{19}ClNO_3P$	1.08	2.33	1.3	4.2	C	NR	NR	6–15–50	—
Cyanazine	$C_9H_{13}ClN_6$	0.89	4.9	1.48	—	C	NR	—	6–15–50	—
Cyanofenphos	$C_{15}H_{14}NO_2PS$	3.1	8.2	4.6	14	C	—	—	—	—
Cyanophos	$C_9H_{10}O_3NSP$	0.47	—	0.59	1.73	C	—	—	—	—
Cycloate	$C_{11}H_{21}NOS$	0.3	—	—	0.14	C	V#	S	15 + 50	3
Cyfluthrin*	$C_{22}H_{18}Cl_2FNO_3$	11.7	—	—	—	C	P	—	15	—
		12.5	—	—	—	—	—	—	—	—
		12.8	—	—	—	—	—	—	—	—
Cymiazole	$C_{12}H_{14}N_2S$	0.73	—	0.89	1.18	—	—	—	—	—
Cymoxanil	$C_7H_{10}N_4O_3$	2.5	0.5	0.16	1.24	V	NR	NR	6–15–50	1–2–3
Cypermethrin*	$C_{22}H_{19}Cl_2NO_3$	—	29	—	—	C	C	C	15	2
		14.1	33	23	—	—	—	—	—	—
		15.1	36	25	—	—	—	—	—	—
Cyprazine	$C_9H_{14}ClN_5$	0.64	1.22	0.74	2.8	C	—	—	—	—
Cyproconazol	$C_{15}H_{18}ClN_3O$	2.04	1.61	2.69	—	C	NR	NR	6–15–50	1–2–3
Cyromazine	$C_6H_{10}N_6$	0.58	—	0.68	18.5	S	—	—	—	—
Dazomet	$C_5H_{10}N_2S_2$	0.4	—	0.71	—	S	NR	—	6–15–50	1–2–3
DCPA	$C_{10}H_6Cl_4O_4$	1.06	1.13	1	1.49	C	C	C	15	2

PESTDATA Chemicals in Order by Chemical Name *(contd.)*

Name	Molecular Formula	RRT/c OV-101	RRT/c OV-225	RRT/c OV-17	RRT/c DEGS	Recoveries				
						302	303	304	Ethers	$MeCl_2$
DDE,*o,p′*-	$C_{14}H_8Cl_4$	1.55	1.28	1.51	—	C	C	C	6	1
DDE,*p,p′*-	$C_{14}H_8Cl_4$	1.88	1.59	1.86	1.63	C	C	C	6	1
DDM	$C_{13}H_{10}Cl_2$	0.72	—	—	—	—	—	—	—	—
DDMS	$C_{14}H_{11}Cl_3$	1.65	—	1.65	—	—	R	—	6	—
DDMU	$C_{14}H_9Cl_3$	1.47	—	—	—	—	—	—	—	—
DDNS	$C_{14}H_{12}Cl_2$	0.83	—	—	—	—	—	—	—	—
DDNU	$C_{14}H_{10}Cl_2$	0.83	—	—	—	—	—	—	—	—
DDT,*o,p′*-	$C_{14}H_9Cl_5$	2.55	2.27	2.7	—	C	C	C	6	1
DDE,*p,p′*-	$C_{14}H_9Cl_5$	3.13	3.6	3.5	4.2	C	C	C	6	1
DEF	$C_{12}H_{27}OPS_3$	1.95	1.65	1.88	1.25	C	C	P	15 + 50	3
Deltamethrin*	$C_{22}H_{19}Br_2NO_3$	17.1	—	21	26.5	C	S#	P	15	2
		27	—	35	42	—	—	—	—	—
		29	19.9	38	50	—	—	—	—	–
Deltamethrin, *trans*-*	$C_{22}H_{19}Br_2NO_3$	17	—	6.2	26	—	P#	NR	15	2
		29	—	20	41	—	—	—	—	—
		31	19.7	38	51	—	—	—	—	—
Demeton-*O* oxygen analog	$C_8H_{19}O_4PS$	0.22	0.32	0.21	0.34	—	—	—	—	—
Demeton-*O* sulfone*	$C_8H_{19}O_5PS_2$	—	—	0.28	—	C	—	—	—	—
		0.71	2.95	0.96	4.5	—	—	—	—	—
Demeton-*O* sulfoxide	$C_8H_{15}O_4PS_2$	0.87	—	1.05	3.6	C	—	—	—	—
Demeton-*O**	$C_8H_{19}O_3PS_2$	—	—	0.2	0.22	C	NR	—	6–15	—
		0.28	—	0.36	0.53	—	—	—	—	—
Demeton-*S*	$C_8H_{19}O_3PS_2$	0.41	0.56	0.41	0.59	C	NR	—	6–15–50	—
Demeton-*S* sulfone	$C_8H_{19}O_5PS_2$	1.15	5.8	1.75	11.4	C	—	—	—	—

Demeton-*S* sulfoxide	$C_8H_{19}O_4PS_2$	—	—	—	8.2	C	—	—	—	—
Des-*N*-isopropyl isofenphos oxygen analog	$C_{12}H_{18}NO_5P$	0.93	—	1.43	—	—	—	—	—	—
Des-*N*-isopropyl isofenphos	$C_{12}H_{18}NO_4PS$	1.21	2.73	1.5	6.5	C	S	—	50	—
Desdiethyl simazine	$C_7H_{12}ClN_5$	0.2	0.86	0.61	—	—	NR	NR	6–15–50	1–2–3
Desethyl simazine	$C_5H_8ClN_5$	0.3	0.8	0.53	—	—	NR	NR	50	1–2–3
Desmedipham	$C_{16}H_{16}N_2O_4$	0.44	0.29	0.45	10	—	—	—	—	—
Desmethyl dephenamid	$C_{15}H_{15}NO$	0.98	—	1.73	—	—	—	—	—	—
Desmethyl norflurazon	$C_{11}H_7ClF_3N_3O$	3.38	1.41	4.9	—	V	NR	NR	6–15–50	1–2–3
Di-allate	$C_{10}H_{17}ClNOS$	0.42	0.26	0.33	—	C	C	—	6	—
Di-*n*-octyl phthalate	$C_{24}H_{38}O_4$	12	—	—	—	—	C	C	15+50	—
Dialifor	$C_{14}H_{17}ClNO_4PS_2$	6.5	—	14.3	33.9	C	C	P	15	2
N,N-Diallyl dichloroacetamide	$C_8H_{11}Cl_2NO$	0.19	0.22	0.14	0.2	C	S	S	15+50	2+3
Diazinon	$C_{12}H_{21}N_2O_3PS$	0.51	0.4	0.44	0.39	C	C	C	15	3
Diazinon oxygen analog	$C_{12}H_{21}N_2O_4P$	0.5	0.53	0.47	0.6	C	NR	NR	6–15–50	1–2–3
Dibromochloropropane	$C_3H_5Br_2Cl$	0.04	—	0.03	—	—	—	—	—	—
Dibutyl phthalate	$C_{16}H_{22}O_4$	0.88	0.92	0.84	—	—	C	C	15+50	—
Dicamba methyl ester	$C_8H_6Cl_2O_3$	0.19	0.18	—	—	—	—	—	—	—
Dichlobenil	$C_7H_3Cl_2N$	0.11	—	0.1	—	C	P	C	15	2
Dichlofenthion	$C_{10}H_{13}Cl_2O_3PS$	0.67	0.64	0.56	0.63	C	C	V	6	2
Dichlofluanid	$C_9H_{11}Cl_2FN_2O_2S_2$	0.9	1.71	1.01	—	C	C#	—	15+50	2+3
Dichlone	$C_{10}H_4Cl_2O_2$	0.55	0.92	—	—	P	S#	S#	6–15–50	2+3
Dichlorobenzene, *p*-	$C_6H_4Cl_2$	0.03	—	0.02	—	—	C	C	6	1
Dichlorobenzophenone, *o,p'*-	$C_{13}H_8Cl_2O$	0.82	1.07	0.92	—	—	C	C	15	2
Dichlorobenzophenone, *p,p'*-	$C_{13}H_8Cl_2O$	0.99	1.25	1.08	—	—	C	C	15	2
Dichlorprop methyl ester	$C_{10}H_{10}Cl_2O_3$	0.28	—	—	—	—	—	—	—	—
Dichlorvos	$C_4H_7Cl_2O_4P$	0.07	0.08	0.08	0.09	C	NR	NR	6–15–50	1–2–3
Diclobutrazol	$C_{15}H_{19}Cl_2N_3O$	2.02	3.4	2.03	—	C	NR	NR	6–15–50	1–2–3
Diclofop-methyl	$C_{16}H_{14}Cl_2O_4$	3.57	4.9	4.7	8.8	C	C	C	15	2

PESTDATA Chemicals in Order by Chemical Name *(contd.)*

Name	Molecular Formula	RRT/c OV-101	RRT/c OV-225	RRT/c OV-17	RRT/c DEGS	Recoveries 302	303	304	Ethers	$MeCl_2$
Dichloran	$C_6H_4Cl_2N_2O_2$	0.42	0.96	0.45	1.62	C	S	P	15 + 50	2 + 3
Dicofol, *o,p′*-*	$C_{14}H_9Cl_5O$	0.86	—	—	—	C	V	S	6 + 15	2
		4.1	1.08	0.91	1.2	—	—	—	—	—
Docofol, *p,p′*-*	$C_{14}H_9Cl_5O$	1.04	—	—	—	C	V	P#	6 + 15	1 + 2
		4.4	1.28	1.08	1.5	—	—	—	—	—
Dicrotophos	$C_8H_{16}NO_5P$	0.31	0.96	0.43	1.53	C	NR	—	6–15–50	—
Dieldrin	$C_{12}H_8Cl_6O$	1.91	1.87	1.84	1.84	C	C	C	15	2
Diethatyl-ethyl	$C_{16}H_{22}ClNO_3$	1.78	3.14	2	4	C	NR	NR	6–15–50	1–2–3
Diethyl phthalate	$C_{12}H_{14}O_4$	0.26	—	—	—	—	P	P	15 + 50	—
Difenoxuron	$C_{16}H_{18}N_2O_3$	0.97	—	0.96	—	—	—	—	—	—
Diisobutyl phthlate	$C_{16}H_{22}O_4$	0.65	0.61	0.56	—	—	P	—	15 + 50	—
Diisohexyl phthalate*	$C_{20}H_{30}O_4$	2.45	—	—	—	—	C	—	15 + 50	—
		2.66	—	—	—	—	—	—	—	—
		2.9	—	—	—	—	—	—	—	—
		3.27	—	—	—	—	—	—	—	—
Diisooctyl phthalate*	$C_{24}H_{38}O_4$	0.91	—	—	—	—	C	C	15 + 50	—
		5.5	—	—	—	—	—	—	—	—
		6.2	—	—	—	—	—	—	—	—
		6.7	—	—	—	—	—	—	—	—
		7.5	—	—	—	—	—	—	—	—
		9	—	—	—	—	—	—	—	—
		10.5	—	—	—	—	—	—	—	—
Dimethachlor	$C_{13}H_{18}ClNO_2$	0.71	1.11	0.71	—	C	—	—	—	—

Dimethametrin	$C_{11}H_{21}N_5S$	—	—	—	2.95	C	—	—	—	—
Dimethenamid	$C_{12}H_{18}ClNO_2S$	0.72	0.98	—	—	—	NR	NR	6–15–50	1–2–3
Dimethoate	$C_5H_{12}NO_3PS_2$	0.4	1.6	0.62	3.27	C	NR	NR	6–15–50	1–2–3
Dimethyl phthalate	$C_{10}H_{10}O_4$	0.15	0.15	0.14	—	—	P	—	6–15–50	—
Dinitramine	$C_{11}H_{13}F_3N_4O_4$	0.52	0.93	0.44	—	C	—	P	15	—
Dinobuton	$C_{14}H_{18}N_2O_7$	1.4	—	1.32	—	C	—	—	—	—
Dinocap*	$C_{18}H_{24}N_2O_6$	—	—	3.5	—	C	P	P	15	2
		4	—	3.9	—	—	—	—	—	—
		4.3	6.9	4.4	—	—	—	—	—	—
		4.8	7.7	4.8	—	—	—	—	—	—
		5.1	9.5	5.6	—	—	—	—	—	—
Dinoseb methyl ether	$C_{11}H_{14}N_2O_5$	0.63	—	—	—	—	—	—	—	—
Dioxabbenzofos	$C_8H_9O_3PS$	0.34	—	0.36	1.06	C	P	—	15	—
Dioxacarb	$C_{11}H_{13}NO_4$	—	—	—	5.8	C	—	—	—	—
Dioxathion	$C_{12}H_{26}O_6P_2S_4$	0.47	—	0.5	0.86	V	NR	—	6–15–50	2
Diphenamid	$C_{16}H_{17}NO$	1.1	—	1.55	3.4	V	NR	—	6–15	—
Diphenylamine	$C_{12}H_{11}N$	0.29	—	0.25	0.59	C	S	—	6 + 15	—
Disul-Na	$C_8H_7Cl_2O_5S{*}Na$	0.23	—	—	—	—	—	—	—	—
Disulfoton	$C_8H_{19}O_2PS_3$	0.54	0.6	0.46	0.54	C	P#	NR	6	1–2–3
Disulfoton sulfone	$C_8H_{19}O_4PS_3$	1.5	6.7	2.39	11	C	NR	—	6–15–50	—
Disulfoton sulfoxide	$C_8H_{19}O_3PS_3$	—	—	—	6.8	C	—	—	—	—
Dithianon	$C_{14}H_4O_2N_2S_2$	4.7	53	11.3	—	NR	—	—	—	—
Diuron	$C_9H_{10}Cl_2N_2O$	0.11	0.09	0.11	—	C	NR	NR	6–15–50	1–2–3
DNOC methyl ether	$C_8H_8N_2O_5$	0.35	—	—	—	—	—	—	—	—
Edifenphos	$C_{14}H_{15}O_2PS_2$	2.87	6.3	5.3	8.5	C	—	—	—	—
Endosulfan I	$C_9H_6Cl_6O_3S$	1.64	1.38	1.47	—	C	C	C	15	2
Endosulfan II	$C_9H_6Cl_6O_3S$	2.21	3.9	2.77	—	C	C	C	15 + 50	2
Endosulfan sulfate	$C_9H_6Cl_6O_4S$	2.83	8.3	4	—	C	C	C	50	2
Endrin	$C_{12}H_8Cl_6O$	2.13	2.22	2.29	2	C	C#	C#	15	2

PESTDATA Chemicals in Order by Chemical Name *(contd.)*

Name	Molecular Formula	RRT/c OV-101	RRT/c OV-225	RRT/c OV-17	RRT/c DEGS	Recoveries				
						302	303	304	Ethers	$MeCl_2$
Endrin alcohol	$C_{12}H_8Cl_6O$	2.55	—	—	—	—	P	C	15 + 50	2 + 3
Endrin aldehyde	$C_{12}H_8Cl_6O$	2.35	—	—	—	C	P	C	15 + 50	—
Endrin ketone	$C_{12}H_8Cl_6O$	3.6	10.3	—	—	—	C	C	50	2
EPN	$C_{14}H_{14}NO_4PS$	4.5	10.6	6.9	20	C	C	C	15	2
Epoxyhexachloronorbornene	$C_7H_2Cl_6O$	—	—	0.2	—	—	—	—	—	—
EPTC	$C_9H_{19}NOS$	0.12	—	—	—	—	P	—	15	—
Esfenvalerate	$C_{25}H_{22}ClNO_3$	22.5	—	—	—	C	C	C	15	2
Etaconazole*	$C_{14}H_{15}Cl_2N_3O_2$	2.36	—	—	—	C	—	—	—	—
		2.43	—	3.17	—	—	—	—	—	—
Ethalfluralin	$C_{13}H_{14}F_3N_3O_4$	0.34	0.27	0.19	0.31	C	C	C	6	2
Ethametsulfuron methyl ester*	$C_{15}H_{18}N_6O_6S$	0.35	2.85	0.4	—	—	NR	NR	6–15–50	1–2–3
		0.55	3.6	0.95	3.14	—	—	—	—	—
Ethephon	$C_2H_6ClO_3P$	3.03	2.74	2.88	1.49	NR	—	—	6 + 15 + 50	1 + 2 + 3
Ethiofencarb	$C_{10}H_{15}NO_2S$	0.6	1.4	0.78	2.38	C	NR	NR	6–15–50	—
Ethiolate	$C_7H_{15}NOS$	0.06	—	—	—	C	—	—	—	—
Ethion	$C_9H_{22}O_4P_2S_4$	2.56	3.93	3.36	4.1	C	C	C	6	2
Ethion oxygen analog	$C_9H_{22}O_5P_2S_3$	1.88	4.1	—	—	C	—	—	—	—
Ethofumesate	$C_{13}H_{18}O_5S$	0.86	1.93	1.02	3.16	C	—	—	—	—
Ethoprop	$C_8H_{19}O_2PS_2$	0.33	0.31	0.25	0.26	C	P#	S#	50	1–2–3
Ethoxyquin	$C_{14}H_{19}NO$	0.6	1.64	0.7	3	C	NR	NR	6–15–50	—
Ethyl *p*-toluene sulfonamide	$C_9H_{13}NO_2S$	—	—	—	2.98	C	—	—	—	—
Ethylenethiourea	$C_3H_6N_2S$	0.5	2.33	0.66	9.5	S	NR	NR	6–15–50	1–2–3
Etridiazole	$C_5H_5Cl_3N_2OS$	0.18	0.12	0.21	—	C	C	P	6	2

Etrimfos	$C_{10}H_{17}N_2O_4PS$	0.58	0.59	0.51	0.63	C	C	C	15	2 + 3
Etrimfos oxygen analog	$C_{10}H_{17}N_2O_5P$	0.51	0.8	0.63	0.88	C	—	—	—	—
Famphur	$C_{10}H_{16}NO_5PS_2$	2.65	14	5	30	C	NR	—	6–15–50	—
Famphur oxygen analog	$C_{10}H_{16}NO_6PS$	2.26	—	4.4	—	C	—	—	—	—
Fenac	$C_8H_5Cl_3O_2$	1.42	3.7	—	—	—	NR	NR	6–15–50	—
Fenac methyl ester	$C_9H_7Cl_3O_2$	0.32	—	—	—	—	—	—	—	—
Fenamiphos	$C_{13}H_{22}NO_3PS$	1.66	3.7	2.41	5.9	C	NR	NR	6–15–50	1–2–3
Fenamiphos sulfone	$C_{13}H_{22}NO_5PS$	4.5	—	8.4	—	C	NR	NR	6–15–50	1–2–3
Fenamiphos sulfoxide	$C_{13}H_{22}NO_4PS$	5.2	—	8.1	—	C	NR	NR	6–15–50	1–2–3
Fenarimol	$C_{17}H_{12}Cl_2N_2O$	6.6	—	10.1	45	C	P#	C#	50	3
Fenarimol metabolite B	$C_{17}H_{14}N_2OCl_2$	4.6	—	—	—	NR	NR	NR	6–15–50	—
Fenarimol metabolite C	$C_{17}H_{14}N_2OCl_2$	4.6	—	—	—	S	—	—	6	—
Fenbuconazole	$C_{19}H_{17}ClN_4$	9.8	—	—	—	C	NR	NR	6–15–50	1–2–3
Fenfuram	$C_{12}H_{11}NO_2$	0.54	1.47	0.62	3.07	C	—	—	—	—
Fenitrothion	$C_9H_{12}NO_5PS$	0.84	1.82	1.05	2.9	C	C	C	15	2
Fenitrothion oxygen analog	$C_9H_{12}NO_6P$	0.72	—	0.83	—	C	—	—	—	—
Feoxaprop ethyl ester	$C_{18}H_{16}NO_5Cl$	8.1	11.3	10.5	—	S	V	V	50	3
Fenoxycarb	$C_{17}H_{19}NO_4$	5	—	7.3	31	C	—	—	—	—
Fenpropathrin	$C_{22}H_{23}NO_3$	4.8	7	5.7	11.9	—	V#	V	15	2
Fenpropimorph	$C_{20}H_{33}NO$	1.09	—	0.62	0.5	C	—	—	50	1–2–3
Fenson	$C_{12}H_{10}O_3ClS$	—	2.03	—	—	—	—	—	—	—
Fensulfothion	$C_{11}H_{17}O_4PS_2$	2.4	—	3.8	16.9	C	NR	NR	6–15–50	1–2–3
Fensulfothion oxygen analog	$C_{11}H_{17}O_5PS$	2.49	—	5	24.6	C	NR	—	6–15–50	—
Fensulfothion oxygen analog sulfone	$C_{11}H_{17}O_7PS_2$	1.99	—	3.8	—	—	—	—	—	—
Fensulfothion sulfone	$C_{11}H_{17}O_5PS_2$	2.8	—	3.6	—	C	NR	—	6–15–50	—
Fenthion	$C_{10}H_{15}O_3PS_2$	0.96	1.46	1.18	2.13	C	S#	NR	6+15	1–2–3
Fenthion oxygen analog	$C_{10}H_{15}O_4PS$	0.78	—	1.12	—	C	NR	NR	6–15–50	1–2–3
Fenthion oxygen analog sulfone*	$C_{10}H_{15}O_6PS_2$	1.77	—	—	—	—	—	—	—	—
		2.29	—	4.1	—	—	—	—	—	—

PESTDATA Chemicals in Order by Chemical Name *(contd.)*

Name	Molecular Formula	RRT/c OV-101	RRT/c OV-225	RRT/c OV-17	RRT/c DEGS	Recoveries 302	303	304	Ethers	$MeCl_2$
Fenthion oxygen analog sulfoxide	$C_{10}H_{15}O_5PS$	0.43	—	0.62	—	C	NR	NR	6–15–50	1–2–3
Fenthion sulfone	$C_{10}H_{15}O_5PS_2$	2.39	—	4.7	—	C	NR	NR	6–15–50	1–2–3
Fenvalerate*	$C_{25}H_{22}ClNO_3$	20.3	44	35	—	C	C	C	15	2
		22.5	51	40	—	—	—	—	—	—
Flamprop-*M*-isopropyl	$C_{19}H_{19}ClFNO_3$	2.46	—	2.81	—	C	—	—	—	—
Flamprop-methyl	$C_{17}H_{15}ClFNO_3$	1.94	—	2.45	—	C	—	—	—	—
Fluazifop butyl ester	$C_{19}H_{20}F_3NO_4$	2.3	2.36	2.31	3.9	C	C	V	15	3
Fluchloralin	$C_{12}H_{13}ClF_3N_3O_4$	0.53	0.76	0.37	0.91	C	C	NR	6	2
Flucythrinate*	$C_{26}H_{23}F_2NO_4$	14.7	36.9	21.4	—	C	C	—	15	2 + 3
		16.1	42	24	—	—	—	—	—	—
Flumetsulam, methylated	$C_{13}H_{11}F_2N_5O_2S$	12.4	—	—	—	—	—	—	—	—
Fluometuron	$C_{10}H_{11}F_3N_2O$	—	—	0.14	—	—	—	—	—	—
Fluridone	$C_{19}H_{14}F_3NO$	16.3	24	—	—	—	NR	NR	6–15–50	—
Fluroxypyr, methylated*	$C_8H_7O_3N_2Cl_2F$	0.61	—	—	—	—	—	—	—	—
		0.79	—	—	—	—	—	—	—	—
Flusilazole	$C_{16}H_{15}F_2N_3Si$	1.97	—	2.33	—	C	—	—	—	—
Fluvalinate*	$C_{26}H_{22}ClF_3N_2O_3$	—	56	35	—	C	C	—	15	2
		25	59	38	—	—	—	—	—	—
FMTU	$C_{10}H_9F_3N_2O_2$	—	—	—	0.98	—	—	—	—	—
Folpet	$C_9H_4Cl_3O_2NS$	1.23	3.01	1.94	4.5	C	C	P	15 + 50	2 + 3
Fonofos	$C_{10}H_{15}O_5PS_2$	0.52	0.56	0.44	0.54	C	C	C	6	2 + 3
Fonofos oxygen analog	$C_{10}H_{15}O_2PS$	0.39	0.53	0.38	0.65	V	NR	NR	6–15–50	1–2–3

Formetanate hydrochloride	$C_{11}H_{16}ClN_3O_2$	0.9	—	0.45	—	—	—	—	—	—
Formothion	$C_6H_{12}NO_4PS_2$	—	—	0.91	3.1	C	NR	NR	6–15–50	1–2–3
Fuberidazole	$C_{11}H_8N_2O$	0.71	—	0.95	—	C	—	—	—	—
Furilazole	$C_{11}H_{13}Cl_2NO_3$	0.46	0.6	0.39	0.95	C	S	—	50	3
G-27550	$C_8H_{12}N_2O$	0.28	—	0.29	0.73	C	—	—	—	—
Gardona	$C_{10}H_9Cl_4O_4P$	1.58	2.72	1.97	4.4	C	NR	NR	6–15–50	1–2–3
GS-31144	$C_8H_{12}N_2O_2$	—	—	—	2.32	—	NR	NR	6–15–50	1–2–3
Haloxyfop methyl ester	$C_{16}H_{13}ClF_3NO_4$	1.55	—	1.4	—	—	—	—	—	—
Heptachlor	$C_{10}H_5Cl_7$	0.83	0.52	0.6	0.5	C	C	C	6	1
Heptachlor epoxide	$C_{10}H_5Cl_7O$	1.29	1.22	1.15	1.26	C	C	C	6	2
Heptachloronorbornene	$C_7H_3Cl_7$	—	—	0.23	—	—	—	—	—	—
Heptenophos	$C_9H_{12}ClO_4P$	—	—	—	0.38	C	—	—	—	—
Hexachlorobenzene	C_6Cl_6	0.45	0.25	0.33	0.31	C	C	P	6	1
Hexachlorobutadiene	C_4Cl_6	—	—	0.04	—	—	V#	P	6	1
Hexachlorocyclopentadiene	C_5Cl_6	0.12	—	0.06	—	—	—	—	—	—
Hexachloroethane	C_2Cl_6	—	—	0.02	—	—	—	—	—	—
Hexachloronorbornadiene	$C_7H_2Cl_6$	—	—	0.12	—	—	—	—	—	—
Hexachlorophene	$C_{13}H_6Cl_6O_2$	13	13	16	—	—	NR	NR	6–15–50	—
Hexachlorophene dimethyl ether	$C_{15}H_{10}Cl_6O_2$	9.7	—	—	—	—	NR	NR	6–15	—
Hexaconazol	$C_{14}H_{17}Cl_2N_3O$	1.86	2.91	1.79	—	C	—	—	—	—
Hexazinone	$C_{12}H_{20}N_4O_2$	2.91	—	—	—	P	NR	NR	6–15–50	1–2–3
HOE-030291	$C_{17}H_{16}Cl_2O_5$	7.9	10.8	12.6	30	—	—	—	—	—
Hydramethylnon*	$C_{25}H_{24}F_6N_4$	2.55	—	—	—	—	—	—	—	—
		32	4.5	—	—	—	—	—	—	—
		44	53	—	—	—	—	—	—	—
Hydroxy chloroneb	$C_7H_6Cl_2O_2$	0.15	0.24	—	—	—	NR	—	6–15	—
Imazalil	$C_{14}H_{14}Cl_2N_2O$	1.76	4	2.08	5.6	C	NR	NR	6–15–50	—
Imazamethabenz methyl ester*	$C_{16}H_{20}N_2O_3$	—	—	2.44	—	C	—	—	—	—

PESTDATA Chemicals in Order by Chemical Name *(contd.)*

Name	Molecular Formula	RRT/c OV-101	RRT/c OV-225	RRT/c OV-17	RRT/c DEGS	Recoveries 302	303	304	Ethers	$MeCl_2$
		1.79	3.5	2.76	—	—	—	—	—	—
Imazethapyr ammonium salt methyl ester	$C_{16}H_{21}N_3O_3$	2.4	4.3	3	—	—	—	—	—	—
Imidacloprid	$C_9H_{10}ClN_5O_2$	1.84	—	—	—	—	NR	NR	6–15–50	1–2–3
IN-A2213	$C_{11}H_{18}N_4O_2$	0.25	0.92	0.28	2.71	C	NR	NR	6–15–50	1–2–3
IN-A3928	$C_{11}H_{18}N_4O_2$	3.06	—	—	—	S	NR	NR	6–15–50	1–2–3
IN-B2838	$C_{10}H_{15}N_3O_3$	0.7	—	0.8	5.3	P	NR	NR	6–15–50	1–2–3
IN-T3935	$C_{11}H_{18}N_4O_3$	4.7	—	—	—	S	—	—	—	—
IN-T3936	$C_{10}H_{15}N_3O_4$	1.41	—	2.55	—	S	NR	NR	6–15–50	1–2–3
IN-T3937	$C_{12}H_{20}N_4O_3$	4.7	—	—	—	S	—	—	—	—
Inoxynil methyl ether	$C_8H_5I_2NO$	0.71	—	—	—	—	—	—	—	—
Iprobenfos	$C_{13}H_{21}O_3PS$	0.6	—	0.54	—	C	—	—	—	—
Iprodione metabolite isomer	$C_{13}H_{13}Cl_2N_3O_3$	5.3	—	7.5	—	C	S	—	50	—
Iprodione*	$C_{13}H_{13}Cl_2N_3O_3$	—	—	5.2	—	C	S#	NR	50	1–2–3
		4.2	18	6.3	37	—	—	—	—	—
Isazofos	$C_9H_{17}ClN_3O_3PS$	0.55	0.8	0.63	0.81	C	C#	—	50	2 + 3
Isocarbamid	$C_8H_{15}N_3O_2$	0.46	—	0.82	—	—	—	—	—	—
Isofenphos	$C_{15}H_{24}NO_4PS$	1.36	1.73	1.38	2.12	C	C	—	15 + 50	—
Isofenphos oxygen analog	$C_{15}H_{24}NO_5P$	1.17	1.74	1.24	3.2	C	—	—	—	—
Isopropalin	$C_{15}H_{23}N_3O_4$	1.14	1.24	1.01	1.22	C	C	—	6	—
Isoprothiolane	$C_{12}H_{18}O_4S_2$	1.6	4.1	3.17	6	C	—	—	—	—
Isoproturon	$C_{12}H_{18}N_2O$	—	—	0.89	—	S	—	—	—	—
Jodfenphos	$C_8H_8Cl_2IO_3PS$	—	—	2.11	2.96	C	—	—	—	—
KWG 1323	$C_{14}H_{16}ClN_3O_3$	0.99	1.91	0.96	4.3	C	NR	NR	6–15–50	1–2–3

KWG 1342	$C_{14}H_{18}ClN_3O_3$	4	15	4.2	—	—	—	—	—	—
Lactofen	$C_{19}H_{15}ClF_3NO_7$	7.3	—	—	—	—	—	C	50	2 + 3
λ-Cyhalothrin	$C_{23}H_{19}ClF_3NO_3$	7.4	—	8	—	C	—	—	—	—
Leptophos	$C_{13}H_{10}BrCl_2O_2PS$	5.8	7.7	8.5	10.6	C	C	C	6	2
Leptophos oxygen analog	$C_{13}H_{10}BrCl_2O_3P$	4.2	7.6	6.5	—	C	—	—	—	—
Leptophos photoproduct	$C_{13}H_{11}Cl_2O_2PS$	2.38	3.24	3.14	4.2	C	—	—	—	—
Lindane	$C_6H_6Cl_6$	0.48	0.69	0.47	0.79	C	C	C	6	1
Linuron	$C_9H_{10}Cl_2NO_2$	0.85	2.13	0.95	—	V	V#	V	50	3
Malathion	$C_{10}H_{19}O_6PS_2$	0.91	1.49	1.05	2.12	C	C	C	15 + 50	3
Malathion oxygen analog	$C_{10}H_{19}O_7PS$	0.68	1.55	0.87	2.64	C	NR	NR	6–15–50	1–2–3
MCPA methyl ester	$C_{10}H_{11}ClO_3$	0.19	—	—	—	—	—	—	—	—
Mecarbam	$C_{10}H_{20}NO_5PS_2$	1.28	2.67	1.58	3.1	C	—	—	50	—
Mecoprop methyl ester	$C_{11}H_{13}ClO_3$	0.19	—	—	—	—	—	—	—	—
Melamine	$C_3H_6N_6$	0.42	—	0.42	19	NR	—	—	—	—
Mephosfolan	$C_8H_{16}NO_3PS_2$	—	—	2.51	9.1	C	—	—	—	—
Merphos*	$C_{12}H_{27}PS_3$	1.34	0.65	1.43	—	—	C	C	6 + 15 + 50	3
		1.95	1.64	1.88	1.22	—	—	—	—	—
Metalaxyl	$C_{15}H_{21}NO_4$	0.81	—	0.9	1.8	C	NR	NR	6–15–50	1–2–3
Metamitron	$C_{10}H_{10}N_4O$	2.24	—	—	—	—	—	—	—	—
Metasystox thiol	$C_6H_{15}O_3PS_2$	0.28	0.49	0.32	—	C	—	—	—	—
Metasystox thiono*	$C_6H_{15}O_3PS_2$	—	—	0.18	—	—	—	—	—	—
		0.31	0.49	0.32	—	—	—	—	—	—
Metazachlor	$C_{14}H_{16}ClN_3O$	1.5	—	1.46	—	C	—	—	—	—
Methabenzthiazuron	$C_{10}H_{11}N_3OS$	0.35	0.7	0.41	1.23	C	NR	NR	6–15–50	1–2–3
Methamidophos	$C_2H_8NO_2PS$	0.07	0.25	0.09	0.71	V	—	—	—	—
Methazole	$C_9H_6Cl_2N_2O_3$	0.97	1.46	—	—	—	—	—	—	—
Methidathion	$C_6H_{11}N_2O_4PS_3$	1.4	3.33	2.28	5.5	C	S	P#	50	3
Methidathion oxygen analog	$C_6H_{11}N_2O_5PS_2$	1.07	—	1.8	7	—	NR	NR	6–15–50	1–2–3
Methidathion sulfone	$C_5H_8N_2O_3S_2$	0.56	2.29	0.82	4.7	—	NR	NR	6–15–50	1–2–3

PESTDATA Chemicals in Order by Chemical Name *(contd.)*

Name	Molecular Formula	RRT/c OV-101	RRT/c OV-225	RRT/c OV-17	RRT/c DEGS	Recoveries 302	303	304	Ethers	$MeCl_2$
Methidathion sulfoxide	$C_5H_8N_2O_4S_2$	0.45	2.25	0.71	5.8	—	NR	NR	6–15–50	1–2–3
Methiocarb	$C_{11}H_{15}NO_2S$	0.88	—	—	3.27	C	—	—	—	—
Methiocarb sulfone	$C_{11}H_{15}NO_4S$	0.8	—	1.17	—	S	NR	NR	6–15–50	1–2–3
Methomyl	$C_5H_{10}N_2O_2S$	0.1	—	—	—	—	NR	NR	6–15–50	1–2–3
Methoprotryne	$C_{11}H_{21}N_5OS$	2.07	—	2.92	—	C	—	—	—	—
Methoxychlor olefin	$C_{16}H_{14}Cl_2O_2$	2.97	3.7	4.2	—	C	C	C	6	2
Methoxychlor, *o,p′*-	$C_{16}H_{15}Cl_3O_2$	3.3	4.5	5	—	—	C	—	6	—
Methoxychlor, *p,p′*-	$C_{16}H_{15}Cl_3O_2$	4.7	7.2	7.2	—	C	C	C	6	2
Methyl 2,3,5-triiodobenzoate	$C_8H_5I_3O_2$	2.28	—	—	—	—	—	—	—	—
Methyl 2,3,6-trichlorobenzoate	$C_8H_5Cl_3O_2$	0.23	0.21	0.18	—	—	—	—	—	—
Methyl 4-chloro-1H-indole-3-acetate	$C_{11}H_{10}ClNO_2$	1.17	—	—	—	R	R#	NR	50	1–2–3
Metobromuron	$C_9H_{11}BrN_2O_2$	0.67	1.44	0.69	2.3	C	NR	NR	6–15–50	1–2–3
Metolachlor	$C_{15}H_{22}ClNO_2$	1.03	1.21	0.93	1.25	C	S#	NR	50	1–2–3
Metolcarb	$C_9H_{11}NO_2$	0.17	—	—	0.48	C	—	—	—	—
Metoxuron	$C_{10}H_{13}ClN_2O_2$	—	—	0.18	—	V	NR	NR	6–15–50	1–2–3
Metribuzin	$C_8H_{14}N_4OS$	0.57	1.47	0.91	4.5	V	NR	NR	50	1–2–3
Metribuzin, deaminated diketo metabolite*	$C_7H_{11}N_3O_2$	0.5	0.79	0.44	4.6	NR	NR	NR	6–15–50	1–2–3
		0.73	1.29	0.52	5.8	—	—	—	—	—
Metribuzin, deaminated metabolite	$C_8H_{13}N_3OS$	0.83	3.77	1.06	22	C	NR	NR	6–15–50	1–2–3
Metribuzin, diketo metabolite	$C_7H_{12}N_4O_2$	0.56	1.41	0.55	5.9	NR	NR	NR	6–15–50	1–2–3

Mevinphos, (*E*)-	$C_7H_{13}O_6P$	0.16	—	0.13	0.33	C	NR	NR	6–15–50	—
Mevinphos, (*Z*)-	$C_7H_{13}O_6P$	0.13	—	0.15	0.4	C	NR	—	6–15–50	—
MGK 264	$C_{17}H_{25}NO_2$	1.6	—	—	1.2	—	—	—	—	—
Mirex	$C_{10}Cl_{12}$	5.8	2.95	5.6	—	P	C	P	6	1
Mirex, 5,10-dihydro-*	$C_{10}H_2Cl_{10}$	2.14	—	—	—	—	—	—	—	—
		2.47	—	—	—	—	—	—	—	—
		3.21	—	—	—	—	—	—	—	—
		4.3	—	—	—	—	—	—	—	—
Molinate	$C_9H_{17}NOS$	—	0.19	—	—	C	—	—	—	—
Monocrotophos	$C_7H_{14}NO_5P$	0.31	1.6	0.5	4.3	C	NR	NR	6–15–50	1–2–3
Monolinuron	$C_9H_{11}ClN_2O_2$	0.48	0.91	0.48	—	C	—	—	—	—
Monuron	$C_9H_{11}ClN_2O$	0.1	—	—	—	—	NR	NR	6–15–50	1–2–3
Myclobutanil	$C_{15}H_{17}ClN_4$	1.9	7.2	2.6	8.7	C	NR	NR	6–15–50	1–2–3
Myclobutanil alcohol metabolite	$C_{15}H_{17}ClN_4O$	3.6	37.1	7.5	—	S	NR	NR	6–15–50	1–2–3
Myclobutanil dihydroxy metabolite	$C_{15}H_{17}N_4O_2Cl$	6.5	—	11.5	—	NR	NR	NR	6–15–50	1–2–3
Naled	$C_4H_7Br_2Cl_2O_4P$	0.34	—	0.32	0.68	C	NR	NR	6–15–50	1–2–3
Napropamide	$C_{17}H_{20}NO_2$	1.7	—	2.12	3.4	C	—	—	—	—
Neburon	$C_{12}H_{16}Cl_{12}N_2O$	0.11	—	—	—	C	NR	NR	6–15–50	1–2–3
Nitralin	$C_{13}H_{19}N_3O_6S$	3.8	24	6.3	—	C	P	P	50	3
Nitrapyrin	$C_6H_3Cl_4N$	0.2	0.18	0.2	—	C	C	V	6	2
Nitrofen	$C_{12}H_7Cl_{12}NO_3$	2.03	3.8	2.71	—	C	C	C	15	2
Nitrofluorfen	$C_{13}H_7ClF_3NO_3$	0.96	1.45	0.86	2.06	C	C	C	15	2
Nitrothal-isopropyl	$C_{14}H_{17}O_6N$	1.1	—	0.68	1.4	C	—	—	—	—
Nonachlor, *cis*	$C_{10}H_5Cl_9$	2.52	3.33	2.61	—	C	C	C	6	1
Nonachlor, *trans*-	$C_{10}H_5Cl_9$	1.75	1.45	1.42	—	C	C	C	6	1
Norea	$C_{13}H_{15}N_2O$	—	—	1.05	1.56	C	—	—	—	—
Norflurazon	$C_{12}H_9ClF_3NO_3$	4.5	—	5.01	—	V	NR	NR	6–15–50	—

PESTDATA Chemicals in Order by Chemical Name *(contd.)*

Name	Molecular Formula	RRT/c OV-101	RRT/c OV-225	RRT/c OV-17	RRT/c DEGS	Recoveries 302	303	304	Ethers	$MeCl_2$
NTN33823	$C_9H_{11}N_4Cl$	3	—	1.18	—	—	NR	NR	6–15–50	1–2–3
NTN35884*	$C_9H_9N_5O_2Cl$	1.59	—	—	—	—	NR	NR	6–15–50	1–2–3
		5	—	—	—	—	—	—	—	—
Nuarimol	$C_{17}H_{12}ClFN_2O$	3.36	7.3	4.8	—	C	NR	C#	50	1–2–3
Octachlor epoxide	$C_{10}H_4Cl_8O$	1.33	0.94	1.05	—	C	C	C	6	1
Octachlorocyclopentane	C_5Cl_8	—	—	0.22	—	—	—	—	—	—
Octhilinone	$C_{11}H_{19}NOS$	0.66	—	0.64	—	C	—	—	—	—
Ofurace	$C_{14}H_{16}NO_3Cl$	2.62	18.6	5.4	—	C	—	—	—	—
Omethoate	$C_5H_{12}NO_4PS$	0.25	1.11	0.39	2.58	C	NR	NR	6–15–50	1–2–3
Oryzalin	$C_{12}H_8N_4O_6S$	4.7	—	—	—	—	NR	NR	6–15–50	—
Ovex	$C_{12}H_8Cl_2O_3S$	1.61	3.04	2.2	—	C	C	C	15	2
Oxadiazon	$C_{15}H_{18}Cl_2N_2O_3$	1.97	2.48	1.96	2.55	C	C	P	15	—
Oxadixyl	$C_{14}H_{18}N_2O_4$	2.5	14	5	26	C	NR	NR	6–15–50	1–2–3
Oxamyl	$C_7H_{13}N_3O_3S$	—	—	—	2.74	—	—	—	—	—
Oxycarboxin	$C_{12}H_{13}NO_4S$	3.28	—	9.4	—	R	—	—	—	—
Oxydemeton-methyl	$C_6H_{15}O_4PS_2$	0.46	0.49	0.31	0.61	C	—	—	—	—
Oxydemeton-methyl sulfone	$C_6H_{15}O_5PS_2$	0.72	—	1.48	13.5	C	—	—	—	—
Oxydeprofos	$C_7H_{17}O_4SP$	—	—	—	0.22	—	—	—	—	—
Oxyfluorfen	$C_{15}H_{11}ClF_3NO_4$	2	4	2.16	5	C	C	—	15	2
Oxythioquinox	$C_{10}H_6N_2OS_2$	1.57	—	1.85	2.19	C	—	—	—	—
Paclobutrazol	$C_{15}H_{20}ClN_3O$	1.52	—	1.59	—	C	—	—	—	—
Parathion	$C_{10}H_{14}NO_5PS$	0.98	1.91	1.07	2.5	C	C	C	15	2
Parathion oxygen analog	$C_{10}H_{14}NO_6P$	0.8	—	0.86	3.23	C	NR	NR	6–15–50	1–2–3

Parathion-methyl	$C_8H_{10}NO_5PS$	0.71	1.64	0.87	2.89	C	C	C	15	2
Parathion-methyl oxygen analog	$C_8H_{10}NO_6P$	0.55	1.71	0.66	3.6	—	NR	NR	6–15–50	1–2–3
Pebulate	$C_{10}H_{21}NOS$	0.17	—	0.1	0.08	C	P	—	15	—
Penconazole	$C_{13}H_{15}Cl_2N_3$	1.24	—	1.32	—	C	—	—	—	—
Pendimethalin	$C_{13}H_{19}N_3O_4$	1.22	1.48	1.21	1.36	C	C	P	15	2
Pentachloroaniline	$C_6H_2Cl_5N$	0.67	0.79	0.66	1.23	C	C	C	6	1
Pentachlorobenzene	C_6HCl_5	0.24	0.13	0.16	0.22	C	C	C	6	1
Pentachlorobenzonitrile	C_7Cl_5N	0.5	0.59	0.45	—	C	C	P	15	2
Pentachlorophenyl methyl ether	$C_7H_3Cl_5O$	0.46	0.3	0.34	—	C	C	C	6	1
Pentachlorophenyl methyl sulfide	$C_7H_3Cl_5S$	0.94	0.69	0.87	0.75	C	C	C	6	1
Permethrin, *cis*	$C_{21}H_{20}Cl_2O_3$	9.4	11.1	13.8	—	C	V#	C	6 + 15	2
Permethrin, *trans*-	$C_{21}H_{20}Cl_2O_3$	10.2	13	15	—	C	V#	C	6 + 15	2
Perthane	$C_{18}H_{20}Cl_2$	2.23	2.01	2.42	—	C	C	C	6	1
Phenmedipham	$C_{16}H_{16}N_2O_4$	0.32	—	0.41	9.5	—	—	—	—	—
Phenothiazine	$C_{12}H_9NS$	1.16	—	1.56	—	—	—	—	—	—
Phenothrin*	$C_{23}H_{26}O_3$	5.4	4.8	6.5	—	—	—	—	—	—
		11.5	10.9	15	—	—	—	—	—	—
Phenthoate	$C_{12}H_{17}O_4PS_2$	1.31	2.05	1.83	2.92	C	C	—	15 + 50	—
Phorate	$C_7H_{17}O_2PS_3$	0.37	0.38	0.32	0.34	C	V#	V#	6	1
Phorate oxygen analog	$C_7H_{17}O_3PS_2$	0.3	0.37	0.29	0.41	C	NR	NR	6–15–50	1–2–3
Phorate oxygen analog sulfone	$C_7H_{17}O_5PS_2$	0.66	—	1.02	5.5	C	NR	NR	6–15–50	1–2–3
Phorate oxygen analog sulfoxide	$C_7H_{17}O_4PS_2$	0.78	—	1.06	4.03	C	NR	NR	6–15–50	1–2–3
Phorate sulfone	$C_7H_{17}O_4PS_3$	0.97	3.26	1.3	4.7	C	S#	S#	6–15–50	3
Phorate sulfoxide	$C_7H_{17}O_3PS_3$	0.89	2.55	1.26	3.44	C	NR	NR	6–15–50	1–2–3
Phosalone	$C_{12}H_{15}ClNO_4PS_2$	5.5	5.5	9.1	28.8	C	C	C	50	2 + 3

PESTDATA Chemicals in Order by Chemical Name *(contd.)*

Name	Molecular Formula	RRT/c OV-101	RRT/c OV-225	RRT/c OV-17	RRT/c DEGS	Recoveries 302	303	304	Ethers	$MeCl_2$
Phosalone oxygen analog	$C_{12}H_{15}ClNO_5PS$	3.8	—	6.2	—	C	—	—	—	—
Phosfolan	$C_7H_{14}NO_3PS_2$	—	—	2.69	12	C	—	—	—	—
Phosmet	$C_{11}H_{12}O_4NPS_2$	4	14.9	8.4	32.7	C	NR	—	6–15–50	3
Phosmet oxygen analog*	$C_{11}H_{12}O_5NPS$	0.5	0.53	0.24	—	—	NR	NR	6–15–50	—
		2	0.93	0.44	—	—	—	—	—	—
		3.1	14.8	7.1	0.22	—	—	—	—	—
Phosphamidon*	$C_{10}H_{19}ClNO_5P$	0.53	—	0.57	1.52	C	NR	NR	6–15–50	1–2–3
		0.67	—	0.76	2.44	—	—	—	—	—
Photodieldrin	$C_{12}H_8Cl_6O$	4.4	15.5	8.5	—	—	C	C	15 + 50	2
Photodieldrin B	$C_{13}H_9Cl_5O$	1.43	—	—	—	—	—	—	—	—
Phoxim oxygen analog	$C_{12}H_{15}N_2O_4P$	0.86	—	—	3.08	C	—	—	—	—
Picloram methyl ester	$C_7H_5Cl_3N_2O_2$	0.75	2.67	—	—	—	—	—	—	—
Picloram*	$C_6H_3Cl_3N_2O_2$	0.25	—	—	—	—	—	—	—	—
		0.67	—	—	—	—	—	—	—	—
Piperophos	$C_{14}H_{28}NO_3PS_2$	4.8	9.7	6.8	9.1	C	—	—	—	—
Pirimicarb	$C_{11}H_{18}N_4O_2$	0.61	—	0.73	0.9	C	—	—	—	—
Pirimiphos-ethyl	$C_{13}H_{24}N_3O_3PS$	1.14	1.03	1.14	0.97	C	C	C	15 + 50	3
Pirimiphos-ethyl oxygen analog	$C_{13}H_{24}N_3O_4P$	1.01	—	1.14	1.68	C	—	—	—	—
Pirimiphos-methyl	$C_{11}H_{20}N_3O_3PS$	0.87	—	0.92	0.96	C	C	C	15	3
PP 890	$C_9H_{10}O_2ClF_3$	1.05	1	—	2.17	—	—	—	—	—
PPG-1576	$C_{19}H_{17}ClF_3NO_5$	6.7	—	—	—	—	—	P	50	2 + 3
PPG-2597	$C_{20}H_{17}ClF_3NO_6$	1.9	3.16	1.86	7.3	—	NR	NR	6–15–50	1–2–3
PPG-847, methylated	$C_{15}H_9ClF_3NO_3$	2.15	5	2.4	—	—	—	—	—	—

PPG-947	$C_{17}H_{11}ClF_3NO_7$	1.04	—	1.13	5	—	NR	NR	6–15–50	1–2–3
PPG-947, methylated*	$C_{18}H_{13}ClF_3NO_7$	0.42	0.97	0.49	—	—	—	—	—	—
		2.14	5	2.4	1.56	—	—	—	—	—
Pretilachlor	$C_{17}H_{26}ClNO_2$	1.88	—	1.99	—	C	—	—	—	—
Probenazole	$C_{10}H_9NO_3S$	—	—	—	6.1	C	—	—	—	—
Prochloraz	$C_{15}H_{16}Cl_3N_3O_2$	10.4	—	15.4	—	C	—	—	—	—
Procyazine	$C_{10}H_{13}ClN_6$	1.5	7.9	—	—	C	—	—	—	—
Procymidone	$C_{13}H_{11}Cl_2NO_2$	1.37	3.04	1.49	3.9	C	C	P	15	—
Prodiamine	$C_{13}H_{17}F_3N_4O_4$	0.94	0.97	0.66	—	C	—	—	—	—
Profenofos	$C_{11}H_{15}BrClO_3PS$	1.8	2.34	2.13	2.68	C	P	P	50	3
Profluralin	$C_{14}H_{16}F_3N_3O_4$	0.53	0.46	0.3	0.4	V	V	—	6	—
Promecarb	$C_{12}H_{17}NO_2$	1.58	—	—	0.8	V	—	—	—	—
Prometryn	$C_{10}H_{19}N_5S$	0.77	—	0.74	1.68	C	P#	P#	50	1–2–3
Pronamide	$C_{12}H_{11}Cl_2NO$	0.51	0.84	0.4	1.52	C	P	—	15 + 50	—
Propachlor	$C_{11}H_{14}ClNO$	0.34	0.37	0.26	0.42	C	NR	NR	6–15–50	1–2–3
Propanil	$C_9H_9Cl_2NO$	0.66	2.82	0.78	4.9	C	NR	NR	6–15	3
Propargite	$C_{19}H_{26}O_4S$	3.8	4.8	4.3	7.9	C	C	—	15	2
Propazine	$C_9H_{16}ClN_5$	0.53	0.65	0.41	1.25	C	S	NR	15 + 50	3
Propetamphos	$C_{10}H_{20}NO_4PS$	0.48	—	0.42	—	C	C#	—	15 + 50	2 + 3
Propham	$C_{10}H_{13}NO_2$	0.13	—	0.12	0.27	C	P	P	15	—
Propiconazole*	$C_{15}H_{17}Cl_2N_3O_2$	3.06	—	—	—	C	NR	NR	6–15–50	1–2–3
		3.21	5.6	4	10	—	—	—	—	—
Propoxur	$C_{11}H_{15}NO_3$	—	—	—	0.83	C	—	—	—	—
Prothiofos	$C_{11}H_{15}Cl_2PO_2S_2$	1.85	1.74	1.82	1.67	C	C	C	6	2
Prothoate	$C_9H_{20}NO_3PS_2$	0.75	1.55	0.79	1.42	C	—	—	—	—
Pyracarbolid	$C_{13}H_{15}NO_2$	1.05	—	1.43	—	—	—	—	—	—
Pyrazon	$C_{10}H_8ClN_3O$	2.67	13	8	—	C	NR	NR	6–15–50	1–2–3
Pyrazon metabolite A	$C_{16}H_8ClN_3O_6$	2.62	—	—	—	—	—	—	—	—
Pyrazon metabolite B	$C_6H_4ClN_3O$	0.42	—	0.9	—	—	NR	NR	6–15–50	1–2–3

PESTDATA Chemicals in Order by Chemical Name *(contd.)*

Name	Molecular Formula	RRT/c OV-101	RRT/c OV-225	RRT/c OV-17	RRT/c DEGS	Recoveries				
						302	303	304	Ethers	$MeCl_2$
Pyrazophos	$C_{14}H_{20}N_3O_5PS$	8.1	—	13	26.9	C	—	—	—	—
Pyrethrins*	$C_{21}H_{27}O_4$	2.12	1.76	1.76	—	—	C	C	50	—
		2.95	2.84	2.7	—	—	—	—	—	—
Pyridaphenthion	$C_{14}H_{17}O_4N_2SP$	4.2	14	8.7	25	C	—	—	—	—
Pyrithiolbac-sodium methyl ester	$C_{14}H_{13}ClN_2O_4$	2.51	4	4.2	6.9	—	—	—	—	—
Quinalphos	$C_{12}H_{15}N_2O_3PS$	1.32	2	1.64	2.5	C	C	—	15	—
Quintozene	$C_6H_{15}NO_2$	0.51	0.46	0.46	—	C	C	C	6	1
Quizalofop ethyl ester	$C_{19}H_{17}ClN_2O_4$	13.6	—	25	—	C	—	—	—	—
R25788	$C_8H_{11}C_{12}NO$	0.19	0.22	0.14	0.2	C	S	S	15 + 50	2 + 3
RH-6467*	$C_{19}H_{15}N_4ClO$	7.9	—	—	—	S	NR	NR	6–15–50	1–2–3
		10.4	—	—	—	—	—	—	—	—
		15	—	—	—	—	—	—	—	—
RH-9129	$C_{19}H_{16}N_3ClO_2$	14	—	—	—	V	NR	NR	6–15–50	1–2–3
RH-9130	$C_{19}H_{16}N_3ClO_2$	12	—	—	—	P	NR	NR	6–15–50	1–2–3
Ronnel	$C_8H_8Cl_3O_3PS$	0.81	0.86	0.76	0.95	C	C	C	6	2
Ronnel oxygen analog	$C_8H_8Cl_3O_4P$	0.64	1.02	0.62	1.27	C	NR	—	6–15–50	—
S-Bioallethrin	$C_{19}H_{26}O_3$	1.37	1.22	1.12	—	—	C	—	50	—
Schradan	$C_8H_{24}N_4O_3P_2$	0.58	—	0.52	1.43	C	NR	—	6–15–50	—
Sethoxydim	$C_{17}H_{29}NO_3S$	3.34	—	—	0.42	—	NR	NR	6–15–50	3
Sethoxydim sulfoxide	$C_{17}H_{29}NO_4S$	0.84	—	—	—	—	NR	NR	6–15–50	3
Silvex	$C_9H_7Cl_3O_3$	0.48	—	—	—	—	—	—	—	—
Silvex methyl ester	$C_{10}H_9Cl_3O_3$	0.45	0.44	—	—	—	—	—	—	—
Simazine	$C_7H_{12}ClN_5$	0.5	0.83	0.5	—	C	NR	NR	50	1–2–3

Simetryn	$C_8H_{15}N_5S$	2.02	1.21	—	—	C	—	—	—	—
Strobane*	$C_{10}H_{11}Cl_7$	1.09	—	—	—	—	C	C	6	1
		1.32	—	—	—	—	—	—	—	—
		1.53	—	—	—	—	—	—	—	—
		1.8	—	—	—	—	—	—	—	—
		1.94	—	—	—	—	—	—	—	—
		2.09	—	—	—	—	—	—	—	—
		2.69	—	—	—	—	—	—	—	—
		3.1	—	—	—	—	—	—	—	—
		3.7	—	—	—	—	—	—	—	—
Sulfallate	$C_8H_{14}ClNS_2$	0.38	0.44	0.36	—	C	C	C	6+15	2
Sulfanilamide	$C_6H_8O_2N_2S$	—	—	2.11	—	NR	NR	NR	6–15–50	1–2–3
Sulfotep	$C_8H_{20}O_5P_2S_2$	0.34	—	0.29	0.37	C	C	P	6+15	2
Sulphenone	$C_{12}H_9ClO_2S$	1.26	3.5	1.92	—	C	—	—	50	3
Sulprofos	$C_{12}H_{19}O_2PS_3$	2.79	—	3.5	4.6	C	—	—	—	—
Sulprofos oxygen analog sulfone	$C_{12}H_{19}O_5PS_2$	5.1	—	10.6	—	C	—	—	—	—
Sulprofos sulfone	$C_{12}H_{19}O_4PS_3$	7.2	—	13.1	—	C	—	—	—	—
Sulprofos sulfoxide*	$C_{12}H_{19}O_3PS_3$	2.78	—	3.6	—	C	—	—	—	—
		6.1	—	11.7	—	—	—	—	—	—
TCMTB	$C_9H_6N_2S_3$	1.5	4.3	2.67	6	C	P	P	15	—
TDE, *o,p′*	$C_{14}H_{10}Cl_4$	1.9	2.46	2.19	—	—	C	C	6	1
TDE, *o,p′*-, olefin	$C_{14}H_9Cl_3$	1.19	1.15	1.2	—	—	—	—	—	—
TDE, *p,p′*-	$C_{14}H_{10}Cl_4$	2.41	3.8	2.87	—	C	C	C	6	1
TDE, *p,p′*-, olefin	$C_{14}H_9Cl_3$	1.45	1.36	1.45	—	C	C	C	6	1
Tebupirimfos (prop)	$C_{13}H_{23}N_2O_3PS$	0.63	—	—	—	—	V	V	6+15	2+3
Tebupirimfos oxygen analog	$C_{13}H_{23}O_4N_2P$	0.51	—	—	—	—	NR	NR	6–15–50	1–2–3
Tebuthiuron	$C_9H_{16}N_4OS$	0.26	—	0.21	0.76	—	—	—	—	—
Tecnazene	$C_6HCl_4NO_2$	0.29	0.26	0.24	—	C	C	C	6	1

PESTDATA Chemicals in Order by Chemical Name *(contd.)*

Name	Molecular Formula	RRT/c OV-101	RRT/c OV-225	RRT/c OV-17	RRT/c DEGS	Recoveries				
						302	303	304	Ethers	$MeCl_2$
TEPP	$C_8H_{20}O_7P_2$	0.21	0.27	0.24	0.15	C	—	—	—	—
Terbacil	$C_9H_{13}ClN_2O_2$	0.54	2.1	0.72	5	C	NR	NR	6–15	2 + 3
Terbufos	$C_9H_{21}O_2PS_3$	0.5	0.44	0.41	0.37	C	P	S	6	—
Terbufos oxygen analog	$C_9H_{21}O_3PS_2$	0.42	—	0.39	0.52	C	—	NR	6–15–50	1–2–3
Terbufos oxygen analog sulfone	$C_9H_{21}O_5PS_2$	0.92	2.9	1.28	4.6	C	NR	NR	6–15–50	1–2–3
Terbufos sulfone	$C_9H_{21}O_4PS_3$	1.2	—	1.58	4	C	C#	C#	6–15–50	2 + 3
Terbumeton	$C_{10}H_{19}N_5O$	0.47	0.42	0.53	—	C	—	—	—	—
Terbuthylazine	$C_9H_{16}N_5Cl$	0.47	0.71	0.48	1.32	C	P	—	15 + 50	—
Terbutryn	$C_{10}H_{19}N_5S$	0.84	1.08	—	—	C	—	—	—	—
Tetradifon	$C_{12}H_6Cl_4O_2S$	5.2	—	8.3	—	C	C	C	15	2
Tetraiodoethylene	C_2I_4	0.55	0.86	1.04	—	—	P	P	6	—
Tetramethrin*	$C_{19}H_{25}NO_4$	4.3	—	—	—	—	—	—	—	—
Tetrasul	$C_{12}H_6Cl_4S$	2.64	2.33	2.8	—	C	C	C	6	1
Tetrasul sulfoxide	$C_{12}H_6Cl_4OS$	4.7	8.6	7.2	—	—	—	—	—	—
Thiabendazole	$C_{10}H_7N_3S$	1.48	—	2.04	20	C	NR	—	6–15–50	—
Thiobencarb	$C_{12}H_{16}ClNOS$	0.94	1	0.98	1.08	C	—	V	15	2 + 3
Thiometon	$C_6H_{15}O_2PS_3$	0.41	0.51	0.4	0.48	C	NR	NR	6–15–50	—
Thionazin	$C_8H_{13}N_2O_3PS$	0.26	—	0.26	0.39	C	P	NR	15 + 50	—
Thionazin oxygen analog	$C_8H_{13}N_2O_4P$	—	—	—	0.54	—	—	—	—	—
Thiophanate-methyl	$C_{12}H_{14}N_4O_4S_2$	—	—	—	0.2	—	—	—	—	—
THPI	$C_8H_9NO_2$	0.21	—	—	1.54	C	NR	NR	6–15–50	—
Tolylfluanid	$C_{10}H_{13}ClFNOS$	1.25	—	1.41	—	C	—	—	—	—

Toxaphene*	$C_{10}H_{10}Cl_8$	—	1.75	—	—	C	C	C	6	1
		—	2.14	—	—	—	—	—	—	—
		—	2.35	—	—	—	—	—	—	—
		—	2.6	—	—	—	—	—	—	—
		1.2	2.74	—	—	—	—	—	—	—
		1.54	3.05	—	—	—	—	—	—	—
		1.8	3.9	—	—	—	—	—	—	—
		2.39	4.3	—	—	—	—	—	—	—
		2.68	4.5	—	—	—	—	—	—	—
		3.12	5.2	—	—	—	—	—	—	—
		3.7	5.6	—	—	—	—	—	—	—
		4.4	6	—	—	—	—	—	—	—
		4.6	6.4	—	—	—	—	—	—	—
		5.1	7	—	—	—	—	—	—	—
Tralomethrin	$C_{22}H_{19}Br_4NO_3$	27	64	44	42	C	V	S	15	2
Tri-allate	$C_{10}H_{16}Cl_3NOS$	0.6	—	0.45	—	C	C	C	6	2
Triadimefon	$C_{14}H_{16}ClN_3O_2$	1.05	1.64	1	2.92	C	S#	S#	50	1–2–3
Triadimenol	$C_{14}H_{18}ClN_3O_2$	1.36	—	1.44	6.5	C	NR	NR	6–15–50	—
Triazophos	$C_{12}H_{16}N_3O_3PS$	2.62	—	5.2	18	C	—	—	—	—
Tributyl phosphate	$C_{12}H_{27}O_4P$	0.3	—	0.23	0.23	—	R	—	50	—
Trichlorfon	$C_4H_8Cl_3O_4P$	0.16	—	0.13	0.77	C	NR	NR	6–15–50	1–2–3
Trichloronat	$C_{10}H_{12}Cl_3O_2PS$	1.13	—	0.97	0.86	C	C	—	6	—
Trichloronat oxygen analog	$C_{10}H_{12}Cl_3O_3P$	0.88	1.05	0.82	0.98	C	—	—	—	—
Triclopyr methyl ester	$C_{10}H_6Cl_3NO_3$	0.36	—	—	—	—	—	—	—	—
Tricylazole	$C_9H_7N_3S$	1.59	—	3.9	—	C	—	—	—	—
Tridiphane	$C_{10}H_7Cl_5O$	0.81	0.85	0.75	—	C	C	—	6	1 + 2
Triflumizole	$C_{15}H_{15}ClF_3N_3O$	1.44	2.23	1.19	3.42	C	—	—	—	—
Trifluralin	$C_{13}H_{16}F_3N_3O_4$	0.34	0.27	0.17	—	C	C	C	6	2
Triflusulfuron methyl ester	$C_{17}H_{19}F_3N_6O_6S$	0.3	—	0.2	0.8	V	NR	NR	6–15–50	1–2–3

PESTDATA Chemicals in Order by Chemical Name *(contd.)*

Name	Molecular Formula	RRT/c OV-101	RRT/c OV-225	RRT/c OV-17	RRT/c DEGS	Recoveries				
						302	303	304	Ethers	$MeCl_2$
Tris(β-chloroethyl) phosphate	$C_6H_{12}Cl_3O_4P$	0.45	—	—	2.79	C	—	—	—	—
Tris(chloropropyl) phosphate	$C_9H_{18}Cl_3O_4P$	0.5	2.02	0.5	1.14	C	NR	NR	6–15–50	1–2–3
Tycor	$C_9H_{16}N_4OS$	0.77	1.62	0.99	4.1	C	S	S	50	3
Tycor DA metabolite	$C_9H_{15}N_3SO$	0.76	3.5	1.47	8	—	NR	NR	6–15–50	1–2–3
Vamidothion sulfone	$C_8H_{18}NO_6PS_2$	2.19	—	—	—	C	—	—	—	—
Vernolate	$C_{10}H_{21}NOS$	0.15	—	0.09	—	—	P	—	15	—
Vinclozolin	$C_{12}H_9Cl_2NO_3$	0.69	1.15	0.64	1.58	C	C	C	15	2
Vinclozolin metabolite B	$C_{12}H_{11}Cl_2NO_4$	0.74	1.2	0.66	1.64	C	P#	C	6 + 15	2
Vinclozolin metabolite E	$C_{11}H_{11}Cl_2NO_2$	0.89	3.02	0.93	—	C	S	NR	15 + 50	—
Vinclozolin metabolite F	$C_{11}H_{13}Cl_2NO_4$	2.87	—	4.6	0.61	R	NR	NR	6–15–50	1–2–3
Vinclozolin metabolite S	$C_{10}H_7Cl_2NO_3$	0.69	2.01	0.79	3.3	V	P	V#	15	2

*Multipeak chemical.

Recovery may vary with choice of florisil elution system; see *PAM*, Volume I.

INDEX